Encyclopedia of Science and Technology

Encyclopedia of Science and Technology

Editor: Amber Hall

NY RESEARCH
P R E S S
New York

Published by NY Research Press
118-35 Queens Blvd., Suite 400,
Forest Hills, NY 11375, USA
www.nyresearchpress.com

Encyclopedia of Science and Technology
Edited by Amber Hall

International Standard Book Number: 978-1-63238-539-0 (Hardback)

Cataloging-in-Publication Data

Encyclopedia of science and technology / edited by Amber Hall.
   p. cm.
Includes bibliographical references and index.
ISBN 978-1-63238-539-0
1. Science. 2. Technology. I. Hall, Amber.
Q113 .E53 2017
500--dc23

Printed in the United States of America.

Contents

Preface

The main aim of this book is to educate learners and enhance their research focus by presenting diverse topics covering this vast field. This is an advanced book which compiles significant studies by distinguished experts. This book addresses successive solutions to the challenges arising in the area of application, along with it; the book provides scope for future developments.

Science and technology is a vast field of study. It is an umbrella discipline that branches out into diverse engineering disciplines. Progress in the domain of science and technology results in advancements in all fields. This book attempts to understand the multiple branches that fall under the discipline of science and technology and how such concepts have practical applications. The ever growing need of advanced technology is the reason that has fuelled the research in this field in recent times. This book will serve as a reference to a broad spectrum of readers. It will help the readers in keeping pace with the rapid changes in the field.

It was a great honour to edit this book, though there were challenges, as it involved a lot of communication and networking between me and the editorial team. However, the end result was this all-inclusive book covering diverse themes in the field.

Finally, it is important to acknowledge the efforts of the contributors for their excellent chapters, through which a wide variety of issues have been addressed. I would also like to thank my colleagues for their valuable feedback during the making of this book.

Editor

Okras (*Abelmoschus Esculentus L.* Moench) Drying Behaviour After Undergoing Blanching or Combined Dewatering-Impregnation-Soaking Process (DISP)/Blanching

Jiokap Nono Yvette[1, *], Aseaku Jude Nkengbeza[1], Desmorieux Helene[2], Degraeve Pascal[3], Kamga Richard[4]

[1]University Institute of Technology (IUT) of the University of Ngaoundere, Department of Chemical Engineering and Environment, Ngaoundere, Cameroon

[2]Process Engineering and Automatic Laboratory of University Claude Bernard – Lyon, Villeurbanne Cedex, France

[3]Food Processing Research Laboratory of Claude Bernard – Lyon 1 University, University Institute of Technology A Lyon 1 – Biological Process Department, Technopole Alimentec – Rue Henri de Boissieu, Bourgen Bresse Cedex, France

[4]National Advanced School of Agro-Industrial Sciences (ENSAI) of the University of Ngaoundere, Department of Applied Chemistry, Ngaoundere, Cameroon

Email address:

jiokapnonoy@yahoo.fr (J. N. Yvette)

Abstract: Traditional (solar) drying of okra fruits (*Abelmoschus esculentus L.* Moench) often gives products of poor storage quality, short shelf life and unpleasant sight. This study assesses the suitability of a dewatering-impregnation-soaking process (DISP) and blanching prior to drying as a means of improving the stability and the appearance of the dried okras. The DISP/Blanching pre-treatment involved immersing whole okra fruits for 12 hrs in a warm concentrated osmotic solution (made of NaCl, ascorbic acid, citric acid and "*kanwa*") that was progressively cooled from 85°C to room temperature (~ 24°C), before drying at 40°C. Compared to simple blanching, the DISP/Blanching pre-treatment yielded dry products that were better appreciated both to the touch and to the sight. The combined DISP/Blanching and drying process gave averages of (78.1 ± 3.6) % weight reduction, (9.1 ± 0.5) % solute gain and (87.2 ± 4.0) % water loss. The total water loss was due at 35.6 % to the DISP/blanching pre-treatment and at 64.4 % to the further drying process. The results obtained here demonstrate the possibility of improving the storage quality of dried okra through the combined DISP/Blanching pre-treatment, reducing post-harvest losses and improving the market quality of dry products. This treatment could be applied industrially to enhance the presentation of tropical vegetables in international markets.

Keywords: Okra, Dewatering-Impregnation-Soaking Process (DISP), Blanching, Air Drying, Kinetics, Quality

1. Introduction

The okra plant (*Abelmoschus esculentus L.* Moench) is a tropical dicotyledonous plant grown throughout the tropics and warm temperate regions for its immature pod used as vegetable, food ingredient, as well as a traditional medicine [1-3]. Okra is not common in most European countries [4] whereas in tropical countries like Cameroon, cultivation of the crop represents an important agricultural and economic activity [3, 5]. Although, scientific studies on okra began in the first half of the 20th century, much of these were dedicated to some of its diseases like the okra virus [6], the rheology and emulsifying properties of its hydrocolloids [7, 8], the medicinal properties of its mucilage [9], the feasibility of using okra gum as fat replacement in food industries [10, 11] or okra as a rich source of high quality edible fat and protein [4, 12, 13,]. Nowadays, research works are mostly directed towards improving okra farming productivity and yields [3, 14, 15, 16], on quality attribute of the fresh and dried pods [17, 18] or on okra dryer design [19, 20].

In Cameroon, okra is often conserved using traditional processing techniques like solar drying in the open. These techniques are however largely inappropriate during the rainy season when solar radiations are low, ambient air relative humidity is high and fruit production is at its peak, hence high post-harvest losses [3]. Furthermore, drying under this time-consuming conditions gives products with pathogenic microorganisms, impurities and a dark colour that reduce the market quality of the slice-dried or milled product [21-23].

Recent publications highlight the need to pre-treat (chemically or physically) fruits and vegetables prior to drying as a means of avoiding any significant structural changes and improve their reconstitution properties [24, 25], ensure maximum inhibition of enzymatic browning reactions and reduce their initial microbial count [26]. Notable amongst these, blanching and/or dewatering-impregnation-soaking process (DISP) generate superior quality marketable products that could be preserved for long [24, 27-31]. The presence of solutes like citric acid reduces the solution's pH, while "*Kanwa*" and ascorbic acid serve as chlorophyll-fixing and folate-retention agents [32], improve the colour or the vitamin retention of processed fruits and vegetables and maintain their texture [29, 33, 34].

In the light of similar research works, some authors [35] found that the optimized condition for okra osmotic dehydration was : sucrose concentration, 49.28(% w/w); solution temperature, 40.79 °C; sample size diameter, 15mm and process time, 4.49hrs. At this optimized condition, water loss and solute gain of 39.78% and 10.16% respectively were observed. Other authors [36] have also tested two levels of sucrose concentrated solution (40 and 60°Brix) to pretreat sliced (7 mm) okra (*Abelmoschus caillei*) before drying in a convective dryer at temperatures ranging from 50 to 80°C. The quality attributes investigated were only: ash content, crude fiber, crude fat, crude protein, bulk density, least gelation concentration and water absorption capacity. In their work, no result is given on the weight reduction and on the solute gain. For the sets of authors [35] and [36], there is no information on the final physical presentation of the product in the market and on the local tastes; that is the acceptability of sweet okra by the local population. Furthermore, they didn't carry out blanching of okra samples.

This work was therefore undertaken to study the behaviour of okra pods during the combined DISP/blanching and drying, with particular emphasis on the water absorption capacity of the fresh fruit and the influence of pre-treatment schedules prior to drying on the transfer kinetics and the product final appearance.

2. Materials and Methods

2.1. Materials and Their Preparation

Four morphologically distinct okra varieties (variety 1, 2, 3 and 4) collected in the locality of Ngaoundere in the Adamawa region of Cameroon were used (Figure 1), with variety 2 being the most cultivated. These varieties were different with respect to their colour (Figure 1), average fruit dimensions, weight and initial moisture content (Table 1). Fresh okra fruits were harvested at maturity from the corresponding okra plant early in the morning, using a sharp knife while leaving a 2 - 3 cm long stalk attached to the pods, put into cartons and promptly transported to the laboratory where they were washed with tap water and drained. Using a razor blade, the stalk was reduced to 1 cm and the sepal whorls hand-removed. Pods with wounds and bruises were discarded and the rest were measured (pod length, diameter using a Mitutoyo digital caliper) and weighed.

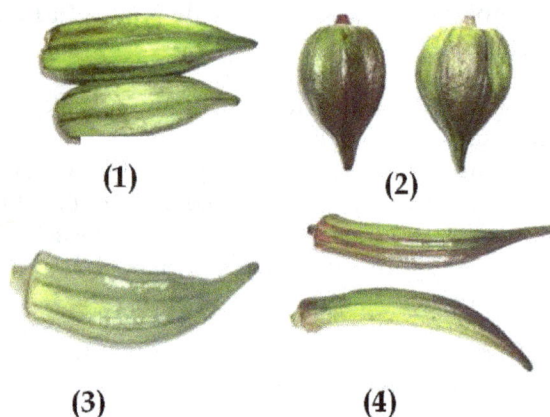

Figure 1. *Morphology of the various okra varieties studied: varieties 1, 2, 3 and 4.*

Table 1. *Physical characterization of okra varieties. Values are means ± standard deviations of measurements from 50 fruits.*

Variety	Length (cm)	Diameter (cm)	Weight (g)	Moisture (w-b) (%)
1	8.0 ± 0.5	3.5 ± 0.2	29.0 ± 2.0	89.2
2	8.0 ± 0.3	3.0 ± 0.2	18.0 ± 1.0	91.2
3	6.5 ± 0.3	1.6 ± 0.1	10.0 ± 0.5	90.2
4	12.0 ± 2.1	1.5 ± 0.1	17.5 ± 1.1	88.2

Fresh okra when stored under ambient conditions (22 - 25°C and 75 - 80% relative humidity) exchanged matter with the atmosphere, resulting in about 9% and 16% weight loss after 12 hrs and 24 hrs of storage, respectively. No appreciable differences were observed following storage for less than 3 hrs. Hence, the duration between harvesting and treatment of fruits was reduced to a maximum of 3 hrs.

Other materials used to prepare the osmotic solution included AnalaR grade ascorbic acid and citric acid, as well as food grade commercial sodium chloride (common salt) and calcium montmorillonite clay locally known as "*Kanwa*".

2.2. Evaluation of the Water Absorption Capacity of Fresh Okra Fruits

The evaluation of the water absorption capacity of okra fruits was conducted in hot water (85°C). This involved immersing the fruits into the hot water, and allowing the whole to cool to room temperature (24°C) over a 48 hrs period. Okras cut into slices (~ 0.5 cm) and uncut okras were studied. The water gain (WG) was calculated as shown in section 2.5. Each experiment was run in triplicate.

2.3. Pre-Treatment of Okra Fruits

The cleaned okra fruits were subjected to either drying without any pre-treatment, or blanching in hot water prior to drying, or subjected to a combined dewatering-impregnation-soaking process/blanching in salt solution (30g/100g) before drying. Each experiment was run in triplicate.

Temperature range of simple blanching in the literature is between 65°C to 100°C with process time varying between 15 sec to 45 minutes [27, 30]. Based on the literature results, our experiment was carried out with three blanching schedules in order to compare their effectiveness: 100°C for 5 s, 100°C for 10 s, and 90°C for 1 min. The blanching consisted of dipping the fresh fruits into hot water (with a mass ratio fruit to water of 1:8) at a predefined temperature and duration, to deactivate and/or destroy all enzymes present while modifying the physical and chemical properties of the fruits. For a given blanching temperature, the fruits were dipped into the hot water using plastic sieve spoons. Once the blanching duration had been exhausted, the spoons were retrieved and the pods were gently rubbed with absorbent tissue, weighed and dried.

The combined process (DISP/blanching) used has been described in detail elsewhere [29]. It consists of immersing the okra fruits into a warm (85°C) osmotic solution (pH 6.5) and the whole allowed to cool to room temperature (~ 24°C). The behaviour of uncut and cut okras (slices) was studied. The osmotic solution was composed of water / common salt / ascorbic acid / citric acid / "kanwa" in ratio (w/w) 100:30:1:1:5. The choice of the concentrated solution was based on literature review and on the popular okra diets which are usually salted [3, 21]. The solution was regularly stirred (every hour) until the end of the experiment. Then, the pods were removed from solution, rubbed gently with absorbent tissue and promptly introduced into the convective dryer. The ratio fruit to solution of 1:8 (w/w) was maintained throughout the DISP.

2.4. Drying of Okra Samples

The pre-treated okra pods were placed on a grilled nylon mesh and introduced into a convective dryer at 40°C, with no-recycling of the hot air. This drying temperature has been shown to be suitable for okra and some leafy vegetable [27, 37, 38]. The relative humidity of the air was estimated using wet and dry bulb thermometers.

During the drying process and on a regular basis, okra samples were retrieved from the oven, rapidly weighed and then returned into the dryer. The drying process ended at a time supposed to be enough to make comparisons between the different processes investigated. Each experiment was run in triplicate.

2.5. Calculations

Water loss, weight reduction, solute gain and water gain during soaking processes relatively to product initial mass were calculated using the following expressions [29]:

$$WL(t) = \frac{M_w(0) - M_w(t)}{M_p(0)} \qquad (1)$$

$$WR(t) = \frac{M_p(0) - M_p(t)}{M_p(0)} \qquad (2)$$

$$SG(t) = \frac{M_p(t) \cdot MS(t) - M_p(0) \cdot MS(0)}{M_p(0)} \qquad (3)$$

$$WG(t) = \frac{M_w(t) - M_w(0)}{M_p(0)} = -WL(t) \qquad (4)$$

We also have:

$$WL(t) = WR(t) + SG(t) \qquad (5)$$

With:

t : Time (sec)

$WL(t)$: Water loss at time t, relatively to product initial mass (g/g)

$M_w(t)$: Mass of water at time t (g)

$M_p(t)$: Mass of product at time t (g)

$WR(t)$: Weight Reduction at time t, relatively to product initial mass (g/g)

$SG(t)$: Solute Gain at time t, relatively to product initial mass (g/g)

$WG(t)$: Water gain at time t, relatively to product initial mass (g/g)

$MS(t)$: Dry matter content (wet basis) at time t (g/g), determined using a standard technique [39]

$X(t)$ is the product moisture content at time t (g/100g d-b).

When water loss or weight reduction is negative, it means that there is respectively a water gain or a gain in weight rather than a loss.

Changes in the volume of okra fruit during drying were equally monitored. The volume of each fruit (fresh or dried) was estimated by submerging the fruit in water contained in a measuring cylinder. The difference between the initial volume (V_i) and the final volume (V_f) gave the shrinkage in volume. The shrinkage ratio was then calculated as follows:

$$Shrinkage(\%) = \left[\frac{(V_i - V_f)}{V_i} \right] \cdot 100 \qquad (6)$$

3. Results and Discussion

3.1. The Behaviour of Okras During Soaking in Solutions

3.1.1. The Behaviour of Okras in Hot Water

The kinetics of water gain during the immersion process, in water at 85°C and allowed to cool till room temperature are presented in Figure 2. This figure shows that the okra absorbs water with a concomitant increase of its mass. The absorption capacity increases progressively from the onset of immersion to a maximum after about 34 hrs for whole okra fruits. The maximum absorption capacities of whole okra are presented in table 2. Okra varieties present different water absorbing capacities, probably due to differences in their specific surface areas (area per unit weight).

Among the four okra varieties tested, variety 2 in its whole form is the one presenting the lowest maximum water absorption capacity. Therefore, the study of the behaviour of sliced forms was done using variety 2 only. When cut into slices, the fruits absorb water much more rapidly, with an initial rate (0.11 kg/kg/s) that is about five times the one obtained with the whole form. The maximum absorption capacity for the sliced okra is reached in just 4 hrs compared to the 34 hrs for the whole okra fruits. For variety 2, the corresponding maximum water gains are respectively 24.8% and 105.2% (Table 2). These peaks should coincide with their respective saturation points when the plant tissues could no longer contain absorbed water. These results could be linked to the surface structures and porosities of these materials.

This high water absorbing capacity of okra could therefore justify its classification as a thickener as proposed by [8].

Table 2. Maximum absorption capacity (in %) of whole and sliced okras.

	Whole			Slices
Variety 1	Variety 2	Variety 3	Variety 4	Variety 2
30%	24,8%	38,0%	29,2%	105,2%

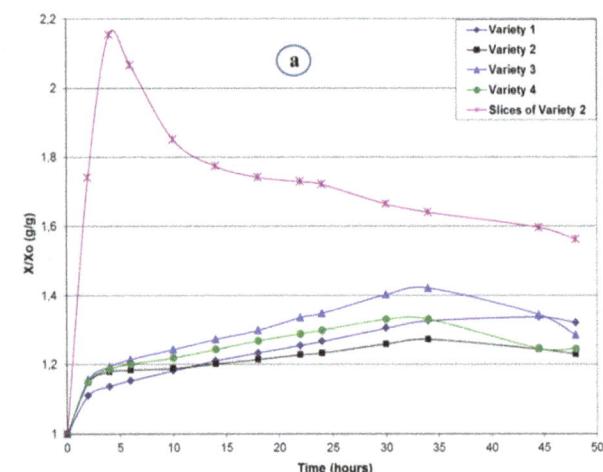

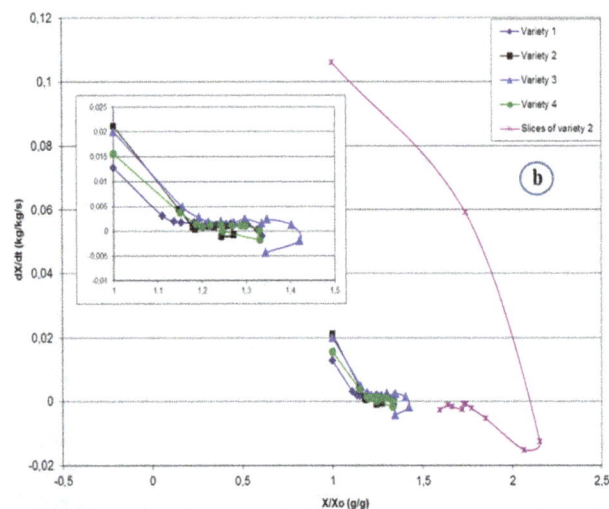

Figure 2. Kinetics of water gain (a) and Water absorption rate (b) of okras during soaking in simple hot water.

After the peaks, the okra fruits progressively lost weight, a phenomenon that was more pronounced for the sliced okra. This could be considered as due to mucilage (gum) dissociation and diffusion into solution. These molecules should mostly involve water-soluble polysaccharides like rhamnose and galacturonic acid often reported to be associated with proteins [2, 9]. From the peak value to the end of soaking (after 48 hrs), material losses for the sliced and whole fruits of variety 2 were respectively 26.3% and 3.1%, relatively to the peak values. Over this time slot, figures 2 (a) and (b) are drawn considering constant dry mass of samples.

3.1.2. The Behaviour of Okras in Concentrated Solutions

Trials to understand okra behaviour in concentrated solutions were carried out with the variety 2 fruits. Okras cut into slices (~ 0.5 cm) and uncut okras were studied. The results collected after 12 hrs and after 24 hrs soaking in concentrated solutions (85°C to ambient temperature) are presented on figure 3. They are compared to the results obtained for the same times while soaking fruits of the same variety in simple hot water (85°C to ambient temperature).

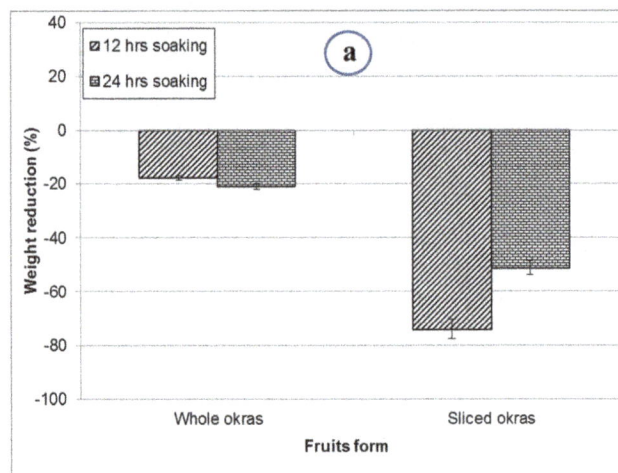

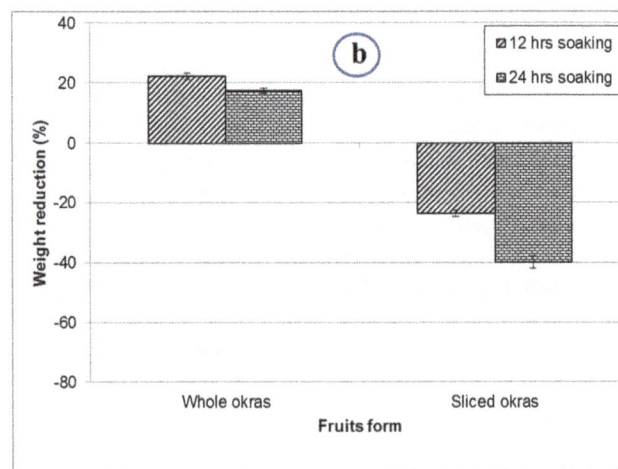

Figure 3. Weight reduction (in %) of okra (variety 2) soaked in different solutions dropping from 85°C to ambient temperature: simple hot water (a) and NaCl solution (30g/100g) (b).

Figure 3 shows that, in sodium chloride concentrated

solution, okra undergoes dehydration only in its entire form. The weight reductions registered are then 22.1% and 17.3% at the end of 12 and 24 hrs soaking respectively. This implies that between 12 and 24 hrs of soaking, the solute gain became more significant than the water loss.

In a salt solution of 30g/100g of water, okras cut into slices (~ 0.5 cm) absorb water as when soaked in simple hot water. In a salt solution, the weight gain is 23.8% at the end of 12 hrs and 40.0% in average at the end of 24 hrs. A test adding to the concentrated solution "*kanwa*" or ascorbic acid at 5g/100g and 1g/100g concentration respectively didn't significantly affect the weight gains during soaking. These results show that sliced okras whether in concentrated solutions or not, undergo a mass increase. This phenomenon could result from the fact that, damaged plant tissues release into solution their mucilage contents, whose interaction with water get it inflated in proportion to the water it is able to absorb. These results are opposite to those obtained by [35] working on osmotic dehydration of sliced okra in sucrose solution. The differences obtained could be explained by the absence of blanching procedure in their method, the variety of okra, the type of the dipping solution or the operating conditions.

Using okra in its cut form equally presents the inconvenience of seeds found in solution. Many authors have shown the potential of okra seed as rich sources of edible fat and protein [12, 13]. Therefore, the treatment of okra in its sliced form reduces its nutritive value if all its elements are not collected at the end of the treatment.

Taking into account these results, further treatments on okra were exclusively conducted using okra in its whole form.

3.2. Convective Drying

Drying was conducted at 40°C either directly on okra pods without treatment or after two different pre-treatments: simple blanching and the combined DISP/Blanching process. The ambient air temperature was 24°C in average and the relative humidity was between 75 and 80%.

3.2.1. The Behaviour of Okra Pods During Simple Drying

Okra pods of the four varieties, freshly harvested, were introduced into the convective dryer. The drying kinetics results obtained are presented on figure 4. It appears that the four okra varieties behave differently during simple drying. The obtained mass transfer rate of variety 3 is greater than those of the three other varieties. It is then followed by variety 2, variety 4 and finally variety 1. The differences observed are largely dependent on the size of the fruits tested and on the okra's variety. Variety 3 has the smallest size of fruits tested. It is therefore advisable not to dry fruits of different varieties or sizes in the same drier to avoid heterogeneity on treated products. Moreover, from global perspectives, drying untreated okras is time demanding, more than 4 days of constant drying, respecting our drying conditions.

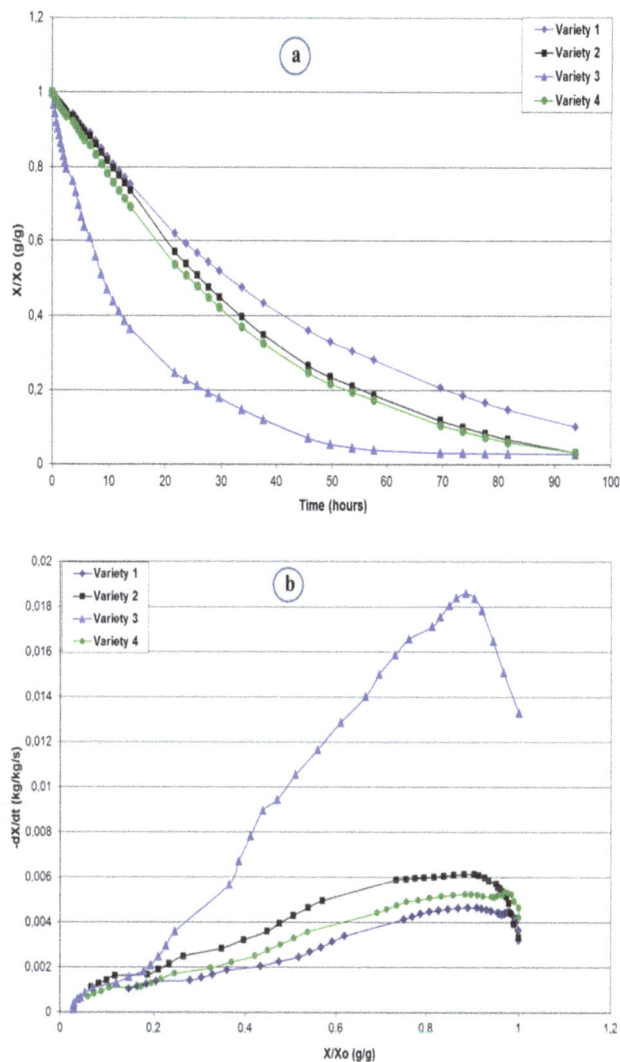

Figure 4. Influence of okra variety on transfer kinetics (a) and on transfer rate (b) during drying at 40°C.

From visual inspection after 93 hrs drying, the colours of the dried okras were not up to our expectations and mindful of the fact that some pre-treatments play a genuine role in ameliorating these colorations, it was then necessary to proceed with pre-treatments which could bring a change.

To rectify these unfavourable economic factors, a pre-treatment which consists in blanching our okra pods was carried out.

3.2.2. Drying of Blanched Okras

(i). Influence of Blanching Schedules

Experiments were carried out using variety 2, with three blanching schedules: 100°C for 5 s, 100°C for 10 s, and 90°C for 1 min. Figure 5 presents the drying kinetics obtained.

A measurement of shrinkage, which is their reductions in volume, was also carried out after drying. These results are reported on figure 6 and show that the greatest value of shrinkage is obtained when okras are blanched for 1min at 90°C.

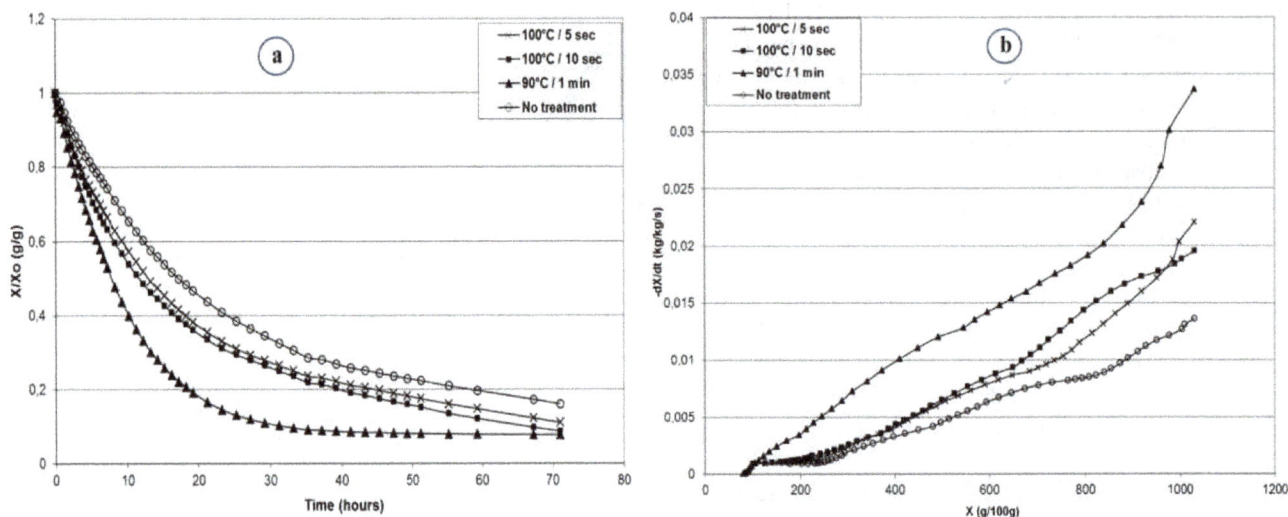

Figure 5. *Influence of blanching schedules on transfer kinetics (a) and on transfer rate (b).*

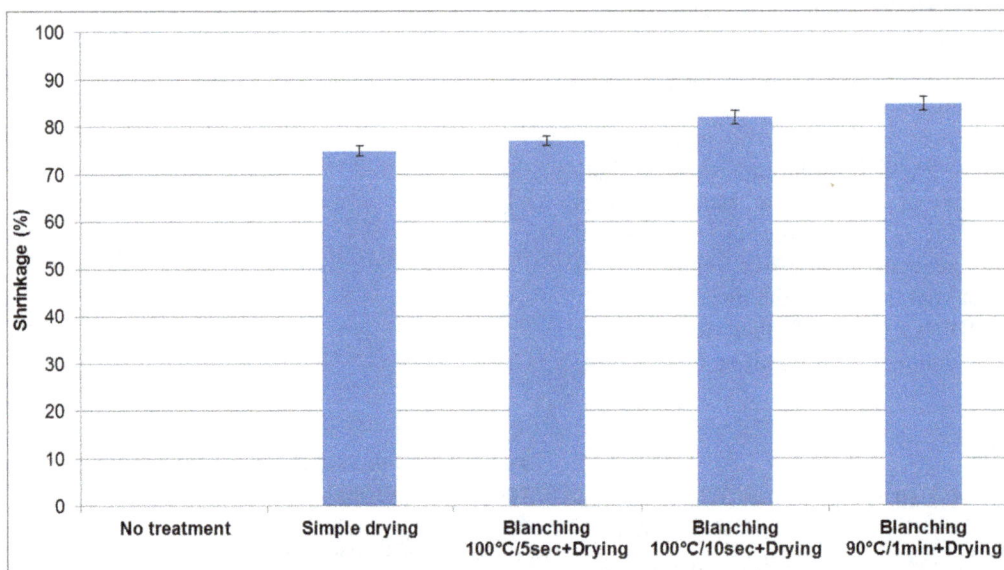

Figure 6. *Shrinkage phenomena of blanched okras compared to the one obtained with simple dried okras.*

On a general note, we remarked that blanched okras dried faster than non-treated okras. Thus, blanching treatments accelerate the drying of the humid products, in the sense that they reduce the duration of drying. Moreover, increasing the time for blanching has the impact of reducing the duration of drying. The drying of blanched okras follows a trend of the order 1 min >> 10 s > 5 s > non-treated.

After 40 hrs drying, okras blanched for 1 min at 90°C were already stable whereas, untreated okras required more than 30 additional hours for a complete drying under the same conditions.

For this variety 2 okra, the colour changed from its green-reddish into a pronounced brown colour, which appeared a few minutes after blanching and which eventually turned dark brown to black. This behaviour reoccurred for other trials and even when "*kanwa*" or ascorbic acid was used in pre-treating solution, implying that, no matter the blanching schedule used, the same results were obtained. Chlorophyll was always

destroyed. Some authors [7] related that chlorophyll degradation in plant tissues with loss of green colour can be linked to the formation of brown pheophytin pigment. Furthermore, other authors [27] reveal contradictory reports on the effect of blanching on vegetable colour. Their works have shown that loss of chlorophyll in leafy vegetables during blanching necessitates a proper combination of time and temperature for blanching. Moreover, according to some [31], blanching of vegetables though makes green leafy vegetables more palatable and less toxic; however it reduces their antioxidant properties drastically.

Taking these into account, we thought it wise to treat our okras with a blanching schedule which was quite short to avoid nutrient loss and which equally guarantees us a short drying process. Hence, a compromise between the nutritional value and the drying duration of blanched okras led us to choose the blanching schedule for 10s at 100°C as the most adequate schedule for further manipulations. This blanching schedule was adopted to treat the four okra varieties, whose drying kinetics are represented on figures 7a and 7b.

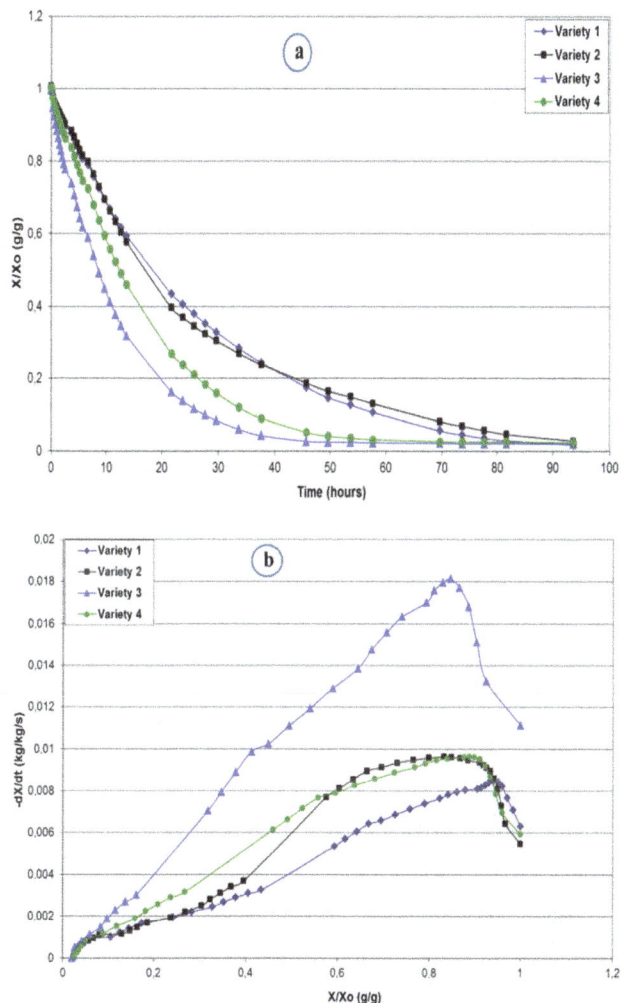

Figure 7. Mass transfer kinetics (a) and transfer rate (b) of blanched (100°C/10s) okra varieties during drying at 40°C.

The blanching procedure reduces drying time by about 20 hrs relatively to non-treated okras. Under the same conditions the drying trend of okra varieties was in the order: variety 3 > variety 4 > variety 2 > variety 1. This result can be linked to okra diameters and explains the radial transfers. The smaller the okra diameter, the higher the transfer rate. Dry mass for blanched okras remained unaffected, implying that blanching was only a propagating factor to water loss.

Furthermore, the observed unattractive colour of variety 2 underwent no colour modifications; however, the colour appearances of the other varieties were much better than that of variety 2.

(ii). Trials to Ameliorate the Colour Properties of Okras

Blanching trials, with or without the presence of simple or mixed solutions of sodium chloride, "*kanwa*", sodium bi-carbonate and citric acid were carried out. The fruit colours were visually appreciated. Among the four okra varieties studied, only variety 2 was reluctant to undergo any colour modifications. No matter the treatment applied, its colour was

always brown after treatments.

For the other three varieties 1, 3 and 4, a solution containing sodium chloride, "*kanwa*", ascorbic acid, gave the best results in terms of colour. The concentrations retained for 100 g of water were 30 g of salt, 5 g of "*kanwa*" and 1 g of ascorbic acid. The pH of the solution was maintained at 6.5 with the help of citric acid. These concentrations were then adopted for use in the combined DISP/Blanching of okras.

3.2.3. Drying of Okras Treated by the Combined DISP/Blanching

A DISP/blanching experiment was conducted using the four okra varieties in their whole forms, soaked in the previously selected DISP solution, having a starting temperature of 85°C falling progressively to room temperature (about 24°C). A summary of the results obtained after DISP/blanching are presented on table 3. The DISP lasted 12 hrs.

Table 3. Combined DISP/Blanching of okras in salt solution (30g/100g at 85°C till room temperature): weight reduction, solute gain, water loss and final moisture content (d-b) obtained after 12 hrs of DISP/Blanching.

Variety	Weight Reduction (% w/w)	Solute Gain (% w/w)	Water Loss (% w/w)	X (g/100g)
1	18.6	8.8	27.4	411.0
2	22.1	7.8	29.9	500.6
3	16.6	9.8	26.4	410.7
4	30.3	9.9	40.2	359.8

Values represent means of triplicate calculations. Standard deviation on each value was less than 5%.

Trials confirmed the dehydration of okra pods in their whole forms. We also noticed a solute intake which could help for conservation means. As shown on table 3, the varieties 1, 2 and 3, were alike in their behaviour and equally presented an average weight reduction of 19.1% as against 30.3% for variety 4. The corresponding average water loss was 27.9% for varieties 1, 2 and 3, versus 40.2% for variety 4. It seems that the ability of dehydration of okra in salt solution depends on its water absorption capacity, i.e. on its polysaccharides content.

The DISP/Blanching was followed by the drying process in the convective dryer at 40°C. Final results are presented on table 4 where values are given in percentage of the initial fresh product for weight reduction and water loss. Differences in tissues structure coupled to differences in okras' initial dimensions could account for the different behaviours observed on the four cultivars.

Table 4. Drying of DISP/Blanched okras at 40°C: weight reduction, solute gain, water loss and final water content (d-b) ratio obtained after 93.7 hrs drying. X_0 is the final X obtained during previous DISP/Blanching (Table 3).

Variety	Weight Reduction (% w/w)	Solute Gain (% w/w)	Water Loss (% w/w)	X/X_0
1	59.7	0	59.7	0.088
2	59.4	0	59.4	0.085
3	59.7	0	59.7	0.110
4	46.0	0	46.0	0.157

Values represent means of triplicate calculations. Standard deviation on each value was less than 5%.

In the course of the soaking process, varieties 1, 3 and 4 retained correctly their colours while variety 2 lost its green colour to light brown. Then, during the drying process it was observed that as drying progressed, the residual salt (NaCl) found on the surface of the pods also underwent drying and produced a whitish colour. This whitish appearance eventually suppressed the visibility of the other colours which were previously present. The dark brown colour of variety 2 fortunately, became masked by the whitish salt colour, giving it a greyish appearance, which in fact modified its colour presentation unlike before.

Table 5. *Contribution (in percentage) of each process to total weight reduction, solute gain and water loss of the combined DISP/blanching and drying process.*

Contribution to	Process	Variety 1	Variety 2	Variety 3	Variety 4	Average
Weight Reduction	DISP	23.8	27.1	21.8	39.7	28.1
	Drying	76.2	72.9	78.2	60.3	71.9
Solute Gain	DISP	100	100	100	100	100
	Drying	0	0	0	0	0
Water Loss	DISP	31.5	33.5	30.7	46.6	35.6
	Drying	68.5	66.5	69.3	53.4	64.4

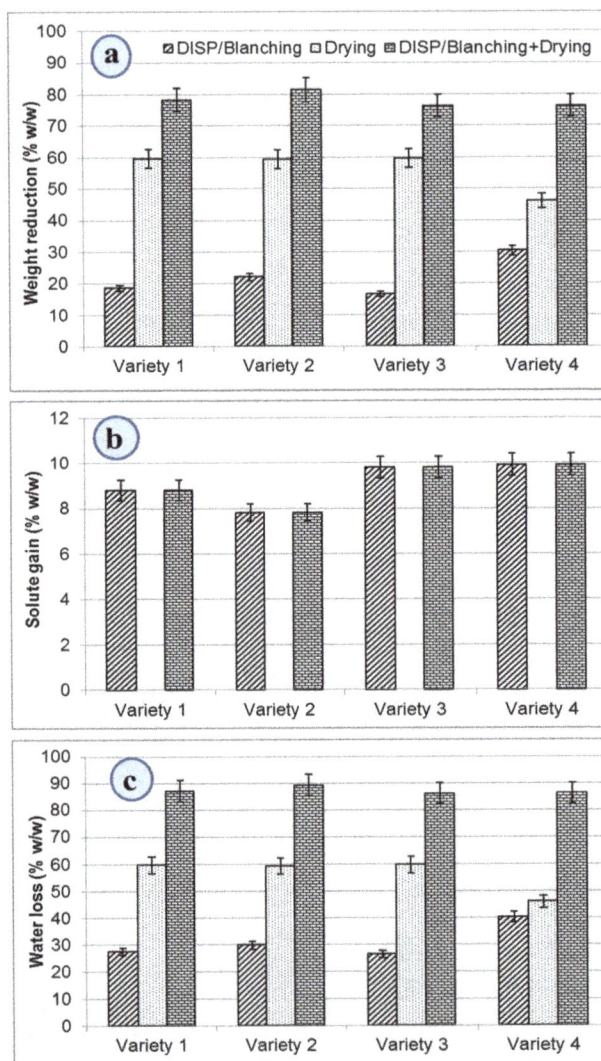

Figure 8. *Okras weight reduction (a), solute gain (b) and water loss (c) during the combined DISP/Blanching and drying process.*

The combined DISP/Blanching and drying of okra varieties gave averages of (78.1 ± 3.6) % weight reduction, (9.1 ± 0.5) % solute gain and (87.2 ± 4.0) % water loss. Details are presented on figures 8a, 8b and 8c. Finally, the DISP/blanching process was responsible for 35.6 % in average of moisture loss over the four okra varieties tested, while drying accounted for a 64.4 % water loss as presented on table 5. These results reflected those of [29], working on tropical fruits. Nevertheless, it can be seen from results presented on table 4 that the moisture contents of the four okra varieties, treated in the whole form, remain high after 93.7 hrs convective drying at 40°C. The final moisture content (w-b) for the tested conditions is respectively 26.6, 29.8, 31.1 and 36.1% for varieties 1, 2, 3 and 4. The air drying conditions (relative humidity and velocity) as well as the load of the dryer could have an important effect on mass transfers as shown in several works [40-42].

3.2.4. The Impacts of Preliminary Okras Treatments on Drying Kinetics

On figure 9 is presented the drying rate of the four okra

varieties, according to the preliminary treatment undergone. This figure shows that mass transfers during the process of drying DISP/Blanched okras is not too significant compared to those of blanched okras. This behaviour and the results

were not surprising because, the okra pods had previously undergone a water loss and a solute intake during the DISP/Blanching.

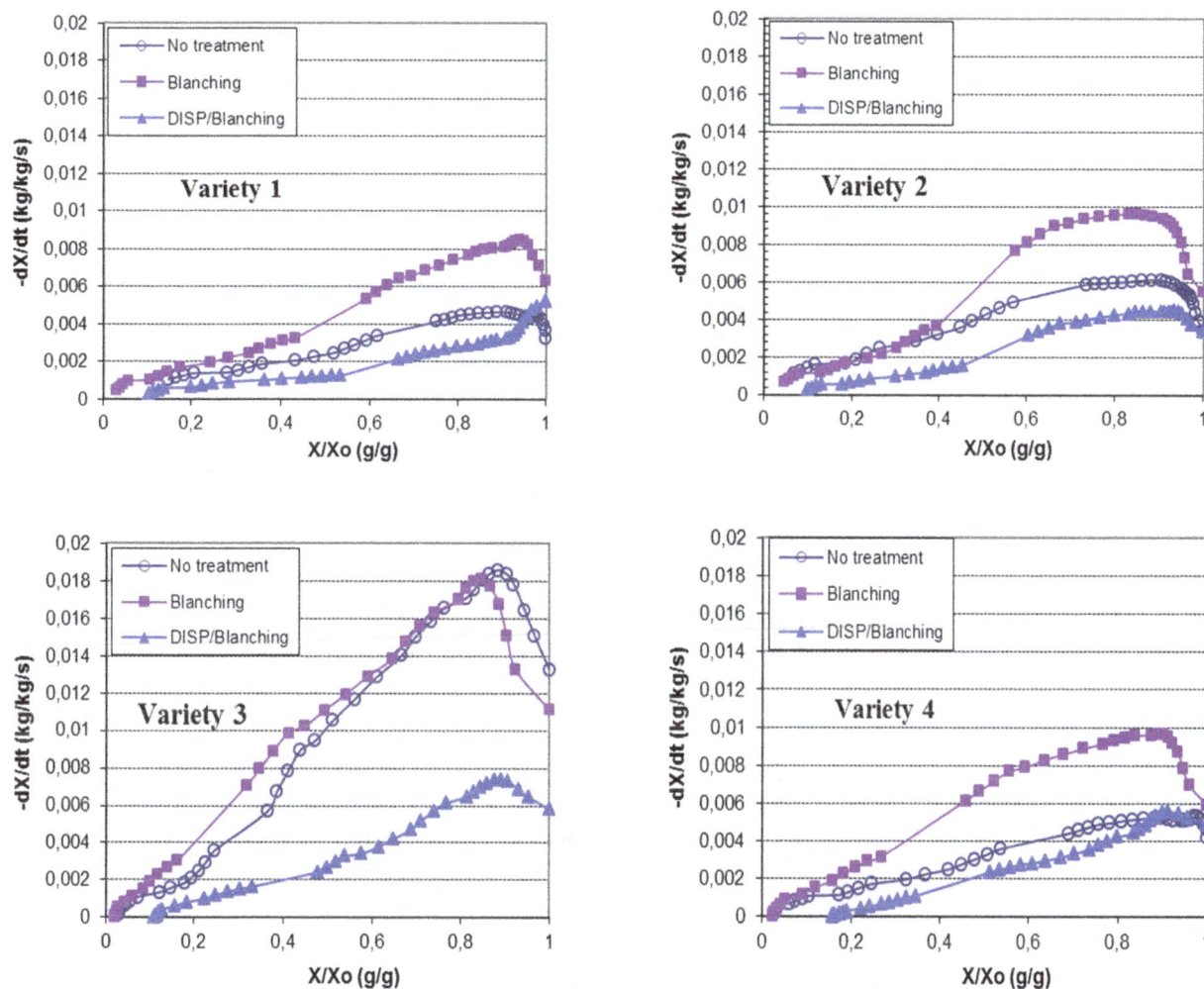

Figure 9. *Drying rate of the four okra varieties: Impacts of the preliminary treatment (blanching or DISP/Blanching) compared to non-treated okras drying kinetics.*

4. Conclusions

The effects of okra pre-treatments on drying kinetics and dried okra final presentation were studied. This work shows that it is possible to improve the dried okra final appearance using the coupled process DISP/Blanching and drying. Results obtained indicate that the only way to register a water loss in the sample during the DISP/Blanching, in a salt solution at 30g/100g with a temperature falling progressively from 85°C to room temperature, is to soak the fruits in their whole forms. The advantage of this process resides on the solute intake (9.1 ± 0.5) % that increases the preservation potentialities of the final product. The drying rates as well as the water loss and solute gains obtained show the importance of taking into account the variety and the size of okra during its treatment by the DISP/blanching and/or the drying processes.

This work is far from being exhaustive. A sensory analysis test on the final product is in project. Nevertheless a study on the possibility of valorising the osmotic dehydration solution and economic aspects should be taken into consideration as well.

Acknowledgement

The authors acknowledge the Process Engineering and Automatic Laboratory (LAGEP) of the University of Lyon 1-France. They also thank The French Embassy in Cameroon for their financial support.

References

[1]   Purseglove J. W, 1987. Tropical crops-dicotyledons. Edition Longman, New York, USA, 368-370.

[2] Sengkhamparn Nipaporn, Verhoef R., Schols Henk A., Sajjaanantakul Tanaboon, Voragen Alphons G. J., 2009. Characterisation of cell wall polysaccharides from okra (Abelmoschus esculentus (L.) Moench). Carbohydrate Research 344, 1824-1832.

[3] Ngbede S. O., Ibekwe H. N., Okpara S. C., Adejumo L., 2014. An Overview of Okra Production, Processing, Marketing, Utilization and Constraints in Ayaragu in Ivo Local Government Area of Ebonyi State, Nigeria. Greener Journal of Agricultural Sciences. 4 (4), 136-143.

[4] Sedat Çalışır, Musa Özcan, Haydar Hacıseferoğulları and M. Uğur Yıldız, 2005. A study on some physico-chemical properties of Turkey okra (Hibiscus esculenta L.) seeds. Journal of Food Engineering, 68 (1), 73-78.

[5] Kumar S., Dagnoko S., Haougui A., Ratnadass A., Pasternak D. and Kouame C., 2010. Okra (Abelmoschus spp.) in West and Central Africa: Potential and progress on its improvement. African Journal of Agricultural Research, 5 (25), 3590-3598.

[6] Stansly P. A., Schuster D. J. and Tong-Xian Liu, 1997. Apparent Parasitism of Bemisia argentifolii (Homoptera: Aleyrodidae) by Aphelinidae (Hymenoptera) on Vegetable Crops and Associated Weeds in South Florida. Biological Control, 9 (1), 49-57.

[7] Adom K. K., Dzogbefia V. P., Ellis W. O. and Simpson B. K., 1996. Solar drying of okra: Effects of selected package materials on storage stability. Food Research International, 29 (7), 589-593.

[8] Ndjouenkeu R., Akingbala J. O. and Oguntimein G. B., 2005. Emulsifying properties of three African food hydrocolloids: okra (Hibiscus esculentus), dika nut (Irvingia gabonensis), and khan (Belschmiedia sp.) Journal of Food Science and Technology. 1-3.

[9] Deters A. M., Lengsfeld C. and Hensel A., 2005. Oligo-and polysaccharides exhibit a structure-dependent bioactivity on human keratinocytes in vitro. Journal of Ethnopharmacology, 102 (3), 391-399.

[10] Tilmon R. W. and Romanchik-Cerpovicz J. E., 2001. Feasibility of using okra exudate as a fat replacer in low fat chocolate dropped cookies. Journal of the American Dietetic Association, 101 (9), A23-A32.

[11] Romanchik-Cerpovicz J. E., Tilmon R. W. and Baldree K. A., 2002. Moisture Retention and Consumer Acceptability of Chocolate Bar Cookies Prepared With Okra Gum as a Fat Ingredient Substitute. Journal of the American Dietetic Association, 102 (9), 1301-1303.

[12] Savello P. A., Martins F. and Hull W., 1980. Nutrient composition of okra seed meals. Journal of Agriculture and Food Chemistry, 28 (6), 1163-1166.

[13] Oyelade O. J., Ade-Omowaye B. I. O. and Adeomi V. F., 2003. Influence of variety on protein, fat contents and some physical characteristics of okra seeds. Journal of Food Engineering, 57 (2), 111-114.

[14] Ijoyah M. O., Atanu S. O. and Ojo S., 2010. Productivity of okra (Abelmoschus esculentus L. Moench) at varying sowing dates in Makurdi, Nigeria. Journal of Applied Biosciences 32, 2015-2019.

[15] Konyeha, S., Alatise, M. O., 2013. Yield and Water Use of Okra (Abelmoschus esculentus L. Moench) under Water Management Strategies in Akure, South-Western City of Nigeria. International Journal of Emerging Technology and Advanced Engineering, 3 (9), 8-12.

[16] Nwaobiala C. U. and Ogbonna M.O., 2014. Adoption Determinants and Profitability Analysis of Okra Farming In Aninri Local Government Area (LGA) of Enugu State, Nigeria. Discourse Journal of Agriculture and Food Sciences. www.resjournals.org/JAFS, 2 (1), 1-10.

[17] Mohammed A. Al-Sulaiman, 2011. Prediction of quality indices during drying of okra pods in a domestic microwave oven using artificial neural network model. African Journal of Agricultural Research Vol. 6(12), pp. 2680-2691.

[18] Olivera Daniela F., Mugridge Alicia, Chaves Alicia R., Mascheroni Rodolfo H. and Viña Sonia Z., 2012. Quality Attributes of Okra (Abelmoschus esculentus L. Moench) Pods as Affected by Cultivar and Fruit Size, Journal of Food Research, 1 (4), 224-235.

[19] Owolarafe O. K., Obayopo S. O., Obayopo S. O., Amarachi O. A., Babatunde O. and Ologunro O.A. 2011. Development and Performance Evaluation of an Okra Drying Machine. Research Journal of Applied Sciences, Engineering and Technology 3 (9), 914-922.

[20] Eke, Ben Akachukwu, 2013. Development of Small Scale Direct Mode Natural Convection Solar Dryer for Tomato, Okra and Carrot. International Journal of Engineering and Technology 3 (2). IJET Publications UK. 199-204.

[21] UNIDO, 2004. "Small-scale fruit and vegetable processing and products. Production methods, Equipment and quality assurance practices". UNIDO Technology Manual, United Nations Industrial Development Organization, Vienna, 106p.

[22] Arise A.K., Arise R.O., Akintola A.A., Idowu O.A. and Aworh O.C., 2012. Microbial, Nutritional and Sensory Evaluation of Traditional Sundried Okra (Orunla) in Selected Markets in South-Western Nigeria. Pakistan Journal of Nutrition 11 (3): 231-236. ISSN 1680-5194. Asian Network for Scientific Information.

[23] Adegbehingbe Kehinde Tope, 2014. Microbial analysis of sun-dried okra samples from some Akoko areas of Ondo state, Nigeria. IMPACT: International Journal of Research in Applied, Natural and Social Sciences (IMPACT: IJRANSS). 2 (5), 87-96.

[24] Shams El-Din M. H. A. and Shouk A. A., 1999. Comparative study between microwave and conventional dehydration of okra. Grasas y Aceites, 50 (6), 454-459.

[25] Aguilera J. M., Chiralt A. and Fito P., 2003. Food dehydration and product structure. Trends in Food Science and Technology, 14 (10), 432-437.

[26] Moreno J., Chiralt A., Escriche I. and Serra J. A., 2000. Effect of blanching/osmotic dehydration combined methods on quality and stability of minimally processed strawberries. Food Research International, 33 (7), 609-616.

[27] Negi P. S. and Roy S. K. 2000. Effect of Blanching and Drying Methods on β-Carotene, Ascorbic acid and Chlorophyll Retention of Leafy Vegetables. Lebensmittel-Wissenschaft und-Technologie, 33 (4), 295-298.

[28] Jiokap Nono Y., Nuadje G. B., Raoult-Wack A.L. and Giroux F., 2001. Comportement de certains fruits tropicaux traités par déshydratation-imprégnation par immersion dans une solution de saccharose. Fruits, (France), 56 (2), 75-83.

[29] Jiokap Nono Y., Reynes M., Zakhia N., Raoult-Wack A. L. and Giroux F., 2002. Mise au point d'un procédé combiné de déshydratation-imprégnation par immersion et séchage de bananes (Musa acuminata groupe cavendish). Journal of Food Engineering, 55, 231-236.

[30] Passo Tsamo C.V., Bilame A.-F., Ndjouenkeu R., Jiokap Nono Y., 2005. Study of material transfer during osmotic dehydration of onion slices (Allium cepa) and tomato fruits (Lycopersicon esculentum), Lebensmittel-Wissenschaft und-Technologie, 38: 495-500.

[31] Ganiyu Oboh, 2005. Effect of blanching on the antioxidant properties of some tropical green leafy vegetables. LWT - Food Science and Technology, 38 (5), 513-517.

[32] Sotiriadis P. K. and Hoskins F. H., 1982. Vitamin retention during storage of processed foods. I. Effect of ascorbic acid on folates in cowpeas, okra and tomatoes. Scientia Horticulturae, 16 (2), 125-130.

[33] Aderiye B. I., 1985. Effects of ascorbic acid and pre-packaging on shelf-life and quality of raw and cooked okra (Hibiscus esculentus). Food Chemistry, 16 (1), 69-77.

[34] Chen J. P., Tai C. Y. and Chen B. H., 2006. Effects of different drying treatments on the stability of carotenoids in Taiwanese mango (Mangifera indica L.). Food Chemistry, 100, 1005-1010.

[35] Agarry S. E. and Owabor C. N., 2012. Statistical optimization of process variables for osmotic dehydration of okra (Abelmoschus esculentus) in sucrose solution. Nigerian Journal of Technology (NIJOTECH), 31 (3), 370-382.

[36] Olaniyan A. M. and Omoleyomi B. D., 2013. Characteristics of Okra under Different Process Pretreatments and Different Drying Conditions. Food Processing & Technology, J Food Process Technol, 4:6. http://dx.doi.org/10.4172/2157-7110.1000237.

[37] Wankhade P. K., Sapkal R. S. and Sapkal V. S., 2012. Drying Characteristics of Okra Slices using Different Drying Methods by Comparative Evaluation. Proceedings of the World Congress on Engineering and Computer Science, Vol II, WCECS 2012, October 24-26, San Francisco, USA.

[38] Famurewa J. A. V. and Olumofin K. M., 2015. Drying kinetics and influence on the chemical characteristics of dehydrated okra (Abelmoschus esculentus) using cabinet dryer. European Journal of Engineering and Technology, 3 (2), Progressive Academic Publishing, UK. www.idpublications.org. 7-19.

[39] AOAC (1980). Official methods of analysis, 13th ed., Association of Official Analytical Chemists, Washington D.C.

[40] Belghit A., Kouhila M., Boutaleb B. C., 1999. Experimental study of drying kinetics of sage in a drying tunnel working in forced convection. Rev. Energ. Ren. 2, 17-26.

[41] Tzempelikos D. A., Vouros A. P., Bardakas A. V., Filios A. E., Margaris D.P., 2014. Case Studies in Thermal Engineering, 3, 79-85.

[42] Taheri-Garavand A., Rafiee S. and Keyhani A., 2015. Effect of temperature, relative humidity and air velocity on drying kinetics and drying rate of basil leaves. Electronic Journal of Environmental, Agricultural and Food Chemistry, 10 (4), 2075-2080.

2

Real Time System Based Monitoring of Turbine Parameters and Protection System in Thermal Power Plant

M. Surekha[1], N. Suthanthira Vanitha[2], K. Yadhari[3]

[1]Student, Embedded System Technologies, Knowledge Institute of Technology, Salem, India
[2]Head of the Department, Department of Electrical & Electronics Engineering, Knowledge Institute of Technology, Salem, India
[3]Assistant Professor, Department of Electrical & Electronics Engineering, Knowledge Institute of Technology, Salem, India

Email address:
surekha.ece14@gmail.com (M. Surekha), hod.eee@kiot.ac.in (N. Suthanthira Vanitha), kyeee@kiot.ac.in (K. Yadhari)

Abstract: Currently in thermal power plant the turbine system parameters like temperature, speed, lubrication oil level and vibrations can be monitored by using the MATLAB. If any problem occurs in the turbine parameters, that can be controlled manually and the protection system is based on the relay mechanism which causes failure in the action of switch. In the proposed system, the drawbacks are eliminated with the support of high speed embedded processor. Real Time processor can be easily altered according to the ports and number of devices can also be connected. The parameters in the turbine system can be measured, monitored and controlled automatically. The parameters variations can be graphically represented by using the LabVIEW.

Keywords: ARM8Processor, MATLAB, LabVIEW, Automation System and Sensors

1. Introduction

Normally, the turbine system is a rotary mechanical device that extracts energy from fluid flow and it generates the power. If the turbine parameters move to an abnormal condition, it will cause a very high damage. In this paper, it mainly focuses on monitoring and controlling of the turbine system parameters. In the existing system, all the parameters in the turbine systemcan be monitored by using the MATLAB and it can be done by using the manual process. In order to overcome the drawbacks in the existing system the proposed system is employed.

The advanced ARM8 processor is used which is the heart of the system that controls all sub devices connected across it. The advanced processor is a flash type reprogrammable memory which has some peripheral devices to play this system as efficient. For the monitoring and controlling the parameters, temperature sensor, speed sensor, level sensor and vibration sensor is used to sense the temperature, speed, lubrication oil level and vibrations in the turbine system respectively. By using the embedded processor these four parameters can be monitored and it can be controlled automatically and graphically represented by using LabVIEW.

2. Literature Survey

Krishna Prasad Dasari, Dr.A.M.Prasad[1] described the main aim of the research paper is to evaluate the method of environmental impact of power plant discharge by reducing the temperature difference between effluent and costal water and flow control. Water temperature control and flow control measurement have been designed in advance technology of industrial control area for thermal discharge model test. Digital temperature sensors, level sensors, Flow meters, different modulated circuits, dedicated interface are used in the test and controlling of the system is adopted in software designing and programming. Measurement procedure, data processing and controlling are done by Proportional – Integral – Derivative (PID) controller. This technology can be implemented where the thermal effluent are discharged in coastal areas.

ShiyamSundar[2] described the paper deals with the Laboratory Virtual Instrumentation Engineering Workbench (LabVIEW) are widely used in industry for supervisory control and data acquisition of industrial processes. This paper described simulation of system for laboratory based thermal power plant generator setup using LabVIEW data logging and supervisory control (DSC) module. Using the input, output and functional parameters the simulation of generator unit and alarm handling technique is achieved.

Peng Guo and David infield[3] describedthe condition monitoring can greatly reduce the maintenance cost for a wind turbine. In this paper, a new condition- monitoring method based on the nonlinear state estimate technique for a wind turbine generator is proposed. The technique is used to construct the normal behavior model of the electrical generator temperature. Generator incipient failure is indicated when the residuals between model estimates and the measured generator temperature become significant. Moving window averaging is used to detect statistically significant changes of the residual mean value and standard deviation in an effective manner; when these parameters exceed predefined thresholds, an incipient failure is flagged. It is demonstrated that the technique can identify dangerous generator over temperature before damage has occurred that results in complete shutdown of the turbine.

Mohamed Zahran and Ali Yousef[4] described the purpose of the work is to investigate a monitoring of autonomous photovoltaic battery wind turbine hybrid system (PVBWHS). In this paper an intelligent graphical user interface (GUI) is built in sub-menus for system characterization. The PVBWHS modeling, simulation, and performance monitoring are also introduced. A LabVIEW model is designed whereby the hybrid system components are simulated as virtual instruments [VI] interacted with functional blocks. The developed monitoring system measures continuously the available power generated from the solar array and wind turbine, and the functional VI compare this with the actual load demand on real time estimates the storage battery operation mode.

Alberto Borghetti and MauroBosetti[5] described the paper presents a procedure for parameter identification along with its application to the model of a combined cycle power plant that includes the surrounding electrical network, built for the analysis of islanding manoeuvres transients. The paper illustrates both the power system computer model, implemented within the EMTP-RV simulation environment, and the developed identification procedure based on the interface between the developed model and MATLAB. The parameter identification procedure is applied to experimental transient recordings that make reference to a similar power plant.

3. Existing System

In the existing system, the turbine parameters like temperature, speed, lubrication oil level and vibrations in the thermal power plant can be monitored and controlled by using the MATLAB. It is a sophisticated data structures, contains built-in editing and debugging tools and supports object-oriented programming. All these parameter can be amplified and controlled the output that match with the input signal. If any problem occurs in the turbine parameters, it can

be monitored and controlled by using the manual process. The protection system is based on the relay mechanism. The relays are connected with the pressure switches and it has lot of drawbacks such as occurrence of drift in the switch, failure in the action of the switch. There is no waveform representation and automation process was available in the existing system.

4. Proposed System

In order to overcome the problem in the existing system, the conventional method of the proposed system is employed. The proposed system is to monitor and control the turbine parameters by using the ARM8 processor. The advanced processor operates at very low voltage and it consumes less power and it can be easily altered and number of devices can be connected according to the input and output ports. The embedded processor is used for real time monitoring of data. The parameters in the turbine system can be monitored and controlled by using the advanced processor.For the monitoring of temperature, speed, vibration and lubrication oil level, the temperature sensor, speed sensor, vibration sensor and level sensor can be used. All these parameter output is given to the amplifier unit. The output of the amplifier is then given to the advanced processor. PC is connected to the processor via RS 232 and it is a serial communication cable. So the parameters can be monitored and controlled by using the embedded processor.Whenever lubrication oil level becomes low, automatically it activates the relay to turn on the DC pump. Speed can be set as constant of 3000 RPM and according to the speed, the vibration can be controlled. Temperature is controlled with the help of cooling fan. By using the advanced ARM8 processor all these four parameters are monitored and it can be controlled automatically. The parameters variations can be graphically represented by using LabVIEW.

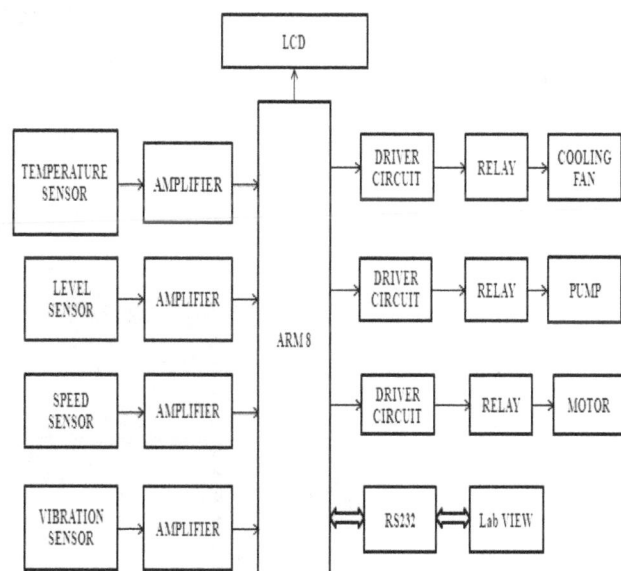

Fig. 1. Block Diagram of Proposed System.

The system mainly operated by ARM8 processor. This paper consists of level sensor, temperature sensor, speed sensor and vibration sensor and they are used to measure the various parameters. The sensors can be connected to the amplifier which amplifies the signal.In the advanced processor, it converts the analog to digital signal through inbuilt ADC. It can be given to the driver circuit and it is used to convert the signal in the required form of processor.The signals can be connected to the relay and it is an electrically operated switch and current flowing through the coil of the relay creates a magnetic field which attracts a lever and changes the switch contacts.So by using the relay, the parameters can be controlled. If the temperature can be raised to higher level, it can be reduced by using the cooling fan. When the oil level is decreased to lower level, it can take oil automatically from the tank. When the speed increases, the turbine will off automatically. According to the speed, the vibration of turbine can be maintained. These parameters are implemented by using the advanced processor. It can be graphically represented by using the LabVIEW.

ARM8

ARM8 processor is used to monitor and control of the parameters automatically. According to the input and output ports, the devices can be connected to the processor. The advanced processor consists of the 64 pins. By using the pin diagram the sensors can be connected and control the parameters. The port1 is connected to the temperature, speed, vibrations and lubrication oil level sensor. The port0 is connected to the cooling fan, pump, and motor. According to the connections, the operations can be performed. Power supply is connected to the pin 10 and 11. The parameters at normal and abnormal conditions can be displayed by using the LCD.

5. LabVIEW

All the parameter variations like temperature, speed, lubrication oil level and vibrations can be done by using the comparator. We have to set the limit value for these parameters. If the parameters can be raised above to that particular limit value, it can be represented as high in the front panel. Otherwise, it can be mentioned as normal.If the speed can be raised above 3000 RPM, it indicates the abnormal condition. It also indicates the speed variations by using the graphical method.

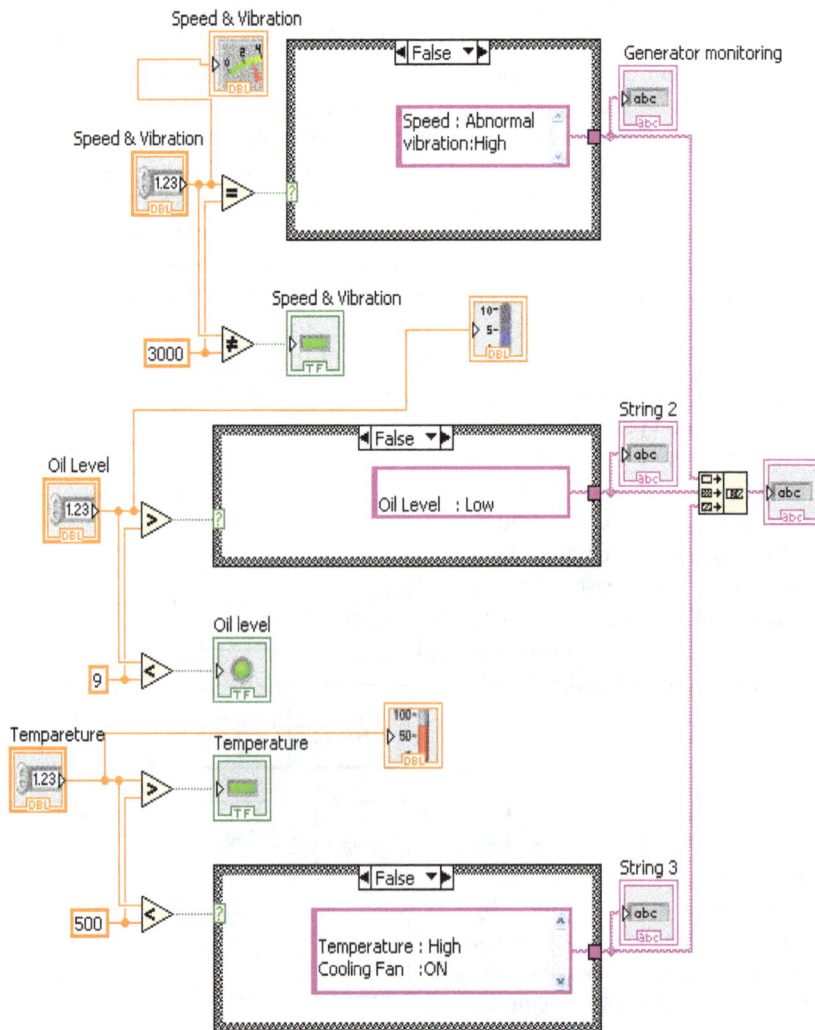

Fig. 2. Proposed Block Diagram of LabVIEW.

If the speed is maintained as 3000 RPM, then it shows as normal condition. So according to the speed, the vibrations can be varied. If the speed can be raised to higher level, vibrations also can be raised. If the speed can be maintained as constant, vibrations will be normal in the turbine system. So we have to maintain the speed as constant. The lubrication oil level can be represented by using the tank as graphical method. If the lubrication oil level can be decreased to lower level, it can be mentioned as low and also the tank level will be represented as lower level. If the lubrication oil level can be maintained as constant, it can be mentioned as normal and the tank level will be represented as constant level in the front panel. The temperature can be represented by using the thermistor. If the temperature can be raised to higher level, it can be displayed as temperature is high. Then automatically the cooling fan will be ON. These variations can be represented in the graphical of thermistor. If the temperature cannot be raised to higher level, it can be displayed as temperature is normal. Then automatically the cooling fan will be OFF. All these conditions of the parameters can be graphically monitored and controlled by using LabVIEW.

6. Flow Chart

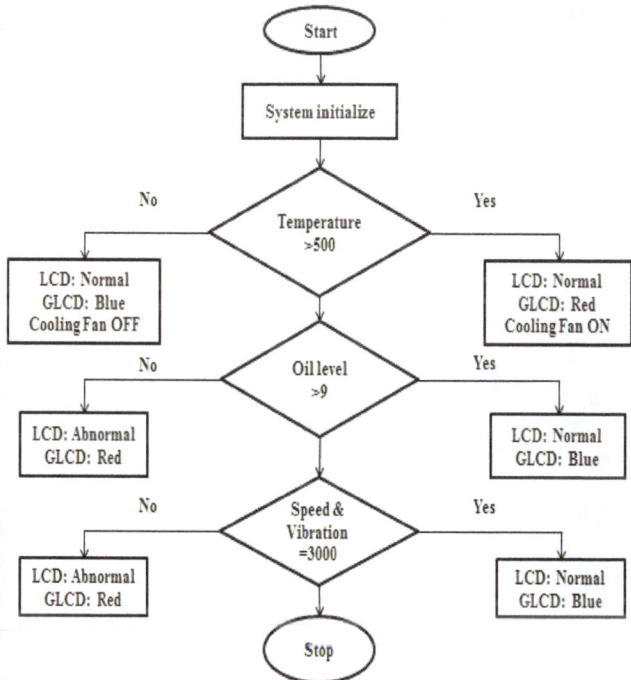

Fig. 3. *Flow Chart of Proposed System.*

Algorithm
Step 1: Start the process.
Step 2: Initialize the system.
Step 3: Representation of the temperature and according to the temperature it indicates the normal and abnormal conditions.
Step 4: The cooling fan will be automatically activated when the temperature moves to higher level.

Step 5: Representation of the speed, vibrations and lubrication oil level according to that the normal and abnormal conditions will be displayed.
Step 6: Stop the process.

7. Results

7.1. Hardware Results

It represents the monitoring and control of the parameters in the turbine system by using the ARM processor. In the ARM8 processor, the control panel consist of four options are temperature sensor, speed sensor, level sensor and vibration sensor. When selecting the option 1, it represents the temperature. If the temperature can be raised to higher level, automatically the cooling fan will be moved to ON condition. When selecting the option 2, it represents the speed. If the speed can be raised to higher level, it automatically represents as abnormal condition and it will reduce the speed. When selecting the option 3, it represents the level of lubrication oil. If the oil level decreases to lower level, it automatically represents as abnormal condition. When selecting the option 4, it represents the vibration level. If the vibration is in normal condition, it represents as normal.

Fig. 4. *ARM8 Processor Output.*

7.2. Software Results

It indicates the parameters are in normal or abnormal condition by LED and graphical representation by using LabVIEW. In the fig.5, it represents the normal condition of the parameters. The speed can be maintained as 3000, according to that vibration also maintained. It indicates the oil level as 10 and temperature as 400. It can be graphically represented and displayed the parameters is in normal condition. In the fig.6, it represents the abnormal condition of

parameters. The speed can be exceeding to 5000 and according to that the vibration can be varied. The oil level decreases to low and the temperature can be exceeding to 700. It can be represented by using the graphical method and displayed the parameters is in abnormal condition.

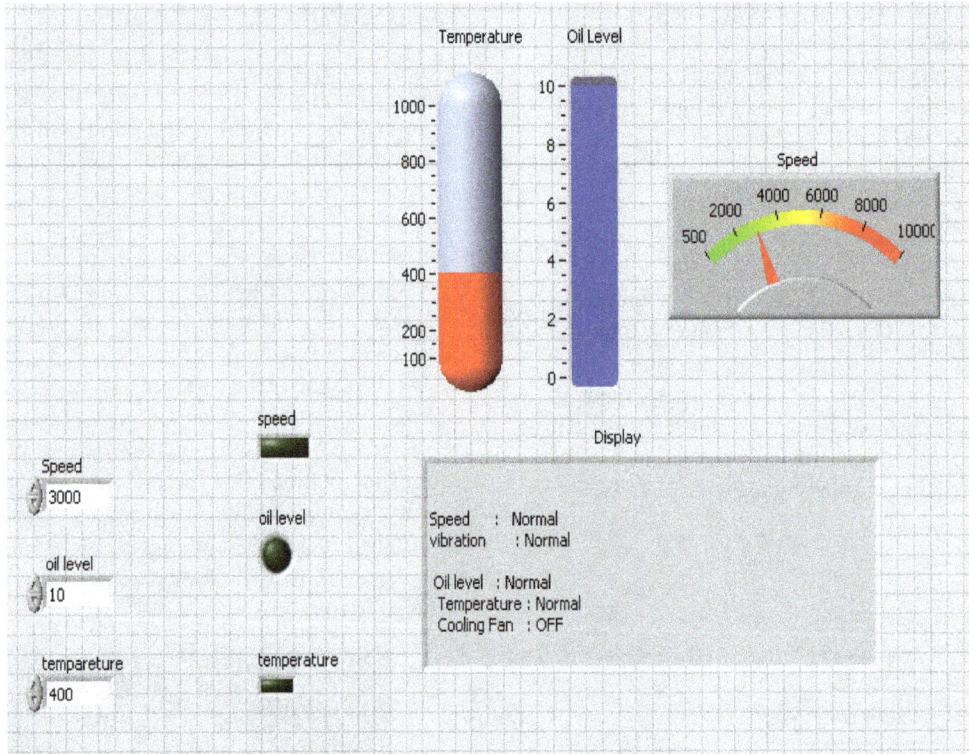

Fig. 5. Normal Condition of Parameters.

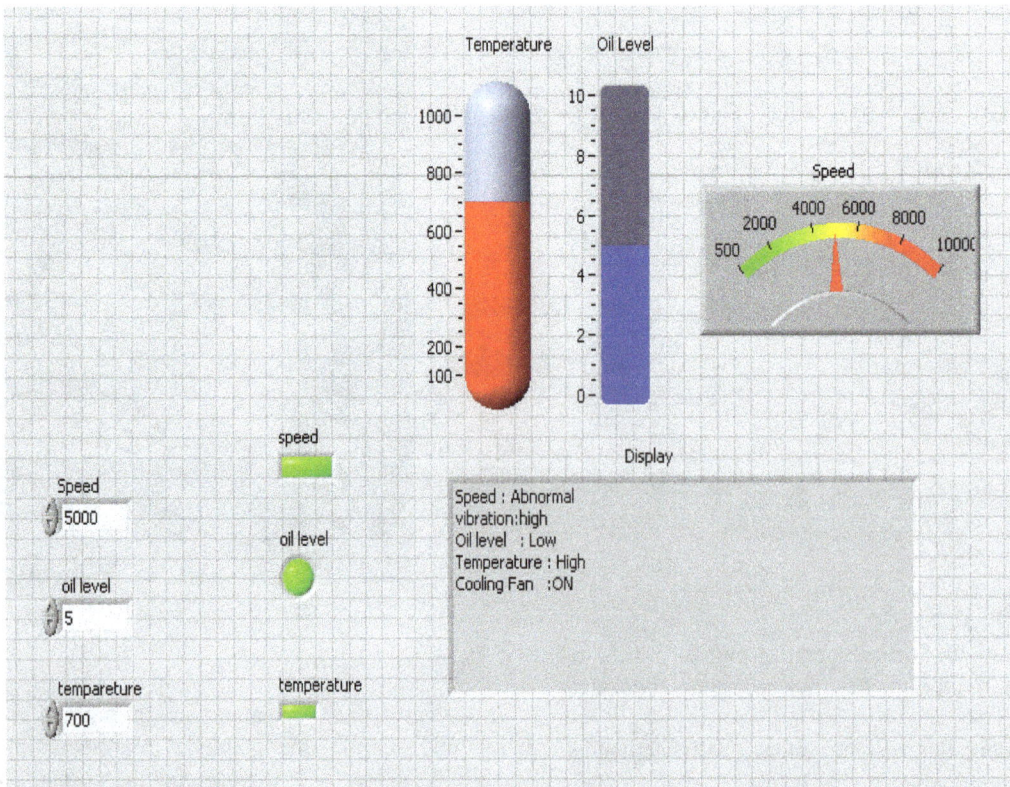

Fig. 6. Abnormal Condition of Parameters.

7.3. Results Comparison

By comparing the hardware and software results, the parameters like temperature, speed, lubrication oil level and vibration can be monitored and controlled automatically. By using the ARM processor of eclipse software, the parameters at normal and abnormal condition can be represented by using the LCD and GLCD. GLCD (Graphical Liquid Crystal Display) is used to represent the variations in the color format.If the parameter is in normal condition, it indicates the blue color in GLCD and the message is displayed in LCD. If the parameter is in abnormal condition, it indicates the red color in GLCD and the message is displayed in LCD. In the LabVIEW, all these parameter variations can be represented by using the graphical method. If the parameters can be raised to higher or lower level, it can be displayed in the front panel. So by comparing both the eclipse software and LabVIEW, LabVIEW is easy to identify the variations of parameters by using the graphical method. By using the eclipse software, it is also very easy to identify the variation of parameters by using the color representation of GLCD.

8. Conclusion

In this project the turbine parameters like temperature, speed, lubrication oil level and vibrations can be monitored and controlled automatically by using ARM8 processor. The parameter variations of normal and abnormal conditions are represented by using the advanced processor and graphically represented by using LabVIEW. The man power is reduced and it reduces the damage that can be occurring in the turbine system. In future, the number of parameters will be added in the system for various applications. The parameters of the turbine system are temperature, speed, lubrication oil level and vibration can be monitored and controlled by using the online system. With the support of the wireless technology like GSM, it helps to prevent the damage occur in the turbine system.

9. Coding

```
#include "macros.h"
#include<ulk.h>
int main(void) PROGRAM_ENTRY;
int main ()
{
int m,n,o,p;
ulk_cpanel_printf("Welcome power plant System\n");
ulk_cpanel_printf("Enter the option 1 or 2 or 3\n");
ulk_cpanel_printf("1.temperature\n");
ulk_cpanel_printf("2.speed\n");
ulk_cpanel_printf("3.level of oil\n");
ulk_scanf_hex(&m);
switch(m)
{
case 1 :
```

```
ulk_fpga_7seg_led_enable();
ulk_fpga_7seg_led_write(1);
ulk_fpga_clcd_init();
ulk_fpga_clcd_display_on();
ulk_scanf_hex(&n);
if(n<4)
{
/***** normal *****/
/***** cooling fan off *****/
}
else
/*****abnormal *****/
/***** cooling fan on *****/
ulk_fpga_clcd_display_string("cooling fan on");
break;
case 2 :
ulk_fpga_7seg_led_enable();
ulk_fpga_7seg_led_write(2);
ulk_fpga_clcd_init();
ulk_fpga_clcd_display_on();
ulk_scanf_hex(&o);
if(o<=5)
{
/***** normal *****/
}
else
/***** automatically reduce speed *****/
break;
case 3 :
ulk_fpga_7seg_led_enable();
ulk_fpga_7seg_led_write(3);
ulk_fpga_clcd_init();
ulk_fpga_clcd_display_on();
ulk_scanf_hex(&p);
if(p=6)
{
/***** normal *****/
/***** no need to add oil *****/
}
else
/***** low level *****/
/***** increase level of oil *****/
break;
default:
ulk_fpga_clcd_init();
ulk_fpga_clcd_display_on();
ulk_fpga_clcd_display_string("Warning!");
}
}
```

References

[1] Mohamed Zahran and Ali Yousef et al published a paper on "monitoring of photovoltaic wind turbine battery hybrid system", (WSEAS-2014).

[2] ShiyamSundar published a paper on "simulation of generator unit in thermal power plant using LabVIEW", (IOSR-2013).

[3] PengGuo and David Infield published a paper on "wind turbine generator condition monitoring using temperature trend analysis", (IEEE-2012).

[4] Krishna Prasad Dasari and Dr.A.M.Prasad et al published a paper on "water temperature and flow control measurement for thermal discharge model using PID controller, (IJERA-2012).

[5] Alberto Borghetti and Mauro Bosetti et al published a paper on "parameters identification of a power plant model for the simulation of islanding transients", (U.P.B.SCI.BULL-2010).

[6] Jaishree.S and Dr.K.Sathiyasekar et al published a paper on "wireless fault detection and preventive system for small wind turbine", (IJAREEIE-2014).

[7] L.SenthilMurugan and K.S.Aswath Rangaraj et al published a paper on "design of monitoring system and fault diagnosis in wind turbine based on can bus using arm", (IJIRSET-2014).

[8] Anish Mathew k et al published a paper on "internal model control of pressure process using arm microcontroller", (ICCEET-2012).

[9] Nie Chun-yan and Xu shan-shan et al published a paper on "data acquisition and realization of communication transmission based on LabVIEW", (IEEE-2012).

[10] V.Rukkumani and D.AngelineVijula et al published a paper on "Multiple parameter Monitor and Control in Wind Mills", (IJCAT-2012).

[11] John Sander et al published a paper on "steam turbine oil challenges", (LELUBRICANTS-2012).

[12] Masaki Kato and Seiichi Asano et al published a paper on "recent technology for reusing aged thermal power generating units", (FUJIELECTRIC-2010).

Line Balancing for Improving Apparel Production by Operator Skill Matrix

Md. Mazharul Islam[1, *], Md. Tanjim Hossain[1], Mohammad Abdul Jalil[2], Elias Khalil[3]

[1]Department of Textile Engineering, Northern University of Bangladesh, Dhaka, Bangladesh
[2]Department of Textile Engineering, Mawlana Bhashani Science and Technology University, Tangail, Bangladesh
[3]Department of Textile Engineering, World University of Bangladesh, Dhaka, Bangladesh

Email address:

mazh999@gmail.com (Md. M. Islam), tanjimseu@yahoo.com (Md. T. Hossain), jalil.mbstu@gmail.com (M. A. Jalil),
eliaskhalil52@gmail.com (E. Khalil)

Abstract: In this modern world, fashion & styles are changing frequently. The emergence of fast changes in fashion has given rise to shorten production cycle time in the garment industry. To meet the dynamic customer demands of momentous quantities in shorten lead time, assembly line production systems are used, where the garment components are assembled into a finished garment through a sub-assembly process. So in the era of product customization, the optimal usage of resources especially the available facilities & operators who are adding the value of product is important. Therefore the assembly line has to be planned in much more flexible way. This paper deals with the maximum utilization of manpower in labor intensive assembly lines. The objective is to accurately delegate workers to the various operations required to complete the product based on their skill & experience so as to achieve the highest level of productivity and delivery as per planned target. The experimental result showed meaningful improvement in productivity as compared to the existing system.

Keywords: Line Balancing, Skill Matrix, Assembly Line, Apparel Production

1. Introduction

As a supply chain of textile industry, garment industry is one of the major industries of the world. The production process of garments is separated into four main phases: designing/ clothing pattern generation, fabric spreading & cutting, sewing and ironing & packing. The most critical phase is sewing phase [1]. As the sewing is the heart of apparel industry, we have to design the sewing line properly so as to achieve the best output at maximum efficiency. Apparel is a mass production system. Assembly line production systems are developed to meet the requirements of mankind, which continue to grow day by day [2]. The demand for greater product variability and shorter life cycles has caused traditional production methods to be replaced with assembly lines [3]. Assembly line is an industrial arrangement of machines, equipments and workers for continuous flow of work pieces in mass production operation. Manufacturing a product in an assembly line requires partitioning the total amount of work into a set of elementary operations called tasks [4]. Tasks are assigned to operators depending on constrains of different labor skill levels. Finally, several workstations in sequence are formed as a sewing line [5]. The sewing process includes a set of work stations, at each of which a specific task is carried out in a restricted sequence, with hundreds of employees and thousands of bundles of sub-assemblies producing different style simultaneously [6]. Therefore this process is of critical importance and needs to be planned more carefully [7]. As a consequence, good line balancing with small stocks in the sewing line has to be drawn up to increase the efficiency and quality [7-9]. So the aim of assembly line balancing in sewing line is to assign tasks to the workstations, so that the machines of the workstation can perform the assigned task with a balanced loading with different labor skill levels [5].

2. Literature Review

A line is defined as a group of operators under the control of one production supervisor [10]. Balancing is the technique of maintaining the same level of inventory at each and every operation at any point of time to meet the production target and to produce garments of acceptable quality [10]. Assembly line balancing (ALB) is a managerial technique and can be applied to balance production flow lines [11-12]. Line balancing is the distribution of work on the line in such a way that everyone gets the same amount of work in terms of time [13]. In practice, a perfect balance could not be achieved but we can improve the situation by using proper technique.

The assembly line balancing problem was first introduced by Bryton in his graduate thesis. In his study, he accepted the amount of workstations as constant, the workstation times as equal for all stations and work tasks as moving among the workstations [14]. The first article was published in 1955 by Salveson [15]. He developed a 0-1 integer programming model to solve the problem. This heuristic method was developed by Helgeson and Birnie of the General Electric Company in 1961 COMSOAL (Computer Method of Sequencing Operations for Assembly Lines) was first used by Arcus in 1966 as a solution approach to the assembly line balancing problem [16,17]. Bartholdi (1993) was the first to address the Two-sided Assembly Line Balancing Problem with the objective of minimizing the number of stations by applying a simple assignment rule. Liu & Chen (2002) presented a Genetic Algorithm approach for assembly planning involving various objectives, such as minimizing cycle time, maximizing workload smoothness, minimizing the frequency of tool change, minimizing the number of tools and machines used and minimizing the complexity of assembly sequences [18-19]. Helgeson ve Birnie (1961) developed the "Ranked Positional Weight Technique" in which operation having the largest ranged weight is assigned to the first workstation, and other operations are assigned to workstations in accordance with their ranked positional weight value [16]. Abdolmajid Yolmeh et al. (2012) proposed a hybrid genetic algorithm to solve the assembly line balancing problem [20].

Operator's Skill Inventory is the database which maintains the record of each operator, who can do what type of operation and at what rating. It is very important to keep this database updated as over the time, operator acquire skills for most of the new operations as well as improve performance in existing operations [21].With the use of skill matrix an engineer's or line supervisors needs very list time to find out and select most efficient operator for a particular operation from the pull of operation. It helps the line supervisor for balancing the line with particular skilled operators according to the work content [21].

There are so many researches going on in the field of

assembly line balancing. Various methods are used for balancing sewing lines which are discussed in the above section. We use operator skill matrix for better allocation of operator throughout the sewing line to get maximum output.

3. Materials and Methodology

In this experiment, 100% cotton Jacket was considered. Total 30 sewing machines were used where number of plain, over lock and flat lock machines was 13, 11, and 6 respectively.

In order to balance a production line in sewing floor a line was chosen & necessary data was accumulated from the line. First the garment was analyzed and operational bulletin or breakdown was created with process sequence, operational description & machine requirements (Table 1). Then workers were placed to different work stations based on operation & machine types and a standard minute for each job was given to the operators (allocated SMV) (Table 1). After one day we calculate the output & found that we didn't get the desired output. To found out the problem, we calculated individual workers performed SMV by work study. After that workers individual efficiency & output at individual efficiency was calculated and then saw that efficiencies varies dramatically due to unplanned layout (Table 1). So to balance the line we have to rearrange the operators within the line. To do this, first machine-wise breakdown (Table 2 to Table 7) was done and workers are reallocated based on two assumptions: a) operators are only be allocated depending on the machine type that he/she can operate; b) allocation is also depend on operation type that he/she can perform. After fulfilling above two conditions we rearranged operators based on SMV. Higher the SMV, higher should be the efficiency % as we know where the workload is high, we need higher efficient worker. After total rearrangement, we again calculated the output (Table 8). We observed that productivity is increased but not up to the theoretical productivity. In some process, huge bottleneck was appeared. So to improve productivity we introduced another worker to the bottleneck operation by dividing the workload between two workers (Table 9). Finally we got satisfactory productivity.

4. Experimental Data

Operational Bulletin of Jacket before and after arrangement, Breakdown of different machines before and after rearrangement along with comparison of productions are shown below in different tables.

Table 1. Operational Bulletin of Jacket before arrangement.

SL	Process Name	Name	M/c Type	Performed SMV	Allocated SMV	Output @ 100% eff	Efficiency %	Output @ performed efficiency %
1	Back part panel join	Jarna	O/L	0.64	0.62	96.77	96	93.26
2	Back panel ts tc	Naher	F/L	0.35	0.3	200.00	87	173.08
3	Left and right panel join	Hasina	O/L	1.72	0.62	96.77	36	34.95
4	Left and right panel join	Eima	O/L	1.28	0.62	96.77	48	46.88
5	Left and right panel ts and tc	Mena	F/L	0.62	0.55	109.09	89	97.30

SL	Process Name	Name	M/c Type	Performed SMV	Allocated SMV	Output @ 100% eff	Efficiency %	Output @ performed efficiency %
6	Front part panel join & tc	Ronju	O/L	0.78	0.5	120.00	64	76.60
7	Front part panel ts & tc	Monzilla	L/S	0.53	0.3	200.00	57	113.92
8	Left & right panel join	Amina	O/L	0.82	0.62	96.77	76	73.47
9	Left and right panel join	Sahina	O/L	0.79	0.62	96.77	78	75.95
10	Left and right panel ts & tc	Nurbanu	F/L	0.53	0.55	109.09	103	112.50
11	Shoulder panel join	Lima	O/L	0.94	0.43	139.53	46	63.60
12	Shoulder panel ts	Alima	F/L	0.88	0.42	142.86	48	68.18
13	Shoulder join	Rojina	O/L	0.92	0.59	101.69	64	65.45
14	Sleeve panel make	Morina	O/L	0.64	0.45	133.33	70	93.26
15	Sleeve panel top stitch & tc	Sokina	F/L	0.47	0.49	122.45	104	126.76
16	Sleeve hem	Momina	F/L	0.42	0.3	200.00	71	141.73
17	Collar make	Pervin	L/S	0.79	0.63	95.24	80	75.95
18	Collar twill tape attach	Rebeka	L/S	0.26	0.22	272.73	85	230.77
19	Collar serving	Roksana	O/L	0.43	0.39	153.85	90	138.46
20	Zipper twill tape make	Sabina	L/S	0.88	0.42	142.86	48	68.44
21	Zipper twill tape attach	Rojina	L/S	0.87	0.5	120.00	57	68.97
22	Zipper holding tuck	Bobita	L/S	0.66	0.3	200.00	45	90.91
23	Zipper patch attach at bottom	Shahanara	L/S	0.45	0.57	105.26	127	133.33
24	Collar join	Sabina	L/S	1.77	0.78	76.92	44	33.90
25	Collar join	Shahila	L/S	1.99	0.78	76.92	39	30.15
26	Zipper tuck with body	Naher	L/S	0.36	0.42	142.86	118	168.22
27	Zipper join with left side	Halima	L/S	0.58	0.5	120.00	86	103.45
28	Zipper join right side	Orchona	L/S	0.76	0.69	86.96	90	78.60
29	Zipper facing join	Beauty	L/S	0.66	0.5	120.00	76	90.91
30	Zipper facing o/l	Sharmin	O/L	0.36	0.3	200.00	83	165.14
				Max Theoretical output		76.92	practical output	30 pcs

Table 2. Breakdown of Plain machines before rearrangement.

SL	Process Name	Name	M/c Type	Performed SMV	Efficiency
7	Front part panel ts & tc	Monzilla	L/S	0.53	57%
17	Collar make	Pervin	L/S	0.79	80%
18	Collar twill tape attach	Rebeka	L/S	0.26	85%
20	Zipper twill tape make	Sabina	L/S	0.88	48%
21	Zipper twill ta[e attach	Rojina	L/S	0.87	57%
22	Zipper holding tuck	Bobita	L/S	0.66	45%
23	Zipper patch attach at bottom	Shahanara	L/S	0.45	127%
24	Collar join	Sabina	L/S	1.77	44%
25	Collar join	Shahila	L/S	1.99	39%
26	Zipper tuck with body	Naher	L/S	0.36	118%
27	Zipper join with left side	Halima	L/S	0.58	86%
28	Zipper join right side	Orchona	L/S	0.76	90%
29	Zipper facing join	Beauty	L/S	0.66	76%

Table 3. Breakdown of Plain machines after rearrangement.

SL	Process Name	Name	M/c Type	Performed SMV	Efficiency
18	Collar twill tape attach	Shahila	L/S	0.26	39%
26	Zipper tuck with body	Sabina	L/S	0.36	44%
23	Zipper patch attach at bottom	Bobita	L/S	0.45	45%
7	Front part panel ts & tc	Sabina	L/S	0.53	48%
27	Zipper join with left side	Monzilla	L/S	0.58	57%
22	Zipper holding tuck	Rojina	L/S	0.66	57%
29	Zipper facing join	Beauty	L/S	0.66	76%
28	Zipper join right side	Pervin	L/S	0.76	80%
17	Collar make	Rebeka	L/S	0.79	85%
21	Zipper twill tape attach	Halima	L/S	0.87	86%
20	Zipper twill tape make	Orchona	L/S	0.88	90%
24	Collar join	Naher	L/S	1.77	118%
25	Collar join	Shahanara	L/S	1.99	127%

Table 4. *Breakdown of Overlock machines before rearrangement.*

SN	Process Name	Name	M/c Type	Performed SMV	Efficiency
1	Back part panel join	Jarna	O/L	0.64	96%
3	Left and right panel join	Hasina	O/L	1.72	36%
4	Left and right panel join	Eima	O/L	1.28	48%
6	Front part panel join & tc	Ronju	O/L	0.78	64%
8	Left & right panel join	Amina	O/L	0.82	76%
9	Left and right panel join	Sahina	O/L	0.79	78%
11	Shoulder panel join	Lima	O/L	0.94	46%
13	Shoulder join	Rojina	O/L	0.92	64%
14	Sleeve panel make	Morina	O/L	0.64	70%
19	Collar serving	Roksana	O/L	0.43	90%
30	Zipper facing o/l	Sharmin	O/L	0.36	83%

Table 5. *Breakdown of Overlock machines after rearrangement.*

SN	Process Name	Name	M/c Type	Performed SMV	Efficiency
30	Zipper facing o/l	Hasina	O/L	0.36	36%
19	Collar serving	Lima	O/L	0.43	46%
14	Sleeve panel make	Eima	O/L	0.64	48%
1	Back part panel join	Ronju	O/L	0.64	64%
6	Front part panel join & tc	Rojina	O/L	0.78	64%
9	Left and right panel join	Morina	O/L	0.79	70%
8	Left & right panel join	Amina	O/L	0.82	76%
13	Shoulder join	Sahina	O/L	0.92	78%
11	Shoulder panel join	Sharmin	O/L	0.94	83%
4	Left and right panel join	Roksana	O/L	1.28	90%
3	Left and right panel join	Jarna	O/L	1.72	96%

Table 6. *Breakdown of Flat lock machines before rearrangement.*

SN	Process Name	Name	M/c Type	Performed SMV	Efficiency
2	Back panel ts tc	Naher	F/L	0.35	87%
5	Left and right panel ts and tc	Mena	F/L	0.62	89%
10	Left and right panel ts & tc	Nurbanu	F/L	0.53	103%
12	Shoulder panel ts	Alima	F/L	0.88	48%
15	Sleeve panel top stitch & tc	Sokina	F/L	0.47	104%
16	Sleeve hem	Momina	F/L	0.42	71%

Table 7. *Breakdown of Flat lock machines after rearrangement.*

SN	Process Name	Name	m/c	Performed SMV	Efficiency
2	Back panel ts tc	Alima	F/L	0.35	48%
16	Sleeve hem	Momina	F/L	0.42	71%
15	Sleeve panel top stitch & tc	Naher	F/L	0.47	87%
10	Left and right panel ts & tc	Mena	F/L	0.53	89%
5	Left and right panel ts and tc	Nurbanu	F/L	0.62	103%
12	Shoulder panel ts	Sokina	F/L	0.88	104%

Table 8. *Operational Bulletin of Jacket after rearrangement.*

SN	Process Name	Name	M/c Type	Performed SMV	Efficiency %	Output @ performed efficiency
1	Back part panel join	Ronju	O/L	0.64	48	46.45
2	Back panel ts tc	Alima	F/L	0.35	48	96.00
3	Left and right panel join	Jarna	O/L	1.72	96	92.90
4	Left and right panel join	Roksana	O/L	1.28	90	87.10
5	Left and right panel ts and tc	Nurbanu	F/L	0.62	71	77.45
6	Front part panel join & tc	Rojina	O/L	0.78	64	76.80
7	Front part panel ts & tc	Sabina	L/S	0.53	44	88.00
8	Left & right panel join	Amina	O/L	0.82	76	73.55
9	Left and right panel join	Morina	O/L	0.79	70	67.74
10	Left and right panel ts & tc	Mena	F/L	0.53	87	94.91
11	Shoulder panel join	Sharmin	O/L	0.94	83	115.81

SN	Process Name	Name	M/c Type	Performed SMV	Efficiency %	Output @ performed efficiency
12	Shoulder panel ts	Sokina	F/L	0.88	89	127.14
13	Shoulder join	Sahina	O/L	0.92	78	79.32
14	Sleeve panel make	Eima	O/L	0.64	64	85.33
15	Sleeve panel top stitch & tc	Naher	F/L	0.47	103	126.12
16	Sleeve hem	Momina	F/L	0.42	104	208.00
17	Collar make	Rebeka	L/S	0.79	86	81.90
18	Collar twill tape attach	Shahila	L/S	0.26	39	106.36
19	Collar serving	Lima	O/L	0.43	46	70.77
20	Zipper twill tape make	Orchona	L/S	0.88	48	68.57
21	Zipper twill tape attach	Halima	L/S	0.87	57	68.40
22	Zipper holding tuck	Rojina	L/S	0.66	45	90.00
23	Zipper patch attach at bottom	Bobita	L/S	0.45	85	89.47
24	Collar join	Naher	L/S	1.77	118	90.77
25	Collar join	Shahanara	L/S	1.99	127	97.69
26	Zipper tuck with body	Sabina	L/S	0.36	57	81.43
27	Zipper join with left side	Monzilla	L/S	0.58	76	91.20
28	Zipper join right side	Pervin	L/S	0.76	90	78.26
29	Zipper facing join	Beauty	L/S	0.66	80	96.00
30	Zipper facing o/l	Hasina	O/L	0.36	36	72.00

Practical output after rearrangement

Table 9. *Final Practical output after sharing of work.*

SN	Process Name	Name	M/c Type	Performed SMV	Efficiency %	Output @ performed efficiency
1	Back part panel join	Raju	O/L	0.64	48	46.45
1	Back part panel join	Ronju	O/L	0.64	48	46.45
2	Back panel ts tc	Alima	F/L	0.35	48	96.00
3	Left and right panel join	Jarna	O/L	1.72	96	92.90
4	Left and right panel join	Roksana	O/L	1.28	90	87.10
5	Left and right panel ts and tc	Nurbanu	F/L	0.62	71	77.45
6	Front part panel join & tc	Rojina	O/L	0.78	64	76.80
7	Front part panel ts & tc	Sabina	L/S	0.53	44	88.00
8	Left & right panel join	Amina	O/L	0.82	76	73.55
9	Left and right panel join	Morina	O/L	0.79	70	67.74
10	Left and right panel ts & tc	Mena	F/L	0.53	87	94.91
11	Shoulder panel join	Sharmin	O/L	0.94	83	115.81
12	Shoulder panel ts	Sokina	F/L	0.88	89	127.14
13	Shoulder join	Sahina	O/L	0.92	78	79.32
14	Sleeve panel make	Eima	O/L	0.64	64	85.33
15	Sleeve panel top stitch & tc	Naher	F/L	0.47	103	126.12
16	Sleeve hem	Momina	F/L	0.42	104	208.00
17	Collar make	Rebeka	L/S	0.79	86	81.90
18	Collar twill tape attach	Shahila	L/S	0.26	39	106.36
19	Collar serving	Lima	O/L	0.43	46	70.77
20	Zipper twill tape make	Orchona	L/S	0.88	48	68.57
21	Zipper twill tape attach	Halima	L/S	0.87	57	68.40
22	Zipper holding tuck	Rojina	L/S	0.66	45	90.00
23	Zipper patch attach at bottom	Bobita	L/S	0.45	85	89.47
24	Collar join	Naher	L/S	1.77	118	90.77
25	Collar join	Shahanara	L/S	1.99	127	97.69
26	Zipper tuck with body	Sabina	L/S	0.36	57	81.43
27	Zipper join with left side	Monzilla	L/S	0.58	76	91.20
28	Zipper join right side	Pervin	L/S	0.76	90	78.26
29	Zipper facing join	Beauty	L/S	0.66	80	96.00
30	Zipper facing o/l	Hasina	O/L	0.36	36	72.00

Final Practical output (after sharing of work)

Table 10. *Comparison of production before & after study.*

Parameter	Before Rearrangement	After Rearrangement	After sharing of work
No of m/c	30	30	31
No of manpower	30	30	31
Output per hr	30	46	68

5. Results & Discussion

Changing from traditional layout to balanced layout model by proper allocation of workers, there are considerable improvements have moved towards us. With final scenario, the best performance results were obtained as summarized in table 10. The average hourly output of the system increased from 30 to 68 pieces. With reference to scenario, it can be said that the balance of sewing line seems appropriate for all performance measures.

6. Conclusion

Skill matrix helps in allocating right person for the right job which helps in achieving desired performance level. It keeps record of all operations an operator had done in the past and efficiency level in each operation. Engineers / line supervisors need minimum time to find and select most efficient operators for an operation from the pull of operators. For line balancing, operators can be selected according to work content. When someone is absent, supervisor can easily find suitable person from the skill matrix table and replace. To analyses the skill availability and distribution throughout the factory. This can be compared with the skill requirement for a particular time period and shortage/excess skill availability to achieve at the training requirement. So productivity can be achieved by allocating skill & semi-skilled workers to the right place and unskilled operator should be trained properly.

References

[1] Chen J.C., Chen C.C., Lin Y.J., Lin C.J., and Chen T.Y. Assembly Line Balancing Problem of Sewing Lines in Garment Industry. International Conference on Industrial Engineering and Operations Management, Bali, Indonesia, 2014.

[2] Eryürük S.H., Clothing assembly line design using simulation and heuristic line balancing techniques, Ege University Textile and Apparel Research & Application Center, 2012.

[3] Eryuruk S. H, Kalaoglu F. and Baskak M. Assembly Line Balancing in a Clothing Company. FIBRES & TEXTILES in Eastern Europe, 2008.

[4] Jithendrababu B. L., RenjuKurian and Pradeepmon T.G. Balancing Labor Intensive Assembly Line Using Genetic Algorithm. International Journal of Innovative Research in Science, Engineering and Technology, 2013.

[5] Jaganathan V. P. Line balancing using largest candidate rule algorithm in a garment industry: a case study. International Journal of Lean Thinking, 2014.

[6] Chan K.C.C, Hui P.C.L., Yeung K.W., Ng F.S.F. (1998). Handling the assembly line balancing problem in the clothing industry using a genetic algorithm, International Journal of Clothing Science and Technology, Vol.10, pp. 21-37.

[7] Tyler D. J. (1991). Materials Management In Clothing Production, BSP Professional Books Press, London.

[8] Cooklin G. (1991). Introduction to Clothing Manufacturing, Blackwell Science, Oxford, p. 104.

[9] Chuter, A. J. (1988). Introduction to Clothing Production Management, Blackwell Science, 1988.Oxford, pp. 60-63.

[10] Babu V.R. (2011), Industrial engineering in apparel production, Woodhead Publishing Series in Textiles, 129.

[11] Robbins S.P., 1985, "Organizational Behavior- Controversies and Applications" (2nd edition), Prentice-Hall of India (Pvt.) Ltd., New Delhi, 288-292.

[12] Tersine R.J. Production/Operations Management: Concepts, Structure and Analysis, pp.352-374, 1985.

[13] Ramdass K. and Kruger D. The effect of time variations in assembly line balancing: lessons learned in the clothing industry in South Africa. Unisa Institutional Repository, South Africa, 2010.

[14] Bryton, B. Balancing of a Continuous Production Line, M.S. Thesis, Northwestern University, Evanson, ILL. 1954.

[15] Salveson M. E. The Assembly Line Balancing Problem, Journal of Industrial Engineering, 6 (3), pp. 18-25, 1955.

[16] Helgeson W. P., Birnie D. P. Assembly Line Balancing Using the Ranked Positional Weight Technique. Journal of Industrial Engineering, Vol. 12 (6), pp. 384-398, 1961.

[17] Arcus A. L. COMSOAL: A Computer Method of Sequencing for Assembly Lines. International Journal of Production Research, 4 (4), pp. 259-277, 1966.

[18] Bartholdi J.J. Balancing two-sided assembly lines: A case study. International Journal of Production Research, Vol.31, 10, pp.2447-2461, 1993.

[19] Liu C.M., Chen C.H. Multi-section electronic assembly line balancing problems: A case study. International Journal of Product Planning & Control. 13 451-461, 2002.

[20] Yolmeh Abdolmajid and Kianfar Farhad. An efficient hybrid genetic algorithm to solve assembly line balancing problem with sequence dependent setup times. International Journal of Computers & Industrial Engineering, Elsevier-Volume 62, Issue 4, Pages 839- 1144, May 2012.

[21] Narkhedkar R. N., Vishnu Dhorugade and Sonavane M.J. Skill matrix: Effective tool to boost productivity. Indian textile journal, 2011.

4

Aquifer Detection and Characterisation Using Material Balance: A Case Study of Reservoirs A, B, C and D

Omoniyi Omotayo Adewale.[1], Iji Sunday[2]

[1]Lecturer, Department of Petroleum Engineering Abubakar Tafawa Balewa, University Bauchi, Bauchi State, Nigeria
[2]Department of Petroleum Engineering Abubakar Tafawa Balewa, University Bauchi, Bauchi State, Nigeria

Email address:
omotosimple4u@gmail.com (O. O. Adewale.)

Abstract: Oil and gas production needs energy, sufficient enough to drive the produced hydrocarbon to the surface of the well. Usually some of this required energy is supplied by nature. The hydrocarbon fluids are under pressure because of their depth. The gas and water in petroleum reservoirs under pressure are the two main sources that help move oil to the well bore and sometimes up to the surface. Depending on the original characteristics of hydrocarbon reservoirs, the type of drive energy is different. The material balance equation has been a very useful tool in analyzing these mechanisms. If none of the terms in the material balance equation can be neglected, then the reservoir can be described as having a combination drive in which all possible sources of energy contribute a significant part in producing the reservoir fluids, and determining the primary recovery factor. For this to happen, the water must be produced from an aquifer. The aquifer water expands slightly, displacing the oil or gas from the reservoir towards the borehole as pressure drops around the borehole. Most literatures have been able to call attention to the analysis of strong and partial water drive. This study was able to bring to light the aquifer characteristics based on weak water drives. Knowledge of the cumulative water influx is also important to the reservoir engineer. This study also goes ahead to add to aquifer detection and characterization, the cumulative water influx of each reservoir. The whole process entailed analyzing reservoirs A,B,C and D using the method proposed by Cole and Campbell. The plots showed a weak water drive for all reservoirs. The water influx for all the reservoirs were calculated and results obtained. The Cole and Campbell plots were proven to be more accurate method of detecting and characterizing aquifer and water drive strength.

Keywords: Aquifer, Hydrocarbon, Material Balance Equation, Water Drive, Water Influx, Reservoir

1. Introduction

Successful reservoir management relies on the ability to generate reliable reservoir performance behavior. The primary questions that reservoir engineers are expected to answer are given in the following, in order of priority:

1. What are the expected quantities of original oil and gas in place (OOIP and OGIP)?

2. How much oil and gas can be economically recovered given the associated probabilities and risks?

3. How can a newly discovered field be developed, followed by implementation of the reservoir management plan and monitoring and evaluation of reservoir performance?

1.1. Natural Producing Mechanisms

There are natural sources of energy in oil reservoirs that control reservoir performance. These include the following:

- Liquid and rock compressibility drive
- Solution gas or depletion drive
- Gascap drive
- Aquifer water drive
- Gravity segregation drive
- Combinations of above'
- Drive mechanisms in gas reservoirs are as follows:
- Gas expansion or depletion drive
- Aquifer water drive
- Combinations of above

1.2. Aquifer Water Drive

When an oil or gas reservoir is in communication with a surrounding (bottom or edge) active aquifer, production from the reservoir results in a pressure drop between the reservoir and the aquifer. This allows influx of water into the reservoir. A producing reservoir is referred to as bottom water drive or edge water drive reservoir, depending on the location of the adjacent aquifer providing energy for production.

Reservoir pressures in water drive reservoirs remain high. Pressure is influenced by the rate of water influx, and by the rate of oil, gas, and water productions. Gas/oil ratios remain low if pressure remains high. Downdip wells produce water earlier, and water production continues to increase. Water drive is usually the most efficient reservoir driving force in oil reservoirs. Recovery efficiencies may vary from 30% to 80%, depending upon the size and strength of the aquifer. Recovery efficiencies for the depletion drive gas reservoirs can be 80% to 90%. However, in the case of water drive gas reservoirs, recovery efficiencies could be in the 50% to 60% range because of the bypassed gas and high reservoir pressures. Recovery from bottom water drive would be substantially affected by a water coning problem. The global average of recovery factor is found to be in the range of 35% or slightly less.

1.3. Reserve Estimation

Many petroleum engineers spend a major part of their professional lives developing estimates of reserves and production capabilities, along with new methods and techniques for improving these estimates. To understand the confidence levels and risks of the estimates, a clear and consistent set of reserve classifications must be used. The confidence level and the techniques implemented by the petroleum engineer depend on the quantity and the maturity of the data available. The data quality, therefore, establishes the classification assigned to the reserve estimates and indicates the confidence one should have in the reserve estimates.

Reserves are classified as proved, proved developed, proved underdeveloped, probable and possible reserves.

Reserve estimation is simply evaluating or assessing a particular reservoir.

One major reason for the estimates of reserves is for management decisions which are seen in the formation of policies for;

1. Exploration and development of oil and gas properties.

2. Design and construction of plants, gathering systems and other surface facilities.
3. Determining and construction of ownership in unitized projects.
4. Establishing sales contracts.

Reserves are frequently estimated before drilling or any subsurface development, during the development drilling of the field, after some performance data are available, and after performance trends are well established.

1.3.1. Reserve Estimation Techniques

Commonly used reservoir performance analysis and reserve estimation techniques are;
- Volumetric
- Decline curves
- Material balance
- Mathematical simulation

For the purpose of this project, attention will be restricted to reservoir estimation based on the material balance method.

1.3.2. Material Balance Method

The material balance equation (MBE) has been used by reservoir engineers for a long time as the basic tool for interpreting and predicting performance. When properly applied, the MBE can be used to;

Estimate initial hydrocarbon volumes in place.

Predict future reservoir performance.

Predict ultimate hydrocarbon recovery under various types of primary driving mechanisms[2]

Schilthuis in 1941 was the first to present the general form of the material balance equation. The equation is derived as a volume balance which equates the cumulative observed production, expressed as an underground withdrawal to the expansion of the fluids in the reservoir resulting from a finite pressure drop[6]

Evaluating the volume balance in reservoir barrels, he obtained;

Underground withdrawal (rb) = Expansion of oil + originally dissolved gas (rb) + Expansion of gascap gas (rb) + Reduction in HCPV due to connate water expansion and decrease in the pore volume (rb)

Mathematically,

$$N_p\left(B_o + \left(R_s - R_{si}\right)B_g\right) = NB_{oi}\left[\frac{\left(B_o - B_{oi}\right) + \left(R_{si} - R_s\right)B_g}{B_{oi}} + m\left(\frac{B_g}{B_{gi}} - 1\right) + \left(1 + m\right)\left(\frac{c_w S_{wc} + c_f}{1 - S_{wc}}\right)\Delta P\right]$$
$$+ \left(W_e - W_p\right)B_w \tag{1}$$

Approximately two decades after the work of Schilthuis, Havlena and Odeh (1963-4) presented two papers describing MBE as a technique of interpreting the MBE as an equation of a straight line, the first paper describes the technique, and the second illustrates the application to reservoir case histories of various fields[6, 7, 8]

One measure of the relative importance of the various drive mechanisms is the intrinsic energy of the different substances, more specifically the compressibility-volume

product, which compensates for reservoir voidage (production) in maintaining reservoir pressure.

Aquifer strength has to be sufficient (size and connectivity) to sweep the oil at elevated pressure (ideally close to initial, bubble point pressure). It is the relative aquifer size, by comparison to the oil leg (and gas cap) that is of importance. Unfortunately, aquifer strength is usually not proven before development takes place but the chance for a strong or sufficient aquifer is accessed based on regional geology. This aspect is particularly important in offshore situations where pre-investment into a water injection plant has to be considered if the chance of a sufficient aquifer is relatively low.

The above material balance equation will be used to detect aquifers as well as to characterise them. The Cole and Campbell plot will be used as important tools for this project. We will see how these plots help us to carry out this project successfully.

1.4. Problem Statement

For proper estimation of reserves, an adequate approach is required so as to be able to gain adequate information about production and production histories. Drive mechanisms are important to the reservoir engineers as well as their strength and drive indices.

This leads us to detecting aquifers that produce through water drives and accurately characterizing them to be able to know the strength under which the reservoir is producing. These include:

Strong, moderate and weak water drives. Reservoir engineers have tried to do this but most works have been less accurate, accounting for mostly strong and moderate water drives. This work presents a more accurate way through which it is done.

1.5. Aim and Objectives of the Study

The aim of this study is to detect and characterize aquifers using four reservoir case histories around the world.

The objectives are to determine:
- Presence of water drive
- Strength of the water drive
- Cumulative water influx, and
- Drive indices

1.6. Significance of the Study

This study will help the reservoir engineer to understand the nature of the aquifer contributing to the production of the hydrocarbon. Failure to account for a weak water drive can result in significant material-balance errors. So the study will show an acceptable method of identifying strong, moderate and weak water drives.

1.7. Limitations

This study is limited to conventional oil and gas reservoirs around the world. Some parameters were assumed to aid full analysis.

2. Methodology

This study will be carried out with respect to gas and oil reservoirs. The two types of reservoirs therefore will involve two different methods of approach to detect and characterize the aquifer.

2.1. Gas Reservoir

2.1.1. The Cole Plot

The Cole plot is a useful tool for distinguishing between water drive and depletion drive gas reservoirs. We can derive the plot from the general gas reservoir material balance.

$$F = G(E_g + E_{f,w}) + W_e \qquad (2)$$

Where F= cumulative reservoir voidage and

$$F = G_p B_w + W_p B_w \qquad (3)$$

$$E_g = B_g - B_{gi} \qquad (4)$$

Eg = Cumulative gas expansion and
Ef,w = cumulative formation and water expansion

$$E_{f,w} = B_{gi} \frac{S_w C_w + C_f}{1 - S_w}(P_i - P) \qquad (5)$$

$$G = OGIP$$

Often in gas reservoirs, Efw is negligible compared to Eg and can therefore be ignored.

$$G_p B_g + W_p B_w = G(B_g - B_{gi}) + W_e \qquad (6)$$

$$\frac{G_p B_g}{B_g - B_{gi}} + \frac{W_p B_w}{B_g - B_{gi}} = G + \frac{W_e}{B_g - B_{gi}} \qquad (7)$$

$$\frac{G_p B_g}{B_g - B_{gi}} = G + \frac{W_e}{B_g - B_{gi}} - \frac{W_p B_w}{B_g - B_{gi}} \qquad (8)$$

$$\frac{G_p B_g}{B_g - B_{gi}} = G + \frac{W_e - W_p B_w}{B_g - B_{gi}} \qquad (9)$$

Cole proposed plotting $\frac{G_p B_g}{B_g - B_{gi}}$ on the Y-axis versus Gp, cumulative gas production, if the reservoir is depletion drive, right-handed term goes to zero and the points plot in a horizontal line with the Y intercept equal to G, the OGIP. If a water drive exists, the right-handed term is not zero and the points will plot above the depletion drive line with a type of slope. So we can say that when a sloping line exists with respect to the horizontal line, it can be used as a diagnostic tool for distinguishing between depletion drive and water drive.

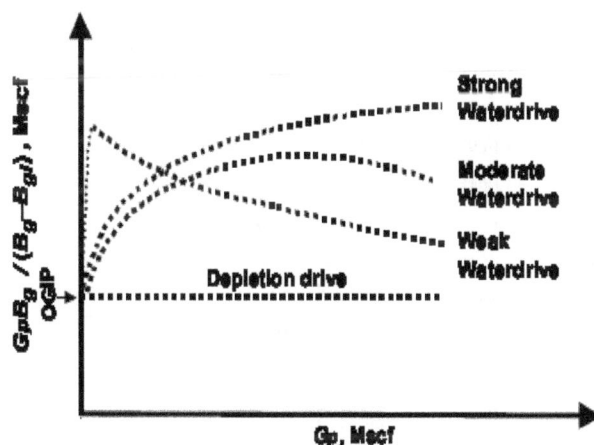

Fig. 1. Cole plot of GpBg/Bg-Bgi vs Gp

2.2. Oil Reservoirs

Campbell extended Cole's theory to oil reservoirs in order to characterize them. He developed a plot called the Campbell plot, used to identify the relative strength of aquifers.

2.2.1. Campbell Plot

For oil reservoirs, the Campbell plot is the counterpart to the modified Cole plot for gas. From the generalized material balance equation;

$$F = N_p[B_o + (R_p - R_s)B_g] + W_p B_w \qquad (10)$$

$$E_o = B_{oi}\left(\left(\frac{B_g}{B_{gi}}\right) - 1\right) \qquad (11)$$

$$E_g = B_{oi}(1 + m)\frac{S_w C_w + C_f}{1 - S_w}\Delta P \qquad (12)$$

Introducing the terms into the MBE, it can be written as

$$F = N(E_O + mE_g + E_{f,w}) + W_e \qquad (13)$$

If we let;

$$E_t = E_o + mE_g + E_{f,w} \qquad (14)$$

We can further simplify equation (13) to

$$F = N(E_t) + W_e \qquad (15)$$

Dividing through by E_t, we have

$$\frac{F}{E_t} = N + \frac{W_e}{E_t} \qquad (16)$$

Where N = OOIP in STB
F = cumulative reservoir voidage

$$F = N_p[B_t + (R_s - R_{si})B_g] + W_p B_w \qquad (17)$$

$$E_o = B_t - B_{ti} \qquad (18)$$

$$E_o = \left(\frac{B_{ti}}{B_{gi}}\right)(B_g - B_{gi}) \qquad (19)$$

$$E_{f,w} = B_{ti}(1 + m)\frac{S_w C_w + C_f}{1 - S_w}(P_i - P) \qquad (20)$$

Et = cumulative total expansion
Eg = cumulative gas expansion
Ef,w = cumulative formation and water expansion
Eo = cumulative oil expansion
m = ratio of initial gas cap volume to initial oil zone volume at reservoir conditions.
Bt = total formation volume factor

$$B_t = B_o + B_g(R_{si} - R_{si}) \qquad (21)$$

Plotting F/Et on the Y axis versus F on the X-axis will yield a plot with one of the characteristic curve shapes as shown below.

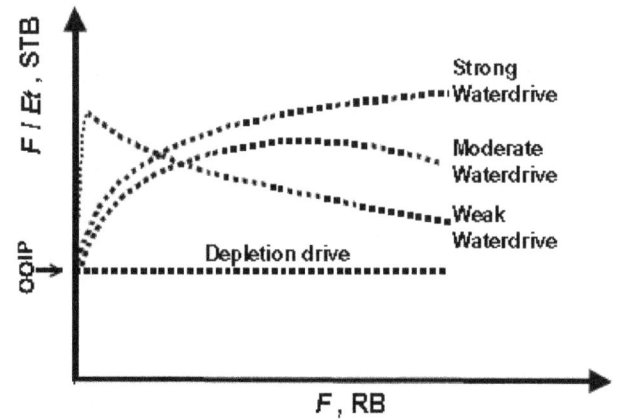

Fig. 2. Campbell plot of F/Et vs F

It should be noted that we assume the reservoir to be a volumetric reservoir, which is not producing under water drive so as to detect whether we have a producing aquifer with water drive or a depletion drive. Therefore we let We=0.

3. Result

Table 1. Reservoir A (Gas Reservoir)

Pressure	Gdf, z	Bg	Bw	Cum gas production	Cum water pproduction	Cum water influx
6411	1.1192	0.6279	1.0452	0	0	0
5947	1.089	0.6587	1.0467	5.475	378	273294
5509	1.0618	0.6933	1.048	10.95	1434	552946
5093	1.0374	0.7327	1.0493	16.425	3056	817481
4697	1.0156	0.7778	1.0506	21.9	5284	1068632
4319	0.9966	0.83	1.0517	27.375	8183	1307702
3957	0.9801	0.891	1.0529	32.85	11864	1535212
3610	0.9663	0.9628	1.054	38.325	16425	1752942
3276	0.9551	1.0487	1.0551	43.8	22019	1962268
2953	0.9467	1.1532	1.056	49.275	28860	2163712
2638	0.9409	1.2829	1.0571	54.75	37256	2359460

Table 2. *Data for the Cole plot for Reservoir A*

GpBg	Bg-Bgi	Gp	GpBg/Bg-Bgi
0	0	0	0
3.606383	0.0308	5.475	117.0903
7.591635	0.0654	10.95	116.08
12.0346	0.1048	16.425	114.8339
17.03382	0.1499	21.9	113.6346
22.72125	0.2021	27.375	112.4258
29.26935	0.2631	32.85	111.248
36.89931	0.3349	38.325	110.1801
45.93306	0.4208	43.8	109.1565
56.82393	0.5253	49.275	108.1742
70.23878	0.655	54.75	107.2348

From the above calculation the above data gave the following plot

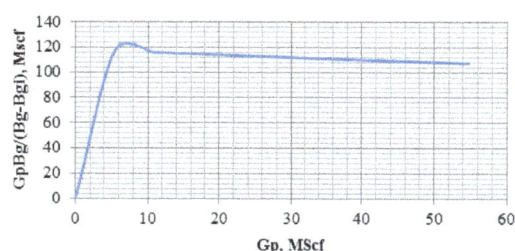

Fig. 3. *Cole plot for Reservoir A*

Table 3. *Reservoir B (oil reservoir)*

Days	Pressure (psia)	Cum oil production.	Cum water production	Cum gas production	Bo(rb/stb)
0	2855	0	0	0	1.2665
305	2779	192821	0	94513	1.2677
700	2627	633942	0	312064	1.2681
1285	2457	1314880	4	710870	1.2554
1456	2402	1524400	7	850934	1.2512
2005	2223	2152960	26	1355720	1.2383
2365	2080	2572000	60	1823250	1.2278
2905	1833	3200560	822	2732860	1.2074
3236	1665	3564680	11135	3397740	1.1949
3595	1460	4003720	97443	4216120	1.1802

Rs	Bg	Bt	Bw
0.501	0.9201	1.2665	1.0222
0.501	0.9637	1.2677	1.0224
0.4973	1.0502	1.272	1.0228
0.4671	1.0977	1.2926	1.0232
0.4574	1.1146	1.2998	1.0233
0.4289	1.201	1.3273	1.0237
0.4024	1.2825	1.3543	1.024
0.3579	1.4584	1.4161	1.0246
0.3277	1.6112	1.4741	1.025
0.2908	1.8526	1.5696	1.0254

EO	Eg	Ef,w	Et	F	F/Et
0	0	0	0	0	0
-0.0012	0.057299	0.482699	0.538799	530447.1	984499.8
-0.0055	0.156895	1.448098	1.599493	1825295	1141171
-0.0261	0.204911	2.52782	2.70663	3646223	1347145
-0.0333	0.221007	2.877141	3.064848	4174010	1361898
-0.0608	0.29622	4.014025	4.249445	5668458	1333929
-0.0878	0.357879	4.922262	5.192341	6483942	1248751
-0.1496	0.467469	6.491034	6.808903	7661163	1125168
-0.2076	0.543246	7.558054	7.8937	8414260	1065946
-0.3031	0.637489	8.860071	9.19446	9711717	1056257

The above calculation gave the following plot:

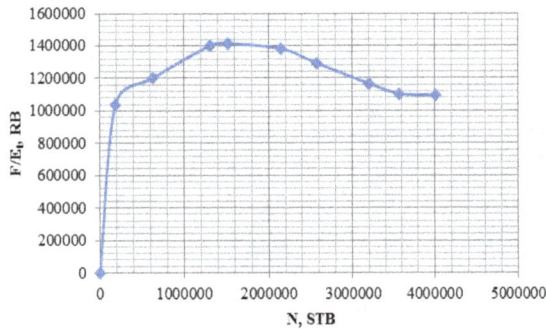

Fig. 4. Campbell plot for Reservoir B

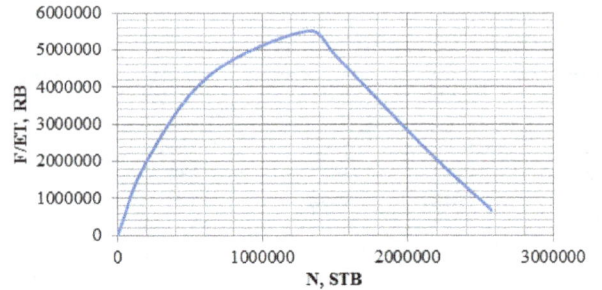

Fig. 5. Campbell plot for Reservoir C

Table 4. *Reservoir C (oil reservoir)*

N	p	Wp	Bo	Bw	Bg	Rs	Bt
0	2000	0	1.467	1.0222	0	838.5	1.467
192821	1800	0	1.472	1.0224	0	838.5	1.472
633942	1700	0	1.475	1.0228	0	838.5	1.475
1314880	1640	4	1.463	1.0232	1.92	816.1	44.471
1524400	1600	7	1.453	1.0233	1.977	798.4	80.7307
2152960	1400	26	1.408	1.0237	2.308	713.4	290.1388
2572000	1200	60	1.359	1.024	2.73	621	595.134
3200560	1000	822	1.322	1.0246	3.328	548	968.106
3564680	800	11135	1.278	1.025	4.163	464	1560.322
4003720	600	97443	1.237	1.0254	5.471	383	2493.278
4200350	400	120500	1.194	1.024	7.786	297.4	4214.199
5002340	200	150220	1.141	1.0245	13.331	190.9	8634.297

The method of calculation for F/Et in the oil reservoir above applies to this oil reservoir to get F/Et

Table 5. *Data for the Campbell plot for Reservoir C*

EO	Eg	F	Ef,w	Et	F/Et
0	-1.467	0	0	-1.467	0
-0.005	-1.467	-1.6E+08	-81.1251	-82.5971	1954022
-0.008	-1.467	-5.3E+08	-121.688	-123.163	4308330
-43.004	0	-1E+09	-146.025	-189.029	5509881
-79.2637	0.043552	-1.2E+09	-162.25	-241.47	4771310
-288.672	0.296456	-1.2E+09	-243.375	-531.751	2210868
-593.667	0.618891	-6.2E+08	-324.5	-917.549	674531.9
-966.639	1.0758	4.25E+08	-405.626	-1371.19	-310289
-1558.85	1.713792	2.59E+09	-486.751	-2043.89	-1266170
-2491.81	2.713186	6.65E+09	-567.876	-3056.97	-2174461
-4212.73	4.481991	1.42E+10	-649.001	-4857.25	-2925923
-8632.83	8.718717	3.91E+10	-730.126	-9354.24	-4176083

The above calculation gave the following plot:

Table 6. *Reservoir D (Gas)*

Time	Average pressure	(Gp)	Bo	Rs	Bg 10E-3
(Days)	(psia)	MMscf	(Rb/stb)	(scf/stb)	(rb/scf)
0	4487	0	1.308	811	0.639
365	4444	1.5	1.301	799	0.644
730	4416	3.1	1.298	793	0.647
1095	4370	5.2	1.297	788	0.65
1460	4332	8.5	1.293	785	0.654
1825	4298	11.8	1.29	779	0.659
2190	4260	16.7	1.287	774	0.653
2555	4228	23	1.285	769	0.666
2920	4230	24.7	1.286	772	0.667
3285	4259	25.4	1.289	778	0.665
3650	4282	25.9	1.299	780	0.665

The method of calculation for GpBg/Bg-Bgi in the oil reservoir above applies to this oil reservoir to get GpBg/Bg-Bgi

Table 7. *Data for the Cole plot for Reservoir D*

GpBg	Bg-Bgi	GpBg / Bg-Bgi
0	0	0
0.000966	0.000005	193.2
0.002006	8E-06	250.7125
0.00338	0.000011	307.2727
0.005559	0.000015	370.6
0.007776	0.00002	388.81
0.010905	0.000014	778.9357
0.015318	0.000027	567.3333
0.016475	0.000028	588.3893
0.016891	0.000026	649.6538
0.017224	0.000026	662.4423

The above calculations and equation gave the following plot:

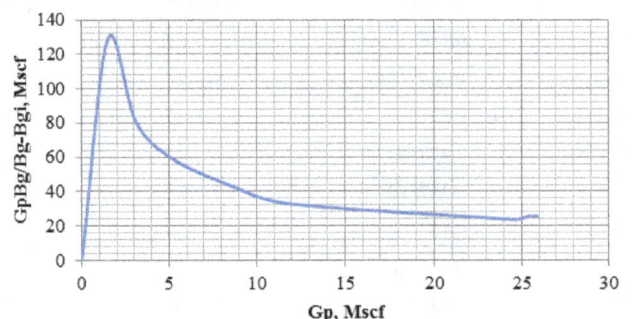

Fig. 6. Cole plot for Reservoir D

3.1. Calculation for Water Influx

Assumptions made:

Aquifers are radial aquifers

All reservoirs are characterized with the following properties:

h = 45ft	μ = 0.2cp
Φ = 0.15	Cw = 6×10-6
K = 100md	Cf = 3×10-6

Subtending angle = 75^0

$$t_D = \frac{2.309 \times 100 \times t}{0.15 \times 0.2 \times 0.000009 \times 5000^2}$$

$$= \frac{230.9t}{6.75}$$

$$= 34.2t$$

$$U = 1.119 f \Phi h c r_0^2$$

$$U = 1.119 \times 0.2083 \times 0.15 \times 45 \times 0.000009 \times 5000^2$$

$$U = 354 bbls/psi$$

Table 8. Reservoir A

T(years)	Pressure(psi)	Plateau pressure Level	ΔP(psi)	t_D	WD(tD) reD = 5
0	6411		451	0	
1	59947	5960	440	34.2	11.80
2	5509	5520	417	68.4	12.02
3	5093	5103	397	102.6	12.10
4	4697	4706	379	136.8	12.20
5	4319	4327	362	171	12.40
6	3957	3965	348	205.2	12.50
7	3610	3617	335	239	12.56
8	3276	3282	325	274	12.60
9	2953	2957	161	308	12.73
10	2638	2796		342	12.8

Table 9. Calculation of the Water Influx for Reservoir A

T(years)	$W_e = U \sum_{j=0}^{n-1} \Delta P_j W_D(T_D - t_{Dj})$	We(bbls)
1	354(451×11.8)	1883917.2
2	354(451×12.02+440×11.8)	3757009.1
3	354(451×12.10+440×12.02+417×11.8)	3833720.9
4	354(451×12.2+440×12.10+417×12.02+397×11.8)	7265191.6
5	354(451×12.4+440×12.2+417×12.10+397×12.02+379×11.8)	8938585
6	354(451×12.5+440×12.4+417×12.2+397×12.10+379×12.02+362×11.8)	9121390.6
7	354(451×12.56+440×12.5+417×12.4+397×12.2+379×12.10+362×12.02+348×11.8)	10477231.8
8	354(451×12.6+440×12.56+417×12.5+397×12.4+379×12.2+362×12.10+348×12.02+335×11.8)	13623428
9	354(451×12.73+440×12.6+417×12.56+397×12.5+379×12.4+362×12.2+348×12.10+335×12.02+325×11.8)	15106507.5
10	354(451×12.8+440×12.73+417×12.6+397×12.56+379×12.5+362×12.4+348×12.2+335×12.10+325×12.02+161×11.8)	14647366

Table 10. Reservoir B

T(years)	Pressure(psi)	Plateau pressure Level	ΔP(psi)	tD	WD(tD) reD = 5
0	2855		38		
0.8	2779	2817	168	33.6	11.4
1.9	2627	2703	161	65.7	11.9
3.5	2457	2542	112	119.7	12.03
4	2402	2430	117	136.8	12.13
5.5	2223	2313	161	188.1	12.3
6.5	2080	2152	195	222.3	12.5
8	1833	1957	208	273.6	12.52
8.9	1665	1749	186	304.4	12.6
10	1460	1563		342	12.8

Table 11. Calculation of the Water Influx for Reservoir B

T(years)	$W_e = U \sum_{j=0}^{n-1} \Delta P_j W_D(T_D - t_{Dj})$	We(bbls)
0.8	354(38×11.4)	153352.8
1.9	354(38×11.9+168×11.4)	838059.6
3.5	354(38×12.03+168×11.8+161×11.4)	1513328.8
4	354(38×12.13+168×12.03+161×11.8+112×11.4)	2003137.3
5.5	354(38×12.3+168×12.13+161×12.03+112×11.9+117×11.4)	2516469.18
6.5	354(38×12.5+168×12.3+161×12.13+112×12.03+117×11.9+161×11.4)	3210564.1
8	354(38×12.52+168×12.5+161×12.3+112×12.13+117×12.03+161×11.9+195×11.4)	4057204.6
8.9	354(38×12.6+168×12.52+161×12.5+112×12.3+117×12.13+161×12.03+195×11.9+208×11.4)	4905797.4
10	354(38×12.8+168×12.6+161×12.52+112×12.5+117×12.3+161×12.13+195×12.03+208×11.9+186×11.4)	5788749.6

Table 12. Reservoir C

T(years)	Pressure(psi)	Plateau pressure Level	ΔP(psi)	tD	WD(tD) reD = 5
0	2000		150	0	
1	1800	1850	130	34.2	11.80
1.5	1700	1720	70	51.3	11.92
2	1640	1650	130	64.4	12.02
3	1600	1520	120	102.6	12.10
4	1400	1400	200	136.8	12.20
4.5	1200	1200	200	154	12.35
5	1000	1000	200	171	12.40
6	800	800	200	205.2	12.50
7	600	600	200	239.4	12.57
8.5	400	400	200	290.7	12.7
9	200	200		307.8	12.75

Table 13. Calculation of the Water Influx for Reservoir c

T(years)	$W_e = U \sum_{i=0}^{n-1} \Delta P_i W_D(T_D - t_{Di})$	We(bbls)
1	354(150×11.8)	626580
1.5	354(150×11.92+130×11.92+70×11.8)	925356
2	354(150×12.02+130×11.92+130×11.8)	1729856.4
3	354(150×12.10+130×12.02+70×11.92+130×11.8)	2034084
4	354(150×12.20+130×12.1+70×12.02+130×11.92+120×11.8)	2101202.4
4.5	354(150×12.35+130×12.20+70×12.1+130×12.02+120×11.92+200×11.8)	3412029
5	354(150×12.4+130×12.35+70×12.20+130×12.1+120×12.02+200×11.92+200×11.8)	4275930.6
6	354(150×12.5+130×12.4+70×12.35+130×12.20+120×12.1+200×12.02+200×11.92+200×11.8)	5146275
7	354(150×12.57+130×12.5+70×12.4+130×12.35+120×12.20+200×12.1+200×12.02+200×11.92+200×11.8)	6023664
8.5	354(150×12.7+130×12.57+70×12.5+130×12.4+120×12.35+200×12.20+200×12.1+200×12.02+200×11.92+200×11.8	6908699.4
9	354(150×12.75+130×12.7+70×12.57+130×12.5+120×12.4+200×12.35+200×12.20+200×12.1+200×12.02+200×11.92+200 ×11.8)	7800177.6

Table 14. Reservoir D

Time, t (Days)	Td	Wd(td)	Pressure (psia)	Pressure drop ΔP (psi)
0	0	0	4487	21.5
365	205.9	74	4444	35.5
730	411.7	140	4416	37.0
1095	617.6	190	4370	42.0
1460	823.4	240	4332	36.0
1825	1029.3	280	4298	36.0
2190	1235.3	328	4260	35.0
2555	1441.0	370	4228	15.0
2920	1646.9	400	4230	-15.0
3285	1852.7	430	4259	-25.0
3650	2058.6	465	4280	

Table 15. Calculation of the Water Influx for Reservoir D

TD	$We = U \sum_{i=0}^{n-1} \Delta Pi Wd(td - Tdi)$ Mrb	We (Mrb)
	21.5 x 0	0
205.9	21.5 x 74	1.59
411.7	(21.5 x 140) + (35.5 x 74)	5.46
617.6	(21.5 x 190) + (35.5 x 140) + (37.0 x 74)	11.79
823.4	(21.5 x 240) + (35.5 x 190) + (37.0 x 140) + (42.0 x 74)	20.19
1029.3	(21.5 x 280) + (35.5 x 240) + (37.0 x 190) + (42.0 x 140) + (36.0 x 74)	30.114
1235.3	(21.5 x 328) + (35.5 x 280) + (37.0 x 240) + (42.0 x 190) + (36 x 140) + (36.0 x 74)	41.56
1441.0	(21.5 x 370) + (35.5 x 328) + (37.0 x 280) + (42.0 x 240) + (36 x 190) + (36 x 140) + (35.0 x 74)	54.51
1646.9	(21.5 x 400) + (35.5 x 370) + (37.0 x 328) + (42.0 x 280) + (36 x 240) + (36 x 190) + (35.0 x 140) + (15 x 74)	67.12
1852.7	(21.5 x 430) + (35.5 x 400) + (37.0 x 370) + (42.0 x 328) + (36 x 280) + (36 x 240) + (35 x 190) + (15 x 140) + (-15.5 x 74)	77.27
2058.6	(21.5 x 465) + (35.5 x 430) + (37 x 400) + (42 x 370) + (36 x 328) + (360 x 328) + (36 x 280) + (35 x 240) + (15 x 190) + (-15.5 x 140) + (-25.0 x 74)	84.72

4. Discussion of Result

4.1. Reservoir A

In the first reservoir, the material balance method was applied to the reservoir history as described by Cole. The process produced a curve similar to one of the curves in the plot presented by Cole. This plot indicates the presence of an aquifer characterized by a weak water drive. From the plot we see an abrupt fall in the Y axis which is GpBg/Bg-Bgi with increasing X axis which is Gp. The abrupt fall can be noticed from around 5.5MScf of cumulative gas production.

In the case of the water influx calculation, the dimensionless time was calculated to be 34.2t with t being the number of years. The aquifer constant for radial geometry was calculated to be 354bbls/psi which was used to calculate the cumulative water influx.

From the cumulative water influx calculated, we see a trend in which the values are gradually increasing within the range of 1.5 to 14MMbbl

4.2. Reservoir B

In the second case, we see a similar trend which was in seen in the Cole plot of the first reservoir. The Campbell plot indicates a weak water drive though with some discrepancies which could be as a result of the nature of the reservoir and aquifer properties. The decrease of the X axis is noticed around 14MMRb of cumulative oil production.

For the cumulative water influx, we also see a trend that increases with large values ranging from 0.1 to 5.8MMbbl.

4.3. Reservoir C

For the third case, we observed a similar but deviated trend in the Campbell plot. The plot also shows the presence of an aquifer characterized by a weak water drive into the reservoir.

A gradual decline was about to be noticed before an abrupt decline on the X axis was noticed around 13MMrb of cumulative oil production.

Regarding the cumulative water influx, the calculated values show an increasing trend with large increment towards the end of the production period ranging from 0.6 to 7.8MMbbls.

4.4. Reservoir D

The last case scenario gave a perfect description of the Cole plot showing a weak water drive. The curve shows an abrupt decline at about 2.7Mscf.

Regarding the water influx calculation, it also shows similar trend to the previously treated cases with We ranging from 1.59 to 84MMbbls

Comparing the results of the three cases, we can summarize:

Table 16. Range of Water Influx values

	Range of We values	Point of decline
Reservoir A	1.5 – 14MMbbls	5.5Mscf
Reservoir B	0.1 – 5.8MMbbls	14Mrb
Reservoir C	0.6 – 7.8MMbbls	13Mrb
Reservoir D	1.59 – 84MMbbls	2.7Mscf

From table 16 above we can infer that reservoir D has the highest range of values for We and the smallest value for the point of decline of the curve.

The steps taken were successful because the Cole and Campbell plot were able to account for the weak water drive.

5. Conclusion and Recommendation

5.1. Conclusion

In this study, the material balance method has proven to be a very useful tool to the reservoir engineer with regards to aquifer detection and characterization. The general material balance equation was re-arranged to come up with an equation which an equation that plots a graph known as Cole and Campbell plot used to characterise the strength of the water drive.

Applying this method to the reservoir data, we were able to come up with plots similar to the modal proposed by Cole using Microsoft excel to aid accurate calculation.

These plots show the presence of a water drive and from the nature of the curves, the plots show weak water drives.

The Cole plot (gas) and Campbell plot (oil) diagnose the presence of a weak waterdrive unambiguously. Depletion-drive plots, such as the p/z, are ambiguous in the presence of a weak water drive and can give OHIP values that are erroneously high by a significant amount. As suggested by previous authors, the weak waterdrive signature on the Cole and Campbell plots is shown to be a negative slope. The study was successful and desirable results gotten.

Generally, the Cole and Campbell method was successful in determining weak water drives in aquifer.

5.2. Recommendation

Production data to be used for material balance analysis should be carefully obtained.

Reasonable assumptions can be made where necessary.

Microsoft excel can be used for calculations regarding material balance to be able to avoid human error and inaccurate result.

The Cole and Campbell plot should be used as a more accurate method for diagnosing aquifer strength.

During diagnoses of water drives, We should be assumed to be neglected in order to ascertain its presence.

References

[1]   Forest, A.G: "Oil and Gas Reserves Classification, Estimation and Evaluation", paper 13946 received in 1985 for SPE's Revision of Petroleum Engineering Handbook, USA.

[2]   Abdus, S. and Ganesh, C.T.: Integrated Petroleum Reservoir Management. 1994, Tulsa, OK, USA.: pp105. PennWell Company.

[3]   Craft, B. C. and Hawkins, M. F.: Applied Petroleum Reservoir Engineering. 1959, NJ, USA.: pp 70-71. Prentice Hall.

[4]   Tarek, A.: Reservoir Engineering Handbook. 2006, 2nd Edition, Houston, Texas, USA.: Gulf Professional Publishing.

[5]   Kewen, L. and Roland, N. H.: "A Decline Curve Analysis Model Based on Fluid Flow Mechanisms", paper 83470 presented at the 2003 SPE Western Region/ AAPG Pacific Section Joint Meeting, Long Beach, California, USA. May, 19-24.

[6]   Dake, L.P.: Fundamentals of Reservoir Engineering. 1994, Amsterdam.: pp310-315. Elsevier.

[7]   Havlena, D. and Odeh, A.S. : "The Material Balance as an Equation of a Straight Line, Part II- Field Cases", paper 559 presented at the 1964 University of Oklahoma- SPE Research Symposium, Dallas, Tx., USA. April, 20-30.

[8]   Havlena, D. and Odeh, A.S. : "The Material Balance as an Equation of a Straight Line, Part II- Field Cases", paper 869 received in 1964 at the Society of Petroleum Engineers' Office, Dallas, Tx., USA. May, 26.

[9]   J.L. Pletcher, Improvements to Reservoir Material-Balance Methods SPE, Marathon Oil Co.

[10] Campbell, R.A. and Campbell, J.M., Sr.; "Mineral Property Economics, Vol.3: Petroleum PropertyEvaluation, Campbell Petroleum Series, Norman, OK, 1978.

[11] Dake, L.P.; "The Practice of Reservoir Engineering," Elsevier, Amsterdam, 1994, page 473.

[12] Wang, B. et al; "OILWAT: Microcomputer Program for Oil Material Balance with Gas Cap and Water Influx," SPE 24437, Petroleum Computer Conference, Houston, Texas, July 19-22, 1992

[13] Arps, J. J.: "Analysis of Decline Curves", Manuscript received in 1944 at the A.I.M.E. Institute, Tulsa, Oklahoma, USA. May, 9.

[14] Ikoku, C. U.: Natural Gas Reservoir Engineering. 1984, New York, USA.: John Wiley & Sons, Inc.

[15] Agarwal, R. G., Gardner, D. C., Kleinsteiber, S. W. and Fussell, D.D.: "Analyzing Well Prodcution Data Using Combined-Type Curve and Decline-Curve Analysis Concepts", paper 57916 presented at the 1998 Annual SPE Technical Conference and Exhibition, New Orleans, USA. September, 27-30.

[16] Okpala, R. I.: "Comparing Various Method of Reserve Estimation", 2005, Project Work.

[17] Main, M. A.: Project Economic and Decision Analysis. Vol. II: Probabilistics Models.

[18] P.Behrenbujch and L.T. Mason, Optimal oilfield development of fields with a small gas cap and strong aquifer,   1993.

[19] R.J. Gajdica, R.A. Wattenberger and R.A. Startzman. A new method of matching Aquifer Performance and determining original gas-in-place. SPE 16935 presented at Dallas Texas September 27-30, 1987.

[20] John McMullan, Material Balance: The forgotten reservoir engineering tool.

[21] Abdus Satter et al, Practical enhanced reservoir engineering; Assisted with simulated software 2008.

[22] Ahmed Y. Abukhamsin June 2009, Optimization of well design and location in a real field,

5

Maturity and Safety of Compost Processed in HV and TW Composting Systems

Kutsanedzie F.[1, *], Ofori V.[2], Diaba K. S.[3]

[1]Research and Innovation Department, Accra Polytechnic, Accra, Ghana
[2]Agricultural Engineering Department, Kwame Nkrumah University of Science and Technology, Kumasi, Ghana
[3]AgriculturalEngineering Department, Anglican University College of Technology, Sunyani, Ghana

Email address:
fkutsanedzie@apoly.edu.gh (Kutsanedzie F.)

Abstract: Two composting systems: passive aerated system, horizontal-vertical system; active aerated system, turned windrow system was designed, constructed and studied for thirteen weeks to compare the maturity and safety of compost processed in them. Waste materials with the following percentage composition: river reed (75%), clay (10%), banana stalk / stem (5%), cow manure / dung (4%), rice chaff (4%), cocoa seed husk (1%), and poultry manure (1%) was processed in the two different systems. Compost materials were weekly sampled from top, bottom and mid portions of the compost masses in each of the systems and bulked to constitute a representative sample for the respective systems. From the weekly representative samples for the two systems, 5g subsamples were used for the determination of the germination index and microbial load for the compost masses. However temperature readings of the compost masses in both systems were recorded insitu daily at three different points using a long stem thermometer and the their respective averages used as the weekly readings for the systems. Temperature and the germination indices of composts processed in the two systems were used as parameters to assess the maturity; and the microbial load and its survival to assess the safety of the compost churned out. There were significant differences in the temperatures recorded in the two different systems during the composting period (p-value = 4.75×10^{-7}, at $\alpha = 0.05$). The total viable counts recorded in HV and TW ranged between 6.90 - 7.75logCFU/g of compost and 7.11–7.79logCFU/g of compost respectively which were significantly different (p-value = 0.027, at $\alpha = 0.05$). The total fungi counts recorded in HV and T-W ranged between 1.11 – 2.32logCFU/g of compost and 1.68 - 2.40logCFU/g of compost respectively. Compost masses in all the systems had germination indices more than 150%.T-W had a higher rate of decomposition and maturity comparatively, hence a better composting system based on compost maturity. Penicillium spp. survived the process and it is known to produces mycotoxins which cause illnesses in humans. It is recommended that compost end-users, farmers, and producers use protectives and observe good hygiene in order to safeguard their health.

Keywords: Passive Aerated Systems, Active Aerated Systems, Germination Index, Maturity

1. Introduction

Composting has been an organic waste management method used for the production of organic fertilizer worldwide. Organic grown foodstuff are highly priced and considered as healthy compared to food grown with chemical fertilizers. Many types of soil-borne pathogens commonly found in soilless potting media are known to be disease causing and also negatively affect yields (Inbar *et al.*, 1993). However compost suppresses these pathogens but the degree of suppression depends on the maturity and type of compost (Ben-Yephet & Nelson, 1999; Hoitink *et al.*, 1997).

Scientists worldwide have been researching on the parameters and methods that yield safe and matured compost within short duration because of its demand for the growing of crops and other horticultural plants.

There are several types of systems used in composting of organic matter but all these different systems are classified broadly into two: passive aeration and active aeration systems. Passive aeration systems are those which are expected to aerate themselves once the process has been set whereas active aerated systems involves using human or machine power to provide aeration for the composting process. According to Sundberg (2005) some common systems includes naturally aerated windrow systems-long rows with triangular cross section; forced aeration static pile systems; tunnel systems-closed rotating cylinders (Sundberg 2005). Sherman (2005) describes passive systems of composting as requiring very little attention while static heaps are maintained through forced aeration or frequent mechanical turning. According to Hao et al. (2001), passive aeration is not as effective as turning in the decomposition of organic matter as observed from the result of the analysis of CO_2 and other gases in composting feedlot manure. Improvement in the aeration in a composting pile has been found to reduce the production of CH_4 production which is a more harmful greenhouse gas compared to CO_2 when a forced aeration system is used (Lopez-Real & Baptista, 1996).

Various parameters that have been used to assess the safety and maturity of composts include the C: N ratio of the finished product, water soluble carbon, cation exchange capacity, humus content, and the carbon dioxide evolution from the finished compost (Garcia et al., 1992; Huang et al., 2001). Germination index, which is a measure of phytotoxicity, has been considered as a reliable indirect quantification of compost maturity (Cunha Queda et al., 2002). Ikeda et al. (2006) compared the germination index method with other methods, using both immature composts and mature composts sold as products. It was observed that the more the decomposition from livestock faeces came to maturity, the higher the germination index became. It was concluded that the germination index method is a highly accurate maturity parameter, under the conditions of setting more than 150% as a maturity standard, using 5 days for seed cultivation. However, degradation of organic materials during composting is done by bacteria, fungi, and actinomycetes, depending on the stage of decomposition, the characteristics of materials, and temperature (Epstein, 1996; USDA, 2000). Bacteria are organisms responsible for compost decomposition and thermophilic bacteria. Pathogens in compost are killed or reduced at temperature above 55 °C (Strom, 1985).

According to Canada Composting Council (2008) at least two of following must be met for compost to be declared mature: C/N ratio is less than or equal to 25; oxygen uptake rate is less than or equal to 150 mg O_2/kg volatile solids per hour; germination of cress (Lepidium sativum) seeds and of radish (Raphanus sativus) seeds in compost must be greater than 90 percent of the germination rate of the control sample, and the growth rate of plants grown in a mixture of compost and soil must not differ more than 50 percent in comparison with the control sample; Compost must be cured for at least 21 days; Compost will not reheat upon standing to 20°C above ambient temperature.

Sequel to the criteria of Canada Composting Council (2008), this study uses the temperature and the germination indices of composts processed in the two systems to assess the maturity; and the microbial load and microbial survival to assess the safety of composts processed in two composting systems: passive aerated system, horizontal-vertical (HV) system; active aerated system, turned windrow (TW) system.

1.1. Problem Statement

Effective and efficient composting systems must churn out composts that are safe and matured earlier if compost is to be prepared on sustainable basis for application on large scale organic farms. Thus the most efficient and effective system to achieve this needs to be selected. This has necessitated the need to compare compost maturity and safety in horizontal-vertical system and turned windrow system for large scale compost production.

1.2. Objective

The main objective of this paper is to design as experiment to compare the compost maturity through the use of germination index and temperature; and safety in terms of microbial loads and microbial survival in the two stated systems.

1.3. Area of Study

Volta River Estate Limited Farms (VREL) is located at Akrade in the Eastern Region of Ghana and owns five farms. They are engaged in banana and pineapple production on large scales. The organic fertilizer used on these farms is prepared via composting using waste materials such as river reed, clay, banana stalk/stem, cow manure/dung, rice chaff, cocoa seed, husk, poultry manure from the farm and environs.

1.4. Description of Systems

1.4.1. Horizontal-Vertical Aeration Technology

The horizontal-vertical aeration technology, known as the T-W aeration is a passive system. It uses of 6 in. diameter uPVC pipes with perforated holes to effect the passive aeration. The pipes are inverted T-shaped, with perforations on the horizontal section allowing ambient air to move into the pile and that of the vertical, allows warm and waste gases to exit compost mass. The dimension of the pile is 6.8m (L)×2.6 (B)m×1.7m (H) and covered with the Toptex (fleece) sheet after mounting the piles. The vertical pipe was perforated to about 1.2m high from the bottom.

Figure 1. Perforated uPVC pipes.

Figure 2. Prepared pile in HV.

1.4.2. Turned Windrows (TW)

With the turned windrow system, the dimension of the TW pile is 35m (L) ×2m (B) ×0.9m (H). Turning was conducted on the condition that temperature or CO_2 levels exceed 650C or 20% respectively and done with the Sandberger ST 300 pulled by a 90 HP tractor. However, turning was done initially four times and a front-loader (165 HP) was used to reshape the pile after which a Toptex (fleece) sheet was used to cover the windrow. On the average, windrow is turned about 6-8 times during the first 2 weeks after which the row was left alone as much as possible.

Figure 3. Sandberger ST 300 turning Feed stock pile.

Figure 4. Covered windrow pile.

2. Materials and Methods

Compost prepared under the two systems consisted of the following in the proportion below: River Reed (RR),75%;

Clay(C), 10%; Banana Stalk/Stem (BS),5%; Cow Manure/Dung (CM), 4%; Rice Chaff (RC), 4%; Cocoa Seed Husk (CSH), 1%; Poultry Manure (PM), 1%. A starter (Soil Tech

Solution) containing genetically modified organisms such as Bacillus spp. and Corynebacterium spp. was mixed in the proportion 500g in 40L water and added to the feedstocks to facilitate the decomposition process.

2.1. Sampling of Compost Mass for Physicochemical and Microbiological Analysis in the Laboratory

Compost masses were sampled at the top, middle and bottom in the two different systems mounted for the research with a forcep every week for laboratory study. The samples taken were bulked to obtain a representative sample. Samples were packed with ice cubes in an ice chest and transported each and every week to the laboratory where they were kept in afreezer at a temperature of 20 °C for a day before microbial analysis was performed.

This experiment was carried out for thirteen (13) weeks, starting from week 0 to week 12 with averaged weekly temperatures; total viable count, coliform count and fungi count; and the germination indices of compost mass in each system determined.

2.2. Temperature Determination

Temperature readings were taken daily from three different sides: top, middle and bottom for 13 weeks in the two composting systems using the long stem thermometer (Salmoiraghi Co. thermometer model 17506) at the site. The daily ambient temperatures were also taken.

2.3. Germination Index Determination

The germination index was determined for compost prepared in both sysyems using the method described by Zucconi (1981), and was computed using the formula expressed as follows:

$$\text{Relative seed germination}(\%)$$
$$= \frac{\text{No. of seeds germinated in litter extract}}{\text{No. of seeds germinated in control}} \times 100$$

$$\text{Relative root growth}(\%)$$
$$= \frac{\text{Mean root length in litter extract}}{\text{Mean root length in control}} \times 100$$

$$\text{GI} = \frac{(\% \text{ Seed germination}) \times (\% \text{ Root elongation})}{100}$$

Where GI = Germination Index

2.4. Serial Dilution for Total Viable, Coliform Count and Fungi Count

1g of representative samples taken from each of the systems mounted for study was weighed into 9ml of 0.1% peptone water contained in 4 different McCartney bottles and incubated at 37°C for 15minutes.They were mixed well and 1ml of the supernatant was drawn from each of the bottles and diluted using 10-fold dilution into 4 other McCartney bottles each containing 9ml of sterile 0.1% blank peptone water. Different pipettes were used for each of the dilution. 1ml of the diluents taken from dilution factors: 1:103,1:104 and 1:105 were transferred into 2 sets of 3 different McCartney bottles one set containing 9ml of molten Plate Count Agar (PCA) and the other set 9ml of Violet Red Bile Agar. Both sets were kept in a water bath at 45 °C to prevent solidification (Collins & Lyne, 1983).

Identification of bacteria was based on the examination of slides of Gram stained microorganisms prepared from pure cultures grown on blood agar, MacConkey agar, Brilliant Green agar and Eosin Methylene Blue agar .The Brilliant Green agar and Eosin Methlyene Blue agar were plated to aid the identification of Salmonella spp. and Escherichia spp. respectively. Slides were observed using the light microscope at x100 with oil immersion. The colonial and cell morphology of microorganisms and reactions to the Gram stain were used in bacteria identification Fungi were identified using their colonial morphology and colour reaction on sabouraud agar in labeled Petri-dishes and incubated at 30°C for 2-7days.

Slides kept in 90% alcohol were dried. The wire loop was flamed and two loopfuls of distilled water was placed at the centre of a cleaned slide. The wire loop was lightly used to touch the identified colony on a plated pure culture, and then gently mixed with the distilled water to dilute their concentration and was spread over the centre of the slide to cover approximately 3cm x 1cm

3. Results and Discussion

3.1. Effects of Temperature on the Factors Determined in the Turned-Windrow System

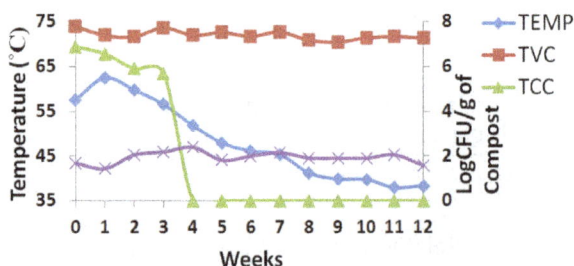

Figure 5. *Variations in Temperature, Total Viable Count, Total Coliform Count and Total Fungi Count in the T-W system.*

3.1.1. Total Viable Count

The total viable count decreased from 7.79logCFU/g of compost to 7.40logCFU/g of compost with a rise in temperature from 57.60°C to 62.56°C during week 0 to week 1 as shown in figure 5. Temperature fell gradually from week 2 to week 7 but was within the thermophilic range. This temperature range caused almost uniform rises and falls in the total viable count during week 2 to week 7. As the temperature fell to 41.17°C (mesophilic condition) during week 8, total viable count decreased to 7.19logCFU/g of compost. There

was a rise in the total viable count from week 10 to 11 and a slight fall in week 12 when temperature fell within the mesophilic range. It was noted also that the percentage frequencies of Corynebacterium spp. increased from week 9 to week 12.

The fall in the total viable count during week 12 might be due to nutrient depletion. The gradual fall in temperature toward the end of the process indicated the stability and maturity of the compost. The bacteria identified during the process were Bacillus spp., Staphylococcus spp., Streptococcus spp., Clostridium spp., Campylobacter spp., Listeria spp., Corynebacterium spp.,Yersinia spp. and Enterobacter spp. Those that survived the process were the Bacillus spp. (28.57%) and Corynebacterium spp. (71.43%).

3.1.2. Total Coliform Count

The total coliform count decreased from 6.88logCFU/g of compost to 6.56logCFU/g of compost with rise in temperature from 57.60°C to 62.56°C during week 0 to week 1 as seen in figure 5. The total coliform count decreased with decreased temperature during the subsequent weeks until week 4 when the total coliform count yielded 0logCFU/g of compost at a temperature of 51.83°C.The total coliform count of 0logCFU/g of compost obtained in week 4 indicated the reduction of faecal contamination. Enterobacter spp. was the only coliform identified during the process.

3.1.3. Total Fungi Count

Total fungi count decreased from 1.68logCFU/g of compost to 1.45logCFU/g of compost with increased in temperature from 57.60°C to 62.56 °C during week 0 to week 1 as shown in figure 5. The total fungi count increased as temperature fell from week 1 to week 4 and subsequently declined in week 5. The total fungi count further rose to values ranging between 1.90logCFU/g of compost to 2.15logCFU/g of compost during week 6 to week 11. There was a fall in the total fungi count in week 12. This distribution of fungi during 6 to 11 confirms the fact that they tolerate low temperature and largely present at the latter stages of composting to decompose cellulose, chitin and lignin. The fungi identified during the process were Penicillium spp., Aspergillus spp., Mucor spp. and Rhizopus spp. The only fungus that survived the process was Penicillium spp.

3.1.4. Germination Index

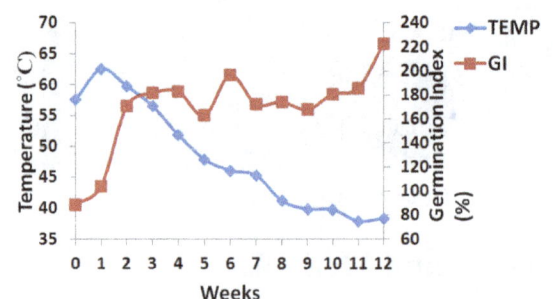

Figure 6. *Variations in Temperature and Germination Index in the T-W systems.*

As temperature increased from 57.60°C to 62.56°C during week 0 to week 1, the germination index increased from 88.58% to 103.89% as illustrated in figure 6. When temperature decreased the subsequent weeks, germination index increased until a decrease was recorded during week 5. There was a fall in the percentage frequency of Bacillus spp., though a rise in the total viable count was recorded during week 5. The rise in the germination index during week 6 could also be attributed to the rise in the level of Bacillus spp., though there was a decrease in the total viable count as seen in Figure 7. The final germination index recorded in the system was 222.64%.

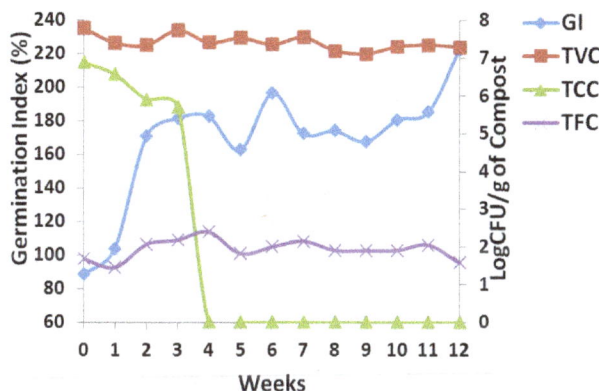

Figure 7. *Variations in Germination Index, Total Viable Count, Total Coliform Count and Total Fungi Count in the T-W system.*

3.2. Effects of Temperature on Factors Determined in the Horizontal-Vertical Technology

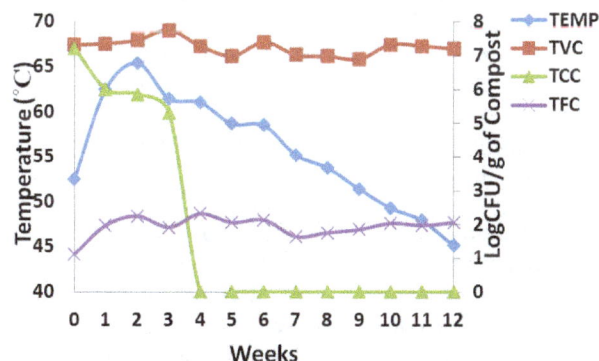

Figure 8. *Temperature variations on Total Viable Count, Total Coliform Count and Total Fungi Count in HV system.*

3.2.1. Total Viable Count

Figure 8 shows that as temperature increased from 52.49°C in week 0 to 65.44°C in week 2, there was a slight rise in the total viable count from 7.31logCFU/g of compost to 7.46logCFU/g of compost. Rise in temperature during week 3 caused a fall in the total viable count. The total viable count ranged between 7.28 logCFU/g of compost and 6.90logCFU/g of compost from week 4 to week 10 with a fall in temperature from 61.07°C to 49.35°C. Temperature fell gradually from the week 3 to week 12. However, the total viable count began to fall from week 10 to week 12.The rise and fall in total viable count between week 2 and week 6 shows that some

microorganisms were being inactivated and favoured at different temperatures during the composting process leading to succession of different organisms at the different stages of the process.

The gradual fall in temperature and total viable count towards the end of process indicate the depletion of nutrients and the process approaching stability, hence less heat was generated. The temperature values recorded in the system during the composting period show that only thermophilic conditions were created.

The bacteria identified during the process were Bacillus spp., Staphylococcus spp., Streptococcus spp., Clostridium spp., Campylobacter spp., Listeria spp., Corynebacterium spp., Yersinia spp and Enterobacter spp. Table I.1 in Appendix I shows that Bacillus spp. (42.86%) and Cornybacterium spp. (57.14%) survived the process and these were the organisms found in the starter used to facilitate the decomposition process.

3.2.2. Total Coliform Count

The total coliform count decreased during week 0 to week 2 from 7.21logCFU/g of compost to 5.85logCFU/g of compost with temperature rise from 52.49°C to 65.44 °C as seen in figure 8. Further fall in temperature from 65.44°C to 61.07°C in week 2 to week 4 led to a fall in the total coliform count from 5.85 logCFU/g of compost to 0, indicating the reduction in faecal contamination. Enterobacter spp. was the only coilform identified.

3.2.3. Total Fungi Count

Figure 8 shows a rise in total fungi count from 1.11logCFU/g of compost to 2.25logCFU/g of compost with rise in temperature from 52.49°C to 65.44°C during week 0 to week 2. The subsequent weeks experienced fluctuations in the total fungi count with a fall in temperature from week 3 to week 7. The total fungi count began increasing during week 8 to week 12 from 1.76logCFU/g of compost to 2.06logCFU/g of compost with a fall in temperature from 53.80°C to 45.19°C. The total fungi count recorded during week 12 was greater than that obtained in week 0. This confirms the fact that fungi are less tolerant to high temperatures and also are responsible for the decomposition of cellulose which are high during the latter stages of composting (Compost Microbiology and Soil Microbiology Web, 2008).

The various fungi identified during the process were Penicillium spp., Aspergillus spp., Mucor spp. and Rhizopus spp., with only Penicillium spp. surviving the process at the end of week 12.

3.2.4. Germination Index

Germination index increased from 82.54% in week 0 with falls in week 1, week 4, week 7 and week 10 and rises during the other weeks till week 12 when 198.11% was recorded for the germination index as shown in fig. 9. The temperature also increased from 52.49°C to 65.44°C from week 0 to week 3 and then fell gradually to 45.19°C in week 12. The rise in germination index from 82.54% to 198.11% denotes the increase in decomposition and maturity of the mass from week

0 to week 12. The gradual fall in temperature toward week 12 denotes the depletion of nutrients, hence less heat generation.

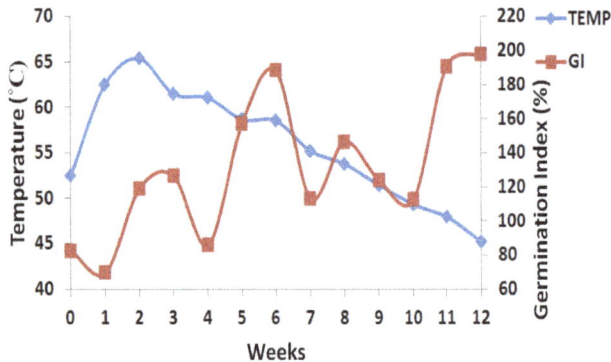

Figure 9. Variations in Temperature and Germination Index in the HV system.

However, when the germination index curve was compared with the total viable count and total fungi count, it was observed from figure 10 that as total fungi count and total viable count rose during week 1, the germination fell. In week 2, rise in total viable count and fungi count gave a rise in germination index. During week 3, the germination index increased with reduction in the total fungi count whereas in week 6, the germination index decreased as the total fungi count increased and total viable count decreased. This indicates that the fungi present were using the nutrient during the early stage of composting hence the low germination index recorded with rises in the fungi count. However, from week 10 to week 12, as the total fungi increased and the total viable count decreased, the germination index increased. This confirm the fact that fungi dominated the process at the end to decompose the cellulose, chitin and lignin coupled with the low temperature that favour their growth as reported in Compost Microbiology and Soil Food Web (2008).

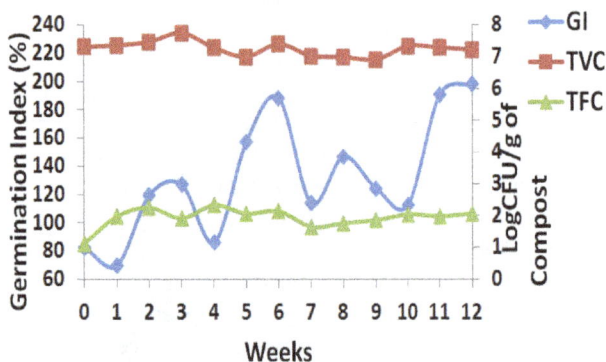

Figure 10. Variations in Germination Index, Total Viable Count and Total Fungi Count in HV system .

3.3. Comparative Analysis of Parameters Determined in the Two Systems

3.3.1. Temperature
Temperature values ranged between 45.19°C - 65.44°C and 37.89°C - 62.56°C in HV and T-W respectively. There were significant differences in temperatures recorded in the different systems during the composting period (p-value =

4.75 x 10-7, at α = 0.05). The T-W system recorded lower temperatures and these can be accounted for by the turning of the compost mass to increase aeration. Different microbes were found in the compost masses in each system at different times a result of the different temperatures recorded in the two systems at various times.

3.3.2. Total Viable Count
The total viable counts recorded in HV, FA and T-W ranged between 6.90 - 7.75logCFU/g of compost and 7.11 –7.79logCFU/g of compost respectively. There was significant differences between the total viable count recorded in the systems (p-value = 0.027, at α = 0.05). The fall in the total viable counts towards the end of the process also suggested the depletion of nutrients.

3.3.3. Total Coliform Count
The total coliform counts ranged between 5.30 – 7.21logCFU/g of compost and 5.70 – 6.88logCFU/g of compost HV and T-W respectively for the initial four weeks. However, during week 4 to week 12, the total coliform counts reduced to 0.00logCFU/g of compost in the various systems studied.

3.3.4. Total Fungi Count
The total fungi counts recorded in HV and T-W ranged between 1.11 – 2.32logCFU/g of compost and 1.68 - 2.40logCFU/g of compost respectively. The total fungi counts recorded in the two systems during the initial week were lower compared to those recorded during the final week except for the T-W system. The decreased moisture contents, temperatures and the presence of large cellulolytic materials in HV system supported the increase in the total fungi count during the latter week of composting while the low total fungi count recorded in T-W system during the final week might have been due to the turnings effected in the compost mass, which consequently resulted in the early decomposition of celluloytic materials.

3.3.5. Germination Index
The germination indices recorded were used as a measure of the rate of decomposition and maturity. Germination indices recorded in the two systems were low but considerably increased during the latter weeks due to the release of nutrients by microorganisms. The germination indices ranged between 82.54% - 198.11% and 88.58% - 222.64% for compost masses in HV and T-W respectively. Compost masses in all the systems had germination indices more than 150% at the end of week 12. Systems T-W had a higher rate of decomposition, hence faster maturity than systems HV; because they had final germination indices of 222.64% and 198.11% respectively.

3.3.6. Microbial Community
Different bacteria were identified in the different systems at different stages of the composting and their frequency decreased at different times due to variation in temperatures recorded in the systems. Bacillus spp., Staphylococcus spp., Streptococcus spp., Clostridium spp., Campylobacter spp., Listeria spp., Corynebacterium spp., Yersinia spp. and

Enterobacter spp. were the bacteria identified during the different phases of the composting process in all the systems.

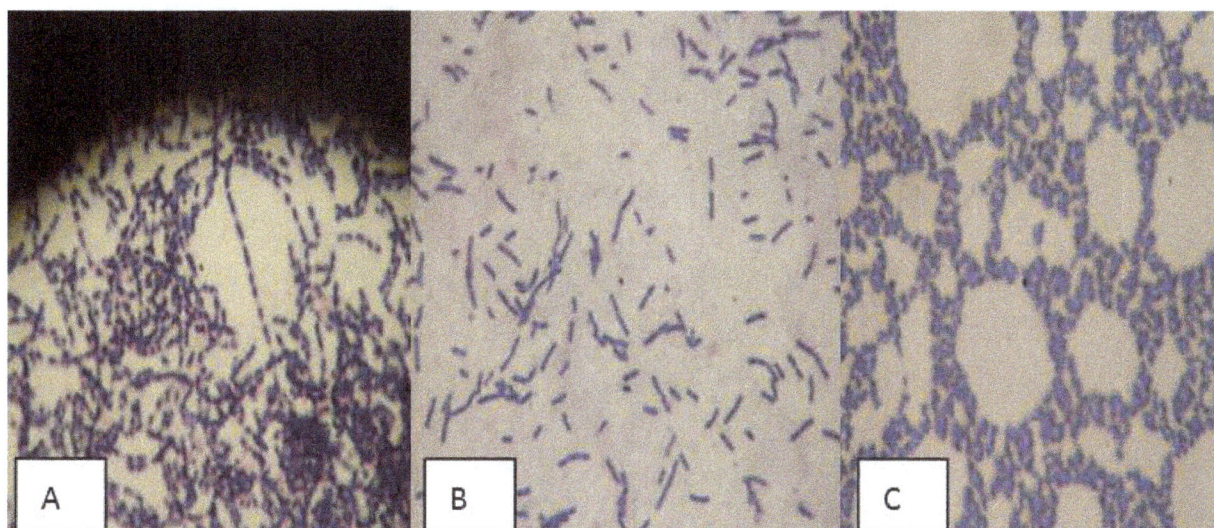

Figure 11. Bacillus spp., A; Corynebacterium spp., B; Staphylococcus spp., C viewed under microscope.

Bacillus spp. (42.86%) and Corynebacterium spp. (57.14%) and (71.43%) survived the process in HV and T-W system respectively. Among the fungi identified in the two systems such as Aspergillus spp., Penicillium spp., Mucor spp. and Rhizopus spp., only Penicillium spp. (100%) survived in both systems.

4. Conclusion

The germination indices obtained from the compost masses in both systems revealed they were higher that 150%, indicating that both systems churned out matured compost. However compost in T-W seems to have a higher rate of decomposition, hence faster maturity comparatively. It therefore serves a better choice in terms of maturity. In addition identical microorganisms survived in both systems; but Bacillus spp. and Corynebacterium spp. were organisms used as starter to facilitate decomposition of compost in both systems. However the only fungi that survived in both systems Penicillium spp. produces mycotoxins such as citrinin, luteoskyrin, ochratoxins and rubratoxin which cause illnesses in humans. Luteoskyrin is associated with high incidence of liver cancer in humans.

It thus can be recommended that compost end-users, farmers, and producers must wear protectives such gloves and observe good hygiene when handling compost in other to safeguard their health.

References

[1] Ben-Yephet, Y. & Nelson, E.B. Differential suppression of damping-off caused by *Pythium aphanidermatum, P. irreulare,* and *P. myriotylum* in composts at different temperatures. Plant Diseases, 83:356-360, 1999.

[2] Collins, C.H. & Lyne, P.M. Microbiological Methods. 5th ed. Butterworth and Co, London,1983, pp.5, 89.

[3] Composting Council of Canada. A Summary of Compost Standards in Canada. Available at: www.compost.org/ standard.html, 2008. (accessed on 31 May 2015).

[4] Compost Microbiology and the Soil Food Web 2008. Available at: http://w ww.ciwmb.ca.gov /publications/Organics/, 2008 (accessed on 5 January 2015).

[5] Cunha Queda, A.C., Vallini, G., Agnolucci, M., Coelho, C.A., Campos, L. & de Sousa, R.B. Microbiological and chemical characterization of composts at different levels of maturity, with evaluation of phytotoxicity and enzymatic activities. In:Insam, H., Riddech, N., Krammer, S. (Eds.), Microbiology of Composting. Springer Verlag, Heidelberg, pp. 345–355, 2002.

[6] Epstein, E. The Science of composting. Technomic Publishing Company, Inc., Lancaster, PA, 1996.

[7] Garcia, C., Hernandez, T., Costa, F. & Ayuso, M. Evaluation of the maturity of municipal waste compost using simple chemical parameters. Commun. Soil Sci. Plant Anal, 23:1501–1512, 1992.

[8] Hao, X.Y., Chang, C., Larney, F.J. & Travis, G.R. Green gas emissions during cattle feedlot manure composting. J. Environ. Qual., 30:376-386, 2001.

[9] Hoitink, H.A., Stone, A.G., and Han, D.Y. 1997,Suppression of plant diseases by compost HortScience, 32:184 -187, 1997.

[10] Hue, N.V. & Liu, J. Predicting compost stability. Compost Science and Utilization, 3:8–15, 1995.

[11] Ikeda, K., Koyama, F. Takamuku, K. & Fukuda, N. The Efficiency of Germination Index Method as a Maturity Parameter for Livestock Faeces Composts. Bulletin of the Fukuoka Agricultural Research Center, 25: 135-139, 2006.

[12] Inbar, Y., Chen, Y., and Hoitink, H.A. 1993, Science and engineering of composting: design, environmental, microbiological and utilization aspects, H.A.J.Hoitink and H. M. Keener (Eds.), Renaissance Publications, Worthington, OH, 669, 1993.

[13] Lopez-Real, J. & Baptista, M. A preliminary comparative study of three manure composting systems and their influence on process parameters and methane emissions. Compost Sci. Util., 4:71-82, 1996.

[14] Sherman, R. Large-scale organic materials composting. North Carolina Cooperative Extension Service. Available at: http://www.bae. ncsu.edu/bae/programs/extension/publicat/vermcompost/ag59 3 .pdf., 2005(accessed on 23 January 2015).

[15] Strom, P.F. Effect of temperature on bacterial species diversity in thermophilic solid-waste composting. Applied and Environmental Microbiology, 50(4): 899-905, 1985.

[16] Sundberg, C. Improving Compost Process Efficiency by Controlling Aeration, Temperature and PH. Doctoral Thesis No.2005:103, Faculty of Natural Resources and Agriculture, Swedish University of Agricultural Sciences, Sweden, pp. 10,11,16-17, 2005

[17] USDA. Composting. Part 637, National Engineering Handbook, NRCS, U.S. Department of Agriculture, Washington, D.C., 2002.

[18] Zucconi, F., Forte, M., Monaco, A. & De Bertoldi, M. Biological evaluation of compost maturity. BioCycle, 22: 27-29, 1981.

6

Experimental and Numerical Study of Free Convective Heat Transfer on the Triangular Fin for the Optimum Condition of Gap's Configuration

Goshayeshi Hamid Reza, Hashemi Bahman

Department of Mechanical Engineering, Mashhad Branch, Islamic Azad University, Mashhad, Iran

Email address:

goshayshi@yahoo.com (G. H. Reza)

Abstract: In this study, the effect of different conditions for making gap in the triangular fin has been investigated on the free convective heat transfer coefficient. Making gap in suitable places on the monolith triangular fins can increase the free convective heat transfer coefficient, so the purpose of this study is to reach the optimum condition of configuration for gaps by which the free convective heat transfer coefficient is increased. For this purpose, eight different conditions of vertical triangular fins located on a vertical plate were made and investigated. One of these conditions doesn't have gap and the other seven conditions had different kinds of gaps configuration. The obtained results showed that the optimum condition is a condition with the configuration of three gaps on the length of the triangular fin. Also for validation of experimental results, the numerical modeling of all the above mentioned conditions was done by using the fluent software.

Keywords: Triangular Fin, Optimum Configuration of Gaps, Monolith Triangular Fin

1. Introduction

Heat transfer is utilized in various industries, cooling and heating systems and controlling desirable condition in various machines such as electronic devices. Hence, studying heat transfer, especially free convective heat transfer in various industrial fields and natural processes are of significance. In fact, the need for decreasing function temperature without using forced convection has resulted in carrying out studies to find free convection improving methods and free convective heat transfer in developed surfaces has been a topic to many laboratory and theoretical studies. In 1963, Starner and McManus [1] studied the free convective heat transfer for four various heat sink models in various dimensions and distances between fins on vertical base surface. In 1965, Welling and Wooldridge [2] carried out laboratory experiences on angled rectangular heat sinks with constant length which were on vertical surface and calculated the fins optimal height for making the highest heat transfer ratio. In 1985, Aziz [3] studied the heat optimization and heat transfer ratio in triangular and rectangular fins under convective boundary conditions.

In 1986, Jofre and Baron [4] studied the free convective

heat transfer by air on vertical surface with triangular groove nesses. In 1988, Kondepudi and O'Neal [5] studied the triangular acicular fins performance under dew layer condition. In 1991, Leung [6] studied the optimal distance for conical fins arrangement and indicated that this distance is relatively independent of the fins arrangement. In 1995, Abratet and Newnham [7] analyzed the triangular fins attached to a thick wall by finite element method. In 1998, Kordyban [8] compared the performance of heat sinks with direct fins and acicular fins. Results to his research suggested that the performance of the heat sink with acicular fins was significantly better than heat sink with direct fins. In 2009, Kobus and Oshio [9] came to this conclusion that the heat performance of an acicular fin heat sink is a weak function of fin diameter and it increases by the increase in fin length. Also, acicular fins have the best heat performance when their diameter and height are not equal. In 2007, Nada [10] carried out a laboratory experiment on horizontal and vertical rectangular finned chambers with fins perpendicular to the chamber surface. He came to this conclusion that the optimal fins distance could be reached for the case that the heat transfer ratio and its performance are maximum. Using numerical modeling in 2008, Edlabadkar et al. [11] studied the effect of angle change in V-shaped fin rays on free convective heat transfer coefficient. They came to this conclusion that, among all studied angles, 90 degree angle had the highest free convective heat transfer coefficient. In the same year, to predict the triangular fins mass and heat transfer, Barman and Debnath [12], Rao [13] and Ji [14]proposed an analytical method. In 2012, Naseriyan and Fahiminiya [15] studied the free convective heat transfer on V-shaped fins and optimal fins arrangement. Research results suggested that, initially, by the increase in fins distance, the heat resistance decreases until it reaches the optimal distance, which has the lowest heat resistance. By the further increase in distance, heat resistance increases. Also, with equal fin height length and constant temperature difference, free convective heat transfer coefficient increases by the increase in fin distance until it reaches the maximum point. Subsequently, by the further increase in distance, heat transfer coefficient decreases.

2. Free Convective Heat Transfer General Equations

Continuity equation and momentum equation in x direction are as following:

$$\frac{\partial u}{\partial x} + \frac{\partial v}{\partial y} = 0 \tag{1}$$

$$\rho\left(u\frac{\partial u}{\partial x} + v\frac{\partial u}{\partial y}\right) = -\rho g - \frac{\partial P}{\partial x} + \mu\left(\frac{\partial^2 u}{\partial x^2} + \frac{\partial^2 u}{\partial y^2}\right) \tag{2}$$

By substituting $-\rho g - \frac{\partial P}{\partial x} = -\beta\rho g(T_\infty - T)$ in equation

(2) and simplification, momentum equation in x direction turns to:

$$u\frac{\partial u}{\partial x} + v\frac{\partial u}{\partial y} = g\beta(T - T_\infty) + v\left(\frac{\partial^2 u}{\partial x^2} + \frac{\partial^2 u}{\partial y^2}\right) \tag{3}$$

In which, β is the fluid volumetric thermal expansion coefficient.

In equation (3), it is clear how the buoyancy force, which forms the flow, is related to the temperature difference. Also, momentum equation in y direction is as following:

$$\rho\left(u\frac{\partial v}{\partial x} + v\frac{\partial v}{\partial y}\right) = -\frac{\partial P}{\partial y} + \mu\left(\frac{\partial^2 v}{\partial x^2} + \frac{\partial^2 v}{\partial y^2}\right) \tag{4}$$

To simultaneous solution of equations (1) and (3) for determination of (u, v, T), energy equation as the following, is needed:

$$u\frac{\partial T}{\partial x} + v\frac{\partial T}{\partial y} = \alpha\left(\frac{\partial^2 T}{\partial x^2} + \frac{\partial^2 T}{\partial y^2}\right) \tag{5}$$

3. Discussion and Results Review

Manufactured triangular fin heat sinks include 8 positions and each heat sink has two fin rows. One of the 8 positions is the position without gaps and with monolith vertical fin, which is considered as the base position (Position 0) and the other 7 positions include various gaps arrangement. The data on fins and base wall dimensions for all manufactured heat sinks is presented in Table (1). Moreover, the distance between the fin rows in each heat sink is 7.9 mm and the angel between fin and base wall in each row is 90 degrees. Figure (1) presents all manufacture triangular fin positions.

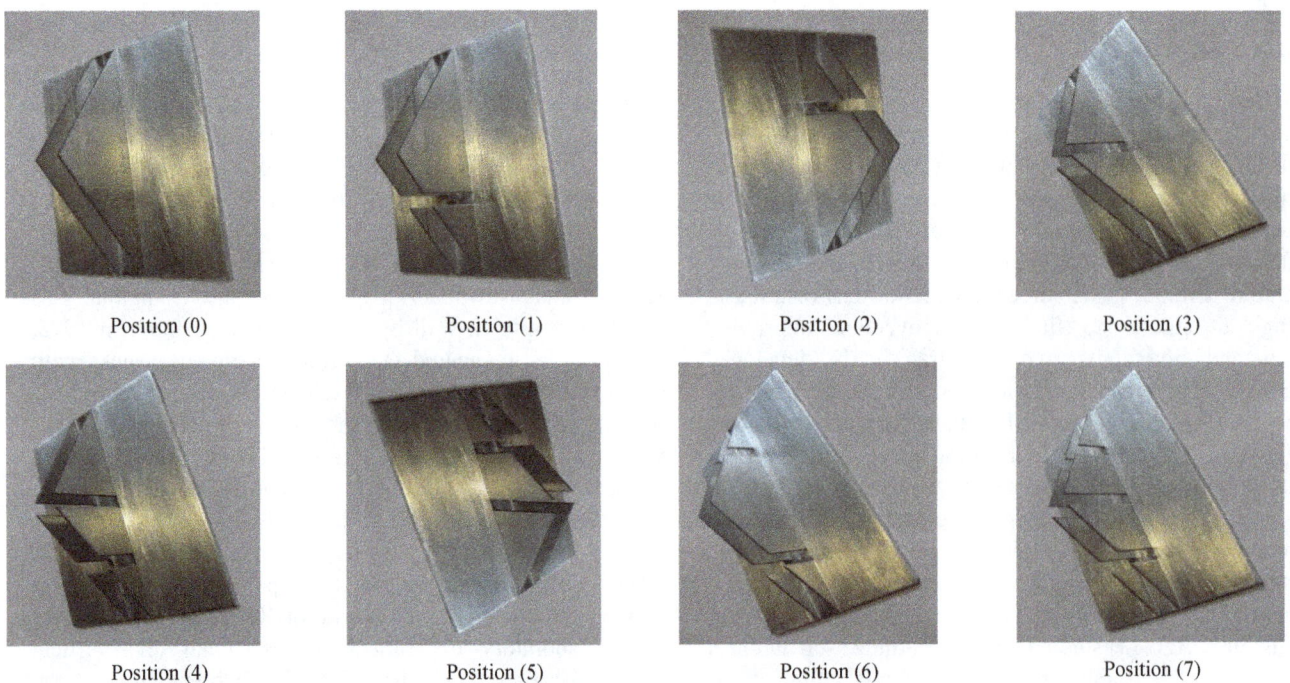

Position (0)   Position (1)   Position (2)   Position (3)

Position (4)   Position (5)   Position (6)   Position (7)

Figure 1. All manufacture triangular fin positions.

Table 1. Fins and Base Wall dimensions.

Base Wall dimensions (mm)			Fins dimensions (mm)			
Length(L)	Width(W)	Thickness (t_b)	Length(L)	Height(H)	Thickness (t)	Gap's Length (S)
80	59.8	1.4	80	20	1	4

The experiments and numerical modeling for all 8 positions were carried out at two temperature differences of 70 to 100 Kelvin, and the results are presented in Diagram (1) and (2). Diagram (1) is related to experimental results and Diagram (2) is related to fluent results in the aforementioned temperature differences.

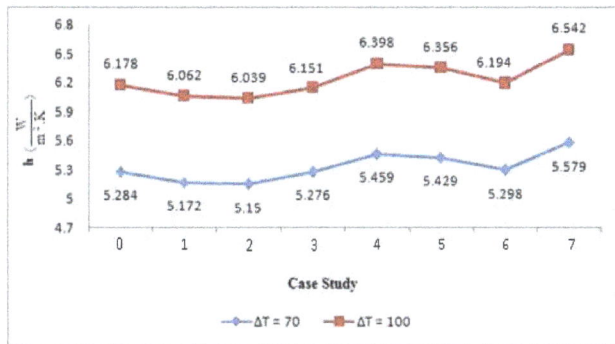

Diagram 1. Experimental results at two temperature differences of 70 to 100 Kelvin.

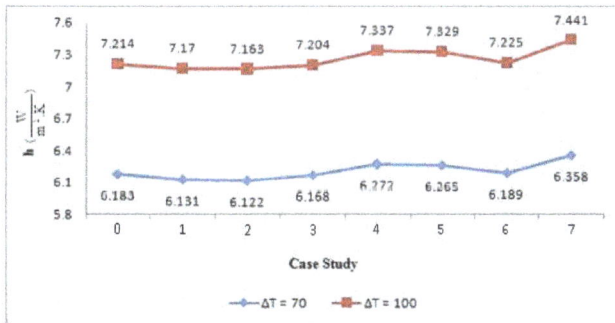

Diagram 2. Fluent results at two temperature differences of 70 to 100 Kelvin.

As it could be observed in diagram (1), it is obvious that the experimental results in the two temperature differences are quite similar. This similarity could also be observed in Diagram (2), for results from Fluent software in two temperature differences. Hence, to prevent the confusion, the results from the temperature difference of 70 Kelvin is going be studied. Also, on diagram (1) and (2), it could be observed that by the increase in temperature difference, free convective heat transfer coefficient increases, for by the increase in temperature difference, Grashof number increases; that is, the buoyancy force which creates free convective flow increases and this leads to increase in flow rate and as a result free convective heat transfer coefficient increases. To verify the experimental results from the 8 positions, the experimental results were compared to the results from Fluent software in diagram (3). As it could be observed from diagram (3), the experimental results comply with the results from Fluent software. Diagram (4), present ratio of the experimental

amounts of free convective heat transfer coefficient for the 7 gapped positions to the experimental amount of free convective heat transfer coefficient for the position without gap (monolith fin), which is considered as the base position (Position 0). Diagram (5), present ratio of the Fluent amounts of free convective heat transfer coefficient for the 7 gapped positions to the Fluent amount of free convective heat transfer coefficient for the position without gap (monolith fin), which is considered as the base position (Position 0).

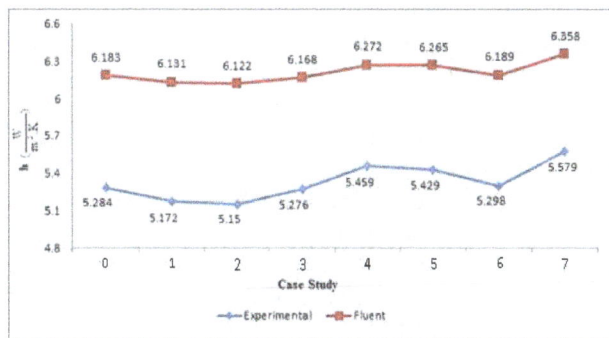

Diagram 3. Comparison of experimental results with results from Fluent at temperature difference of 70 Kelvin .

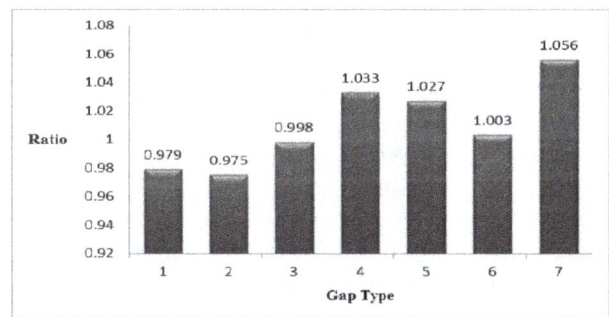

Diagram 4. Ratio of the experimental amounts of free convective heat transfer coefficient for the 7 gapped positions to the experimental amount of free convective heat transfer coefficient for the position without gap at temperature difference of 70 Kelvin.

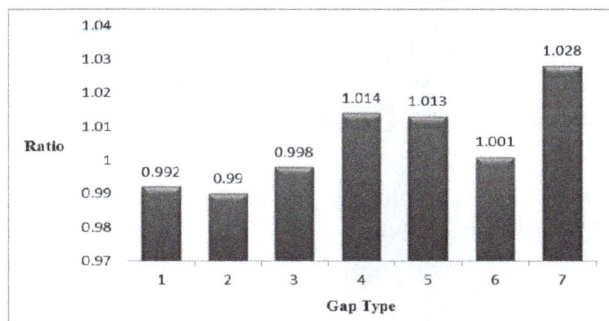

Diagram 5. Ratio of the Fluent amounts of free convective heat transfer coefficient for the 7 gapped positions to the Fluent amount of free convective heat transfer coefficient for the position without gap at temperature difference of 70 Kelvin.

In all studied heat sinks, by the increase in distance from the base surface initial edge, the convective heat penetration in free flow increases and the thermal boundary layers grow in two transverse and longitudinal directions of the base surface, which leads to a decrease in free convective heat transfer coefficient. Also, the thermal boundary layers which are not able to grow unlimitedly in the space between the two fins, crash into each other and develop. Thermal boundary layers development also has a remarkable effect on decreasing the free convective heat transfer coefficient, for this prevents establishing a proper free air flow in the space between the fins and as a result, the temperature steep decreases and by the decrease in the temperature steep, free convective heat transfer coefficient and also convective heat transfer ratio decrease. Forming gaps in proper locations on the fins could lead the air to properly flow between the fins and by merging the air between the fins in gaps places with free air flow, the thermal boundary layer thickness decreases and as a result the boundary layer development weakens or even in some cases it is lost, which in turn, it increases the free convective heat transfer coefficient.

Considering the diagram (4) and (5), it is obvious that positions (1) , (2) , and (3) have a relatively lower free convective heat transfer coefficient comparing to monolith fins position. Hence, in studied heat sinks, forming only one gap on the fins is not helpful, for it does not have a huge impact on decreasing thermal boundary layers thickness in two fins outer space and also, lack of ability in preventing thermal boundary layers development or lack of ability in decreasing the development effects formed in two fins outer space. Hence, forming only one gap on the fins not only does not impact the increase in free convective heat transfer coefficient comparing to the monolith fins position, but also it decreases the free convective heat transfer coefficient due to the omission of the useful fins parts which could lay in the thermal boundary layer outer space and exchange heat with the environment air. Also, experimental and Fluent results for position (6) suggest that the increase in free convective heat transfer coefficient for this position is not much sensible and forming gap on the monolith fins under this position arrangement does not provide a huge difference on free convective heat transfer coefficient, comparing to the monolith fins position. That is due to the fact that in position (6), despite the presence of two gaps on fins length, since the distance between these two gaps is huge, thermal boundary layers have the opportunity to remarkably grow after the first gap and before reaching the second gap. Hence, in addition to the fact that the thermal boundary layersin the two fins outer space increases in thickness, the thermal boundary layers in the two fins outer space become completely developed, so that the formation of the second gap is hugely ineffective and cannot decrease the thermal boundary layers effectively and also, leads to a decrease in development effects in the space between the two fins. Accordingly, forming a gap on the monolith fins under this position arrangement is not helpful.

From diagram (4) and (5), it is obvious that positions (4), (5), and (7) have a higher free convective heat transfer coefficient comparing to monolith fins position. Hence, forming gaps on monolith fins under aforementioned positions could lead to an increase in free convective heat transfer coefficient. Due to the continues and close effects which have on the thermal boundary layers, presence of two gaps on the fins length (Positions 4 and 5) could have a great impact on the decrease in boundary layers thickness, delaying their development, or decreasing their development effects and as a result the increase in free convective heat transfer coefficient. The impact could be highly decreased, especially if one of the gaps is formed before reaching the fins half-length (Position 4), for this prevents the excessive growth of thermal boundary layers thickness in fins initial length, where the thermal boundary layer thickness is low, and thermal boundary layers do not have a high thickness while reaching the second gap. As a result, the development of thermal boundary layers happen in higher longitudinal distance from the base surface initial edge. Ultimately, by forming three gaps in fins length, air properly flows between the fins and as a result by merging the air between the fins with free air flow in gaps places, the thermal boundary layers thickness decreases remarkably. This fact leads to a delay in thermal boundary layers development and also effective decrease of its impacts and as a result free convective heat transfer coefficient increases to the maximum. Hence, the best arrangement for forming gaps in a heat sink with triangular fins is the arrangement with three gaps in fins length (Position 7). Figure (2) , (3) and (4) , present the free convective heat transfer coefficient contours for monolith fins (Position 0), the worst gaps arrangement (Position 2) and position (6) are respectively. Also, Figure (5) , (6) , and (7) present the free convective heat transfer coefficient contours for position (4) , (5) and position (7) which has the best gap arrangement, respectively.

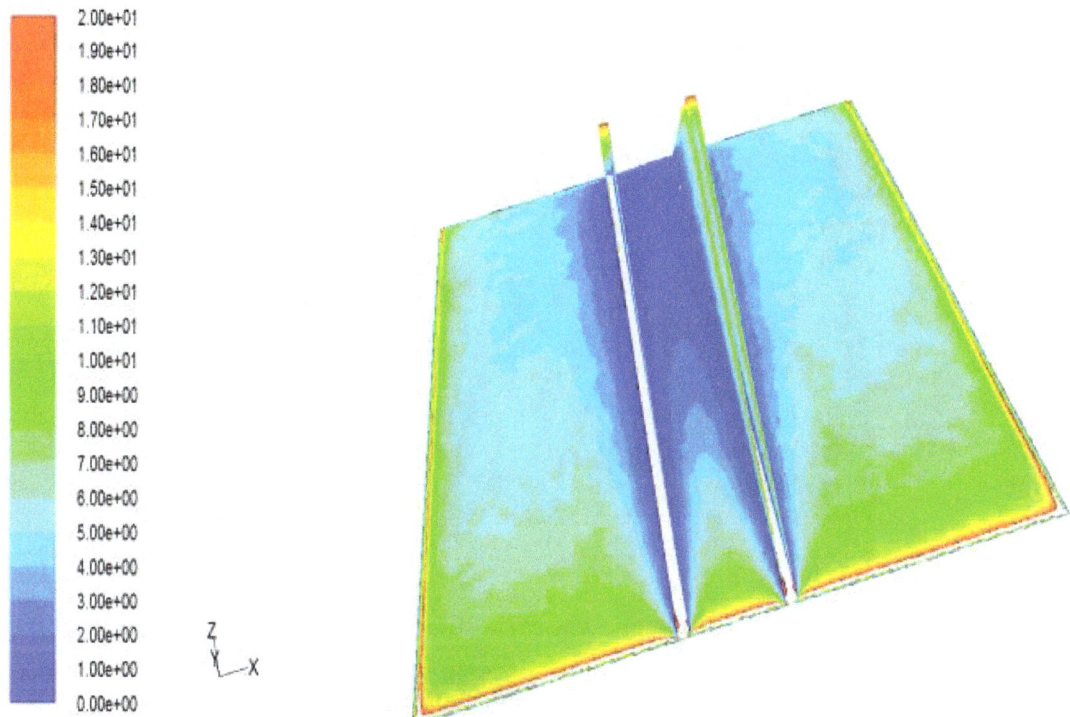

Contours of Surface Heat Transfer Coef. (w/m2-k)

Nov 26, 2013
FLUENT 6.3 (3d, pbns, lam)

Figure 2. Contour of free convective heat transfer coefficient for monolith fin.

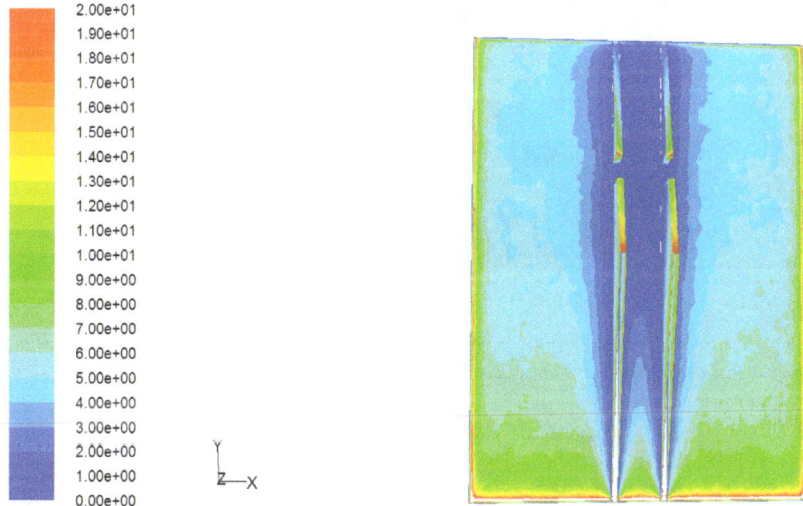

Contours of Surface Heat Transfer Coef. (w/m2-k)

Oct 19, 2013
FLUENT 6.3 (3d, pbns, lam)

Figure 3. Contour of free convective heat transfer coefficient for position 2.

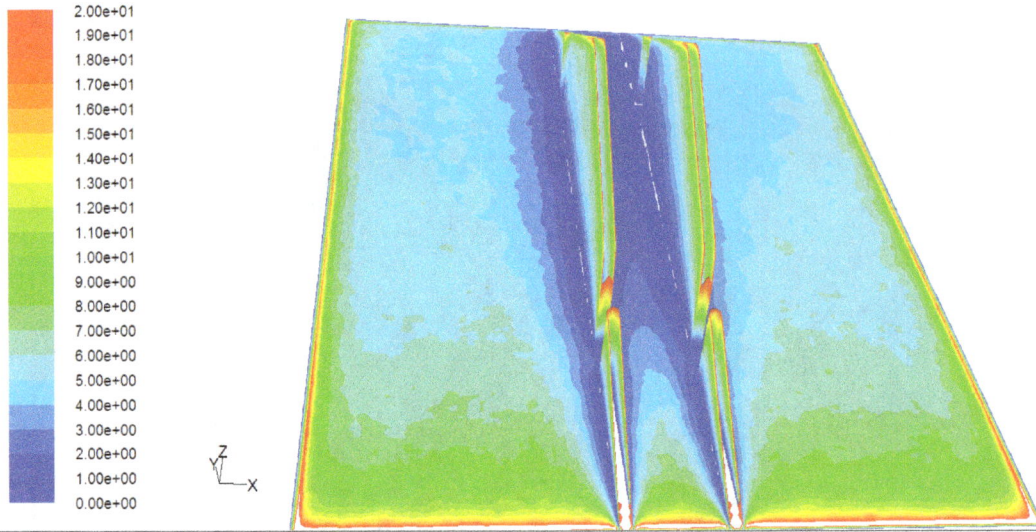

Contours of Surface Heat Transfer Coef. (w/m2-k)

Oct 19, 2013
FLUENT 6.3 (3d, pbns, lam)

Figure 4. *Contour of free convective heat transfer coefficient for position 6.*

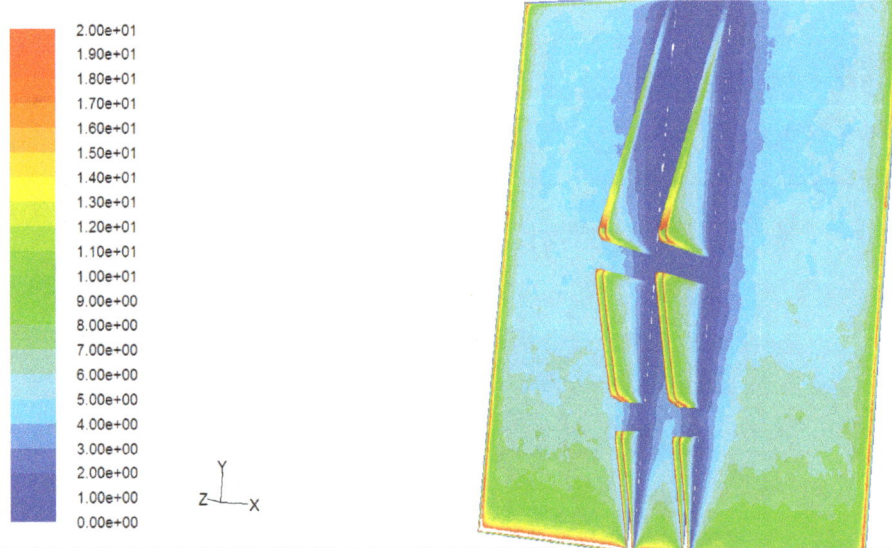

Contours of Surface Heat Transfer Coef. (w/m2-k)

Oct 19, 2013
FLUENT 6.3 (3d, pbns, lam)

Figure 5. *Contour of free convective heat transfer coefficient for position 4.*

2.00e+01
1.90e+01
1.80e+01
1.70e+01
1.60e+01
1.50e+01
1.40e+01
1.30e+01
1.20e+01
1.10e+01
1.00e+01
9.00e+00
8.00e+00
7.00e+00
6.00e+00
5.00e+00
4.00e+00
3.00e+00
2.00e+00
1.00e+00
0.00e+00

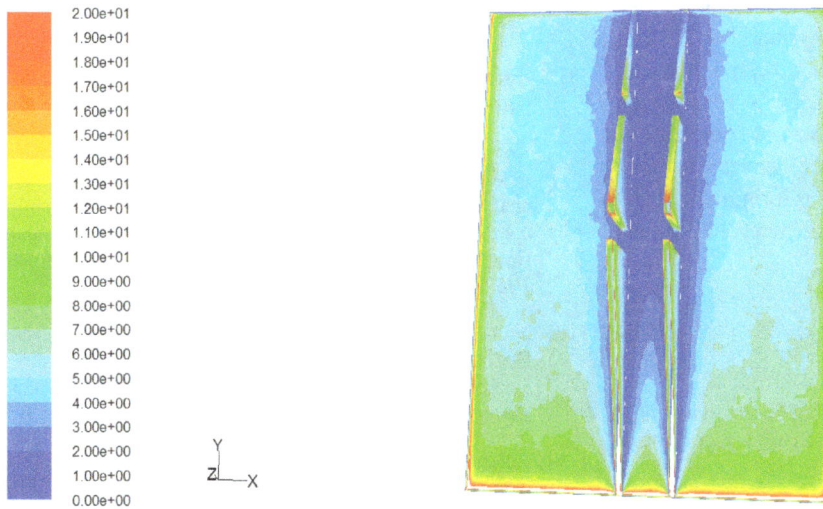

Contours of Surface Heat Transfer Coef. (w/m2-k)   Dec 12, 2013
                                                      FLUENT 6.3 (3d, pbns, lam)

Figure 6. Contour of free convective heat transfer coefficient for position 5.

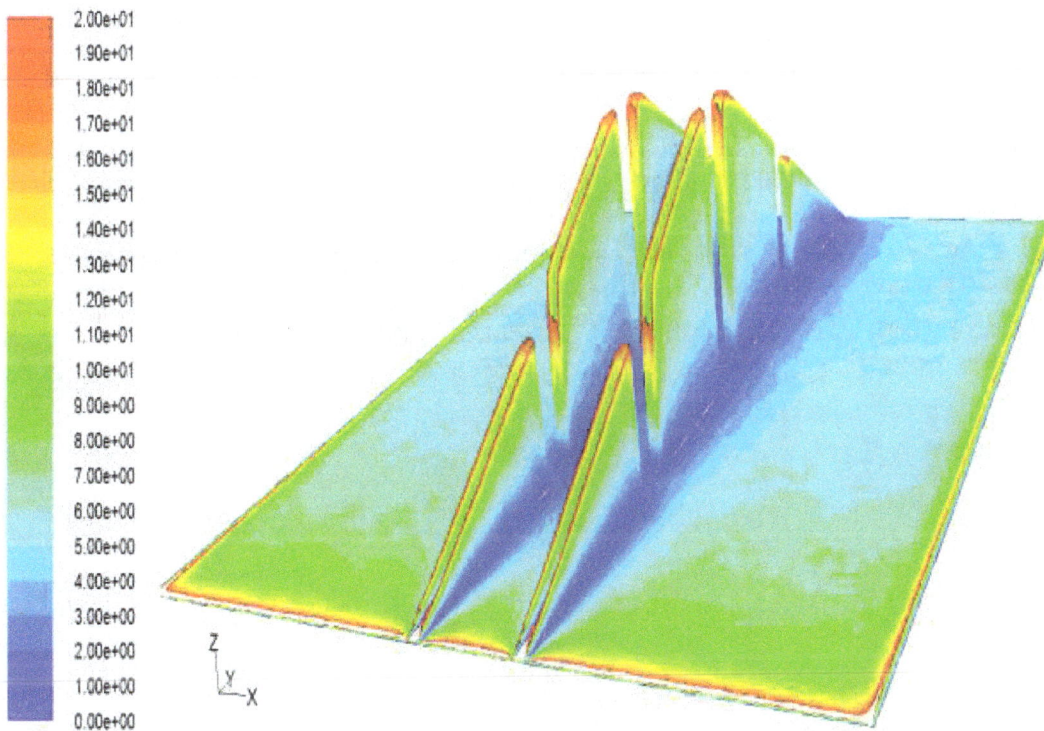

2.00e+01
1.90e+01
1.80e+01
1.70e+01
1.60e+01
1.50e+01
1.40e+01
1.30e+01
1.20e+01
1.10e+01
1.00e+01
9.00e+00
8.00e+00
7.00e+00
6.00e+00
5.00e+00
4.00e+00
3.00e+00
2.00e+00
1.00e+00
0.00e+00

Contours of Surface Heat Transfer Coef. (w/m2-k)   Oct 19, 2013
                                                      FLUENT 6.3 (3d, pbns, lam)

Figure 7. Contour of free convective heat transfer coefficient for position 7.

4. Conclusion

By the increase in the distance from the base surface initial edge in heat sinks with vertical fins, the convective heat penetration in free flow increases and the thermal boundary layers grow. By the increase in thermal layer boundary layers thickness and development formation, the temperature steep decreases and by the decrease in the temperature steep, free convective heat transfer coefficient and also convective heat transfer ratio decrease. By forming proper gaps in heat sink fins, the thermal boundary layer development decreases.

Forming gaps in proper locations on the fins could lead the air to properly flow between the fins and by merging the air between the fins in gaps places with free air flow, the thermal boundary layer thickness decreases and as a result the boundary layer development weakens or even in some cases it is lost, which in turn, it increases the free convective heat transfer coefficient. The best arrangement for forming gaps in a heat sink with triangular fins is the arrangement with three gaps in fins length (Position 7).

References

[1] Starner, K.E., McManus, H.N. (1963) , "An experimental investigation of free convection heat transfer from rectangular fin arrays", J., Heat Transfer 85(2), pp. 273–278.

[2] Welling, J.R., Wooldridge, C.B. (1965) , "Free convection heat transfer coefficients from rectangular vertical fins",Trans. ASME J. Heat Transfer 87 (3), pp. 439–444.

[3] Aziz, A. (1985) ," OPTIMIZATION OF RECTANGULAR AND TRIANGULAR FINS WITH CONVECTIVE BOUNDARY CONDITION," INT. COMM. HEAT MASS TRANSFER, Vol. 12, pp. 479-482.

[4] Jofre, R.J., and Baron, R.F. (1986) , "Free convection heat transfer to a rough plate," ASME Paper No. 67.

[5] Kondepudi, S. N., O'Neal, D. L. (1988) , "PERFORMANCE OF TRIANGULAR SPINE FINS UNDER FROSTING CONDITIONS," Heat Recovery Systems & CHP. Vol. 8, No.1, pp. 1-7.

[6] Leung, S. E., Ho, C. W., and Probert, S. D. (1991) , "Performances of heat exchangers with tapered fins".

[7] Abratet, S., Newnham, P. (1995) , "FINITE ELEMENT ANALYSIS OF TRIANGULAR FINS ATTACHED TO A THICK WALL," Computers & Structures, Vol. 51, No.6, pp. 945-957.

[8] Kordyban, T. (1998) , "Hot air rises and heat sink everything you know about cooling electronice is wrong," ASME Press.

[9] Kobus, C.J., and Oshio, T. (2004) , "Development of a theoretical model for predicting the thermal performance characteristics of a vertical pin-fin array heat sink under combined forced and natural convection with impinging flow," International Journal of Heat Mass Transfer 48, 1053-1063.

[10] Nada, S.A. (2007) ," Natural Convection Heat Transfer in Horizontal and Vertical Closed Narrow Enclosures with Heated Rectangular Finned Base Plate", Int. J. Heat Mass Transfer, Vol. 50, pp. 667-679.

[11] Edlabadkar, R.L., Sane, N.K., Parishwad, G.V. (2008) , " Computational Analysis of Natural Convection With Single V-Type Partition Plate", 5th European Thermal-Sciences Conference, Netherlands.

[12] Kundua, B., Barman, D., Debnath, S. (2008) , " An analytical approach for predicting fin performance of triangular fins subject to simultaneous heat and mass transfer," International journal of refrigeration 31, pp. 1113–1120.

[13] Rao, V. D., Naidu, S.V., Rao, B.G., Sharma, K.V.,(2006) "Heat Transfer from a Horizontal Fin Array by Natural Convection and Radiation a Conjugate Analysis", Int. J. Heat and Mass Transfer. Vol. 49, pp. 3379–3391.

[14] Ji, S.Y. Kim,S, Y, (2007) "Pressure drop and heat transfer correlations for triangular folded fin heat sinks", IEEE Trans. Compon. Packaging Technol. 30 (1).(2007) 3–8.

[15] Naserian, M.M., Fahiminia, M., Goshayeshi, H.R. (2013) , "Experimental and numerical analysis of natural convection heat transfer coefficient of V-type fin configurations " Journal of Mechanical Science and Technology.

7

CFD Analysis on Flow Through a Resistance Muffler of LCV Diesel Engine

Pradyumna Saripalli, K. Sankaranarayana

Mechanical Engineering, Gitam Institute of Technology, Gitam University, Visakhapatnam (A.P), India

Email address:

Pradyumnasaripalli@gmail.com (P. Saripalli)

Abstract: The exhaust pollution has become one of the important problems of environment pollution with applications in automobile industry, and the exhausted muffler has been paid attention to improve the performance of engines. Computational Fluid Dynamics (CFD) method was used to explore the aerodynamic performance of the muffler. Resistance muffler research relates with the fields of acoustics, fluid dynamics, heat transfer and mechanism design. The project report simulates the field by numerical method with Cosmos Flow and analyses the effect which the internal flow field has on the performance of the muffler. With this method the pressure distribution in the muffler is simulated and the pressure loss is predicted for the structure modification. The experiment results verify that the assembly performance of the muffler modified is better than the original muffler.

Keywords: CFD, Muffler, Cosmos Flow, Field Flow, Acoustics, Pressure Loss

1. Introduction

Internal combustion engines are the major power sources for automobiles and also the work developers for the small scale power generation units. They operate by taking in a mixture of air and fuel (petrol engines) or by taking only air (diesel engines) and compressing it followed by combustion. At this stage power is produced while combustion of the fuel and ultimately exhaust gases are produced and have to be pushed out of engine cylinder so that there is ample volume left for the fresh charge to be induced.

The exhaust from the automobiles is at a high pressure and leads to generation of noise while rejection to the atmosphere. To reduce the exhaust pressure and subsequently to reduce the noise a muffler is improvised in the exhaust system of the automobiles.

A muffler is a device that is used for reducing the amount of noise produced by the engines. The basic constructions of muffler usually consist of the tubular metal jacket, perforated tubes and the expansion chamber. The arrangement of these components will guide the exhaust gas to flow from the inlet pipe of the muffler to the outlet (tail pipe). Inside the muffler, the noise from the exhaust gas will be cancelled out by the basic physics principle on noise cancellation before the gas flows out to the atmosphere. The noise cancellation will reduce the noise that radiated by the vehicle to the surroundings. The present day mufflers that are being incorporated in automobiles are majorly of two types:

1. Reflective mufflers
2. Dissipative mufflers

A reflective muffler consists of a number of tubular elements of different transverse dimensions joined together so as to cause, at every junction, impedance mismatch and hence reflection of substantial part of the incident acoustic energy back to the source, whereas dissipative mufflers consist of ducts lined on the inside with acoustically absorptive materials. Both mufflers are having different construction, geometry, and principles in their application.

In resistance or reflective mufflers the noise attenuation is done by the basic physics principle of wave cancellation. The steady or mean flow should be allowed to pass unimpeded through the muffler while the fluctuating flow which is associated with the acoustic pressure fluctuation is impeded. If the steady flow is not significantly impeded the so-called 'back pressure' will be very low and the engine will function more efficiently. Because of the back pressure the volumetric efficiency decreases and specific fuel consumption increases. Hence there is a need for an optimum muffler design.

The various factors that affect the muffler performance are
1. Muffler design
2. Muffler material
3. Restriction to the flow inside the muffler
Muffler consists of four main components
I.   Inlet pipe
II. Outlet pipe
III. Muffler shell
IV. Perforated tube
When the exhaust gases from inlet pipe pass through the perforations inside the shell, the gases get scattered in different directions. After reflection from the inside surface of the shell, the sound cancellation of waves occur. The gases pass through the perforations multiple times and even get reflected from the shell surface. Due to the combined effect of these, the level of sound at the muffler outlet is reduced significantly. After doing a lot of research and experimentation it has been found that the reduction in the noise is inversely proportional to the backpressure, which is not a desired characteristic and hence an optimum muffler design has to be implemented so that there is maximum noise reduction and minimum possible back pressure.

The flow through the muffler and variation of various parameters such as velocity and pressure along the length of the model can be accurately demonstrated with the help of CFD analysis which display accurate results within a short span of time.

2. Literature Survey

Erdem Özdemir et al [1] The model comprised of a perforated inlet and had four pipes between the expansion chambers. The analysis was performed by changing the lengths of the front, middle and expansion chambers as it was found that the noise attenuation can be done by increasing the length of the middle chamber. The air inlet was considered at 473 K and the air was considered to be incompressible ideal gas for the analysis. It is observed that 30% reduction on length of rear chamber did not make any difference on acoustic characteristics with base muffler model. To generate cross flow, inlet and outlet pipe's perforated part stand in the middle chamber. A decrease at the length of middle chamber prevents the cross flow. Thus, It can be concluded that a greater pressure loss occur at this model.

Sileshi Kore et al [2] The analysis was performed in ANSYS FLUENT and the meshing of the geometry was done using GAMBIT. The muffler was designed for the simulation of the Alfa Romeo 1995 model. The solver implemented was a 2-D, segregated implicit solver with 2nd order implicit time stepping. Second order upwind discretization was used for the density, momentum, energy, turbulent kinetic energy and the turbulent dissipation rate equations. The inlet conditions were considered as pressure inlet of air (ideal gas) at a pressure of 101325 Pa and 300 K

temperature. In this case the inlet velocity is normally 80 m/s and the outlet boundary condition has been set at atmospheric pressure. From the above analysis it was observed that the maximum transmission loss is around 14dB with an optimal amount of backpressure for the developed model.

Tutunea et al [3] Here the model was generated using SOLIDWORKS 3D modelling package and the analysis and simulation of the internal flow was carried out using ANSYS FLUENT 14. The observations were made on a muffler which had three main chambers and the middle chamber had a perforated tube. The boundary conditions were incorporated as velocity inlet, where the velocity was considered as 25m/s and at 280 Deg C of temperature. It was observed that there was a considerable decrease in the pressure at the outlet of the muffler and the results were in good agreement.

The flow of the exhaust was assessed by generating a CAD model and simulating them in the virtual environment of ANSYS fluent. The thermal and the acoustic parameters were analysed and the performance of the designed muffler were evaluated. [4, 5, 6, 7, 8, 9, 10, 11, 12, 13].

3. Basic Theory

3.1. Theoretical Computation

Cylinder firing rate (CFR) = Engine Speed in RPM/60 …. For a two stroke engine= Engine Speed in RPM/120 ….For a four-stroke engine

Engine firing rate (EFR) = n X (CFR), where n is the number of cylinders

3.2. Muffler Volume (V_m) Calculation

$$V_m = V_f x [\pi/4 (D^2 xl)] x (no of cylinders /2)$$

Where V_f is the volume factor, d is the bore and l is the stroke of an internal combustion engine.

4. Problem Formulation

The exhaust system is defined as the hardware necessary to vent the exhaust from the vehicle beginning at the exhaust plane defined by the engine manufacturer and necessary to isolate the exhaust thermally from vehicle structures. The virtual design of the exhaust muffler, as a minimum, include an accurate estimate of space required for the exhaust, backpressure to the engine, system weight, gas species distributions, gas temperature distributions, the interaction of the plume with external surfaces both on the vehicle and the ground, and the thermal interaction of the exhaust system with external surfaces through internal convection, conduction, and radiation. The muffler being simulated is being designed as per the engine output of Maruti Ciaz.

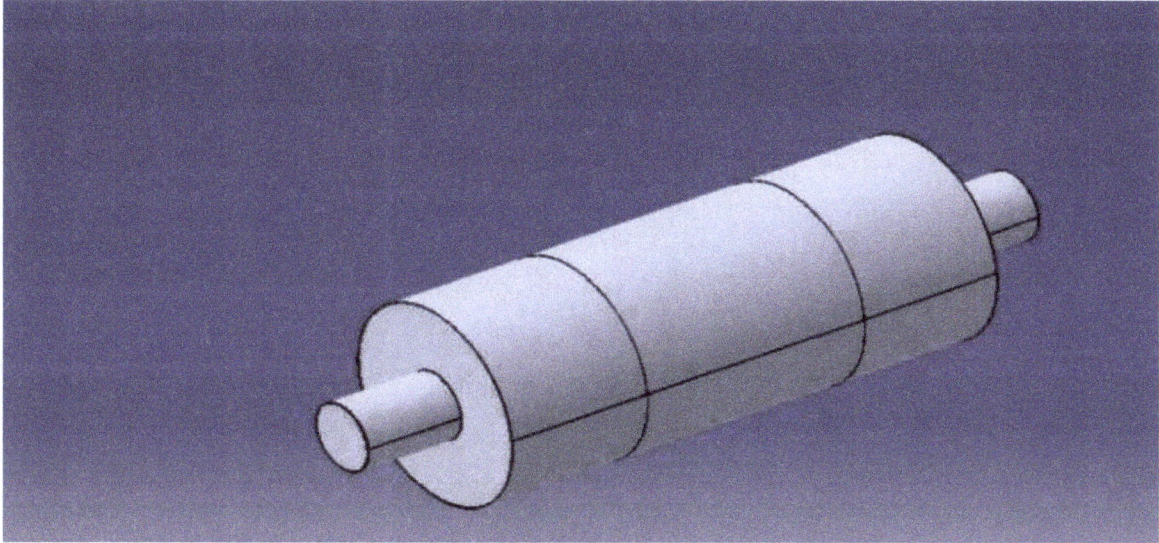

Fig. 1. *CATIA model of the designed muffler.*

Case Study – LCV Diesel engine vehicle (Ciaz)
Engine Data:
Bore (D) = 69.6 mm
Stroke (L) = 82 mm
No. Cylinders (n) = 4
Engine power (P) = 88.7 bhp at 4000 RPM
Muffler Volume Calculations:
Swept volume per cylinder = 0.25 (π xD^2xl)= 0.25 (3.14 x 69.6^2x 82)
(Vs) = 0.3119 lit.
Total swept volume in litres = 4x0.3119= 1.247 Lit.
Volume to be considered for calculation = 0.5 x V_s x n= 0.6239 Lit
Silencer volume: Volume of silencer must be at least 12 to 25 times the volume considered [6]. Volume can be adjusted depending on the space constraint.

Factor considered is = 22
Silencer volume = factor x consider volume = 13.73 Lit
Diameter of Muffler Calculations:
V_m = 0.25x π x d^2 x l
0.01373 = 0.25x π x d^2 x 0.5
d = 0.187 m= 187 mm
Diameter of Pipe Calculations:
As per the standards of the supercritical grade of mufflers, the diameter of the body should be about three times than the exhaust pipe diameter.
d = 3* d $_{exhaust}$
187= 3* d $_{exhaust}$
d $_{exhaust}$ = 62.33 mm

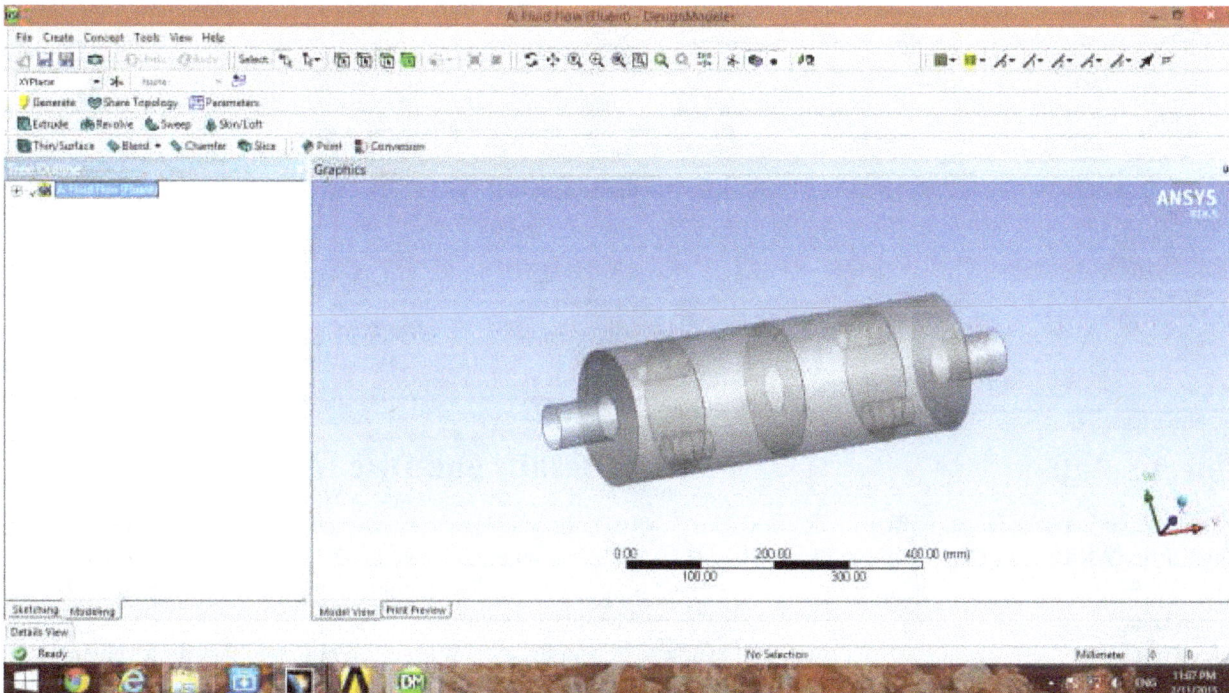

Fig. 2. *Internal view of designed muffler (model 1).*

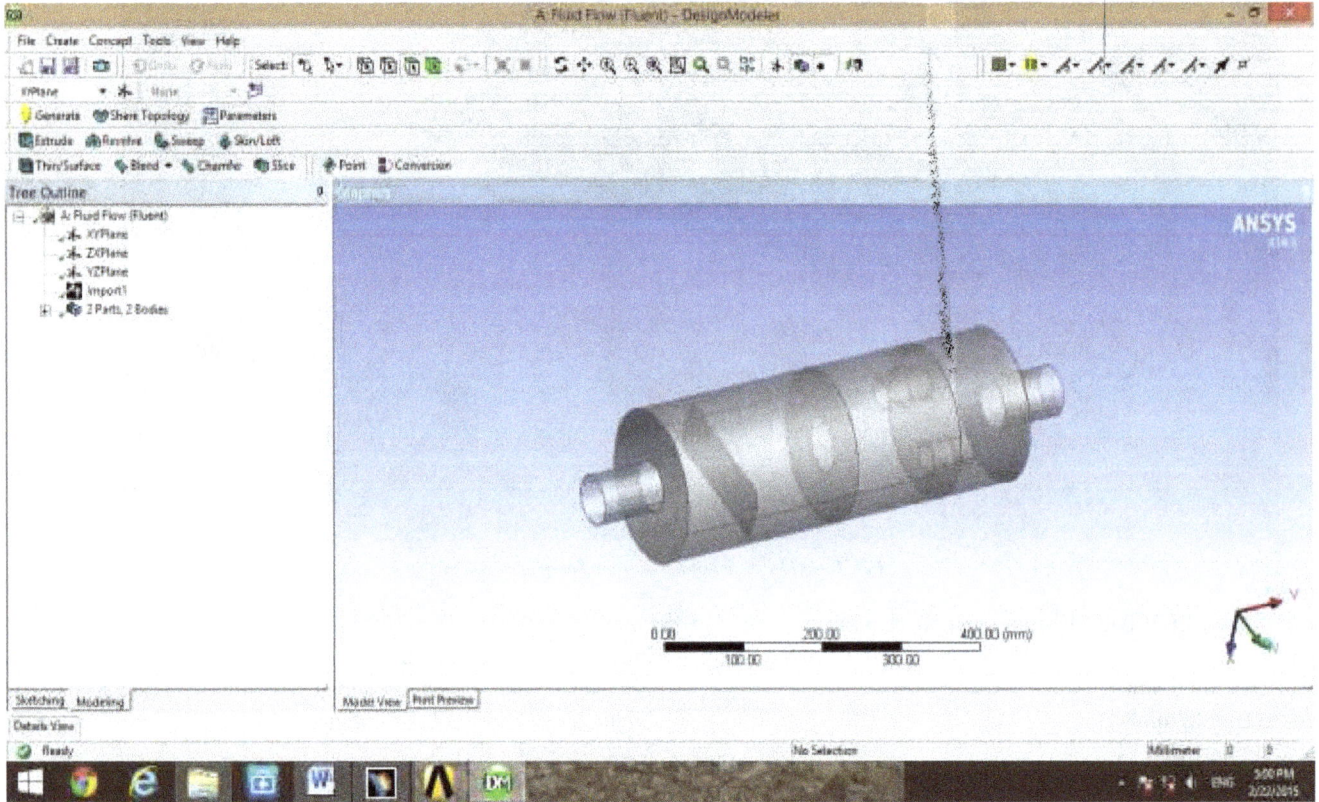

Fig. 3. Internal view of designed muffler (model 2).

***Table 1.** Material properties.*

Property	Units	Value
Density	kg/m3	7900
Specific heat	J/(kg*K)	500
Thermal Conductivity	W/(m*K)	16.2
Conductivity type	isotropic	Isotropic

***Table 2.** Dimensional data.*

Entity	Dimensions (mm)
Shell length	500
Shell Dia	187
Inlet pipe Dia	62.33
Inlet pipe length	85
Outlet pipe Dia	62.33
Outlet pipe length	85

***Table 3.** Initial values for velocity inlet as the inlet boundary condition.*

Area (m2)	0.003051
Temperature (K)	470
Viscosity (kg/ms)	2.7e-05
Enthalpy (J/kg)	749575.3
Density (kg/m3)	0.696
Length (mm)	500
Velocity(m/s)	80
Ratio of specific heats	1.4

***Table 4.** Initial values for pressure inlet as the inlet boundary condition.*

Area (m2)	0.003051
Temperature (K)	470
Viscosity (kg/ms)	2.7e-05
Enthalpy (J/kg)	749575.3
Density (kg/m3)	0.696
Length (mm)	500
Velocity(m/s)	80

5. Problem Solution

After the construction of exhaust muffler model in CATIA it is analysed in ANSYS FLUENT 14.5. The FLUENT analysis is carried out by considering three types of inlet boundary conditions.
1. Velocity inlet
2. Pressure inlet

6. Results and Discussion

The mean flow performance of the muffler considered in the flow analysis has been assessed. The results of the simulated muffler models obtained with the use of CFD modelling are very encouraging.

From Figure 4 it has been observed that for a velocity inlet boundary condition in model 1, the exhaust from the engine enters the muffler at a velocity of 80m/s and increases to a magnitude of about 133 m/s in the expansion chamber once it passes through the opening. This is observed as a result of decrease in the flow area the pressure increases and subsequently the velocity increases. After this the gases get spread in the chamber and they enter the next chamber from the other two splits. Then on hitting the baffle they swirl and come out of the exhaust at an increased speed of about 106m/s.

Also from figure 9 we observe that for a pressure inlet boundary condition in model 1, the gases enter the muffler at a pressure of 3.02 bar and hit the baffle. On making the impact the exhaust recedes a bit and swirls are observed in that particular entrance chamber. The swirls have the pressure magnitude of about 2.47 bar. The gases enter the second chamber from the baffle openings at top and bottom and hence it has been observed that the pressure intensity is more near the walls in this chamber and has a magnitude of about 2.8 bar. This is because of the fact that due to the presence of the slits or openings near the top and the bottom of the baffle, the pressurized gas or exhaust is directed towards the walls of the muffler and on hitting them the exhaust comes away from the walls. From the second chamber the exhaust enters the third chamber through an opening at the center of a baffle and hence it is observed that there is a 2.26 bar pressure region almost throughout the center of the chamber and the rest has a swirl region with a pressure of 1.5bar. This is due to the reason that as the area decreases the pressure increases

and the velocity decreases. So as the pressure is more hence there is a constant or maintained region in the chamber. Again the gases hit the baffle and enter the next chamber with a similar effect as it was observed in the second chamber but with a reduced pressure intensity of about 1.93 bar near the walls. Finally the exhaust gases come out of the outlet pipe at a pressure of 1.61 bar.

In model 2 as evident from figure 6 it has been observed that the gases enter the inlet pipe at a pressure of 3.01 bar and on hitting the baffle there occurs a swirl region in the first chamber which has a pressure intensity of 2.9 bar. The flow of the gases was more or less similar as that in model 1but due to the changed arrangement of the baffles more intensity of pressure was observed near the walls. The velocity magnitude over this region is nearly 310 m/s according to figure 17. Due to the reduction in the flow area the velocity of the gases increase and reach a maximum intensity of 620 m/s. The flow pattern remains similar to that of model 1 further and the gases come out of the outlet pipe at a speed of 372 m/s and a pressure of 1.64 bar.

Even though the second model displayed a similar behaviour, there was a more significant drop in the pressure of the exhaust gases in the first case than the second. The drop in the pressure of the exhaust gases in the first model was about 57% whereas the drop in the second model was nearly 51%.

6.1. Velocity Inlet Boundary Condition for Model 1

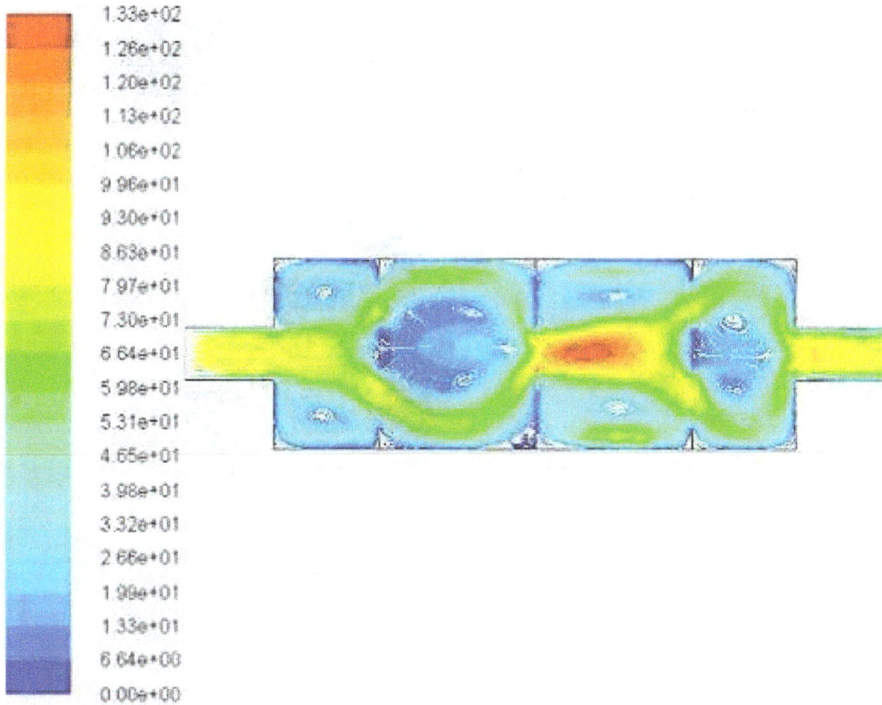

Pathlines Colored by Velocity Magnitude (m/s)

Feb 05, 2015
ANSYS Fluent 14.5 (2d, pbns, rke)

Fig. 4. Velocity field cut plot.

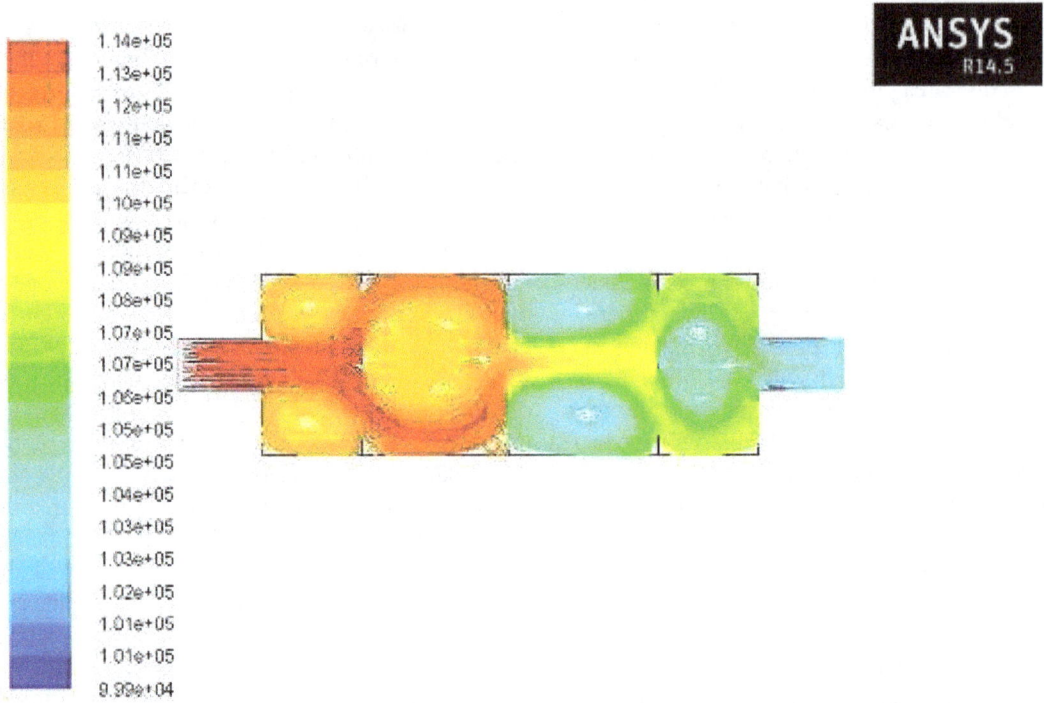

Fig. 5. *Pressure field cut plot.*

XY PLOTS

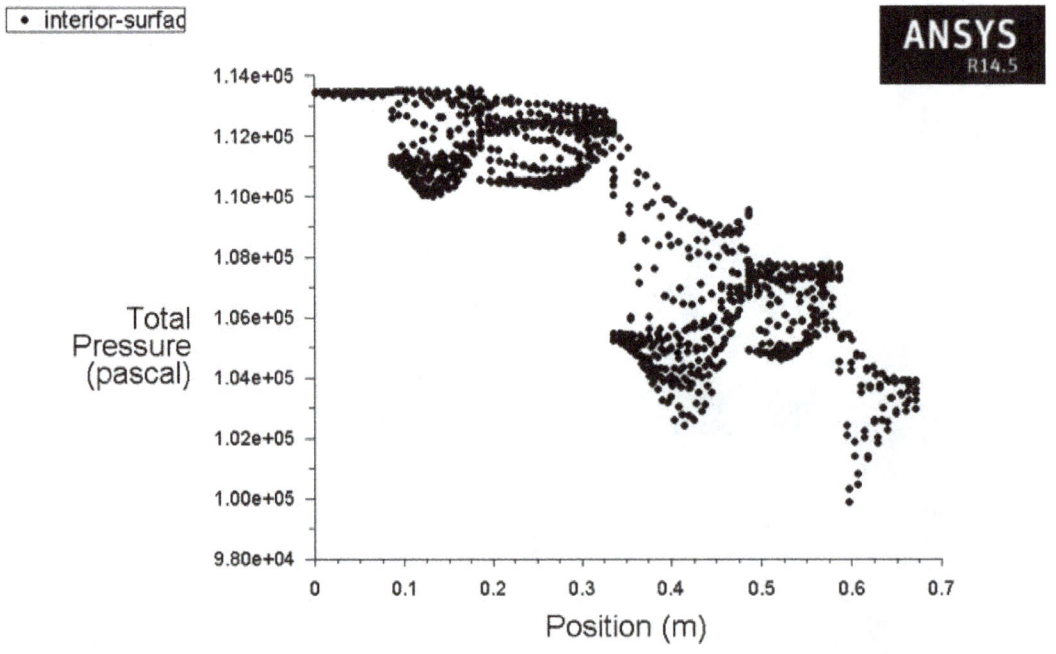

Fig. 6. *XY plot (Total pressure vs position).*

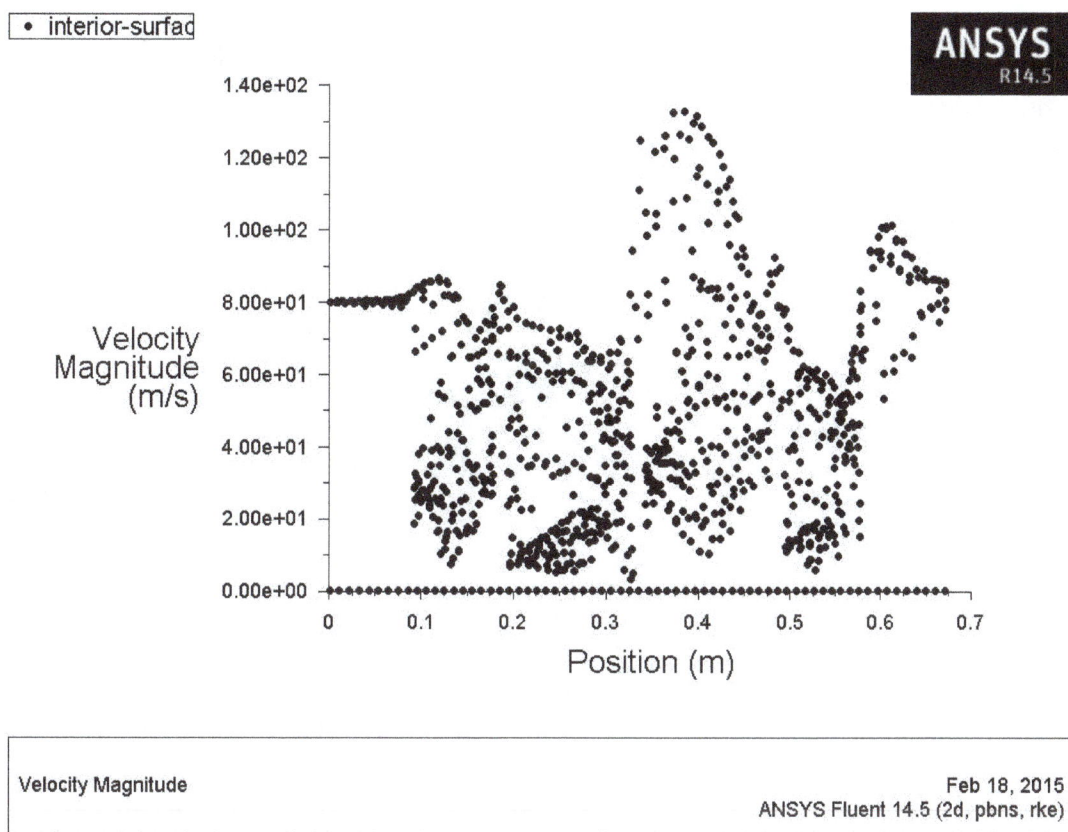

Fig. 7. XY plot (Velocity magnitude vs position).

6.2. Pressure Inlet Boundary Condition for Model 1

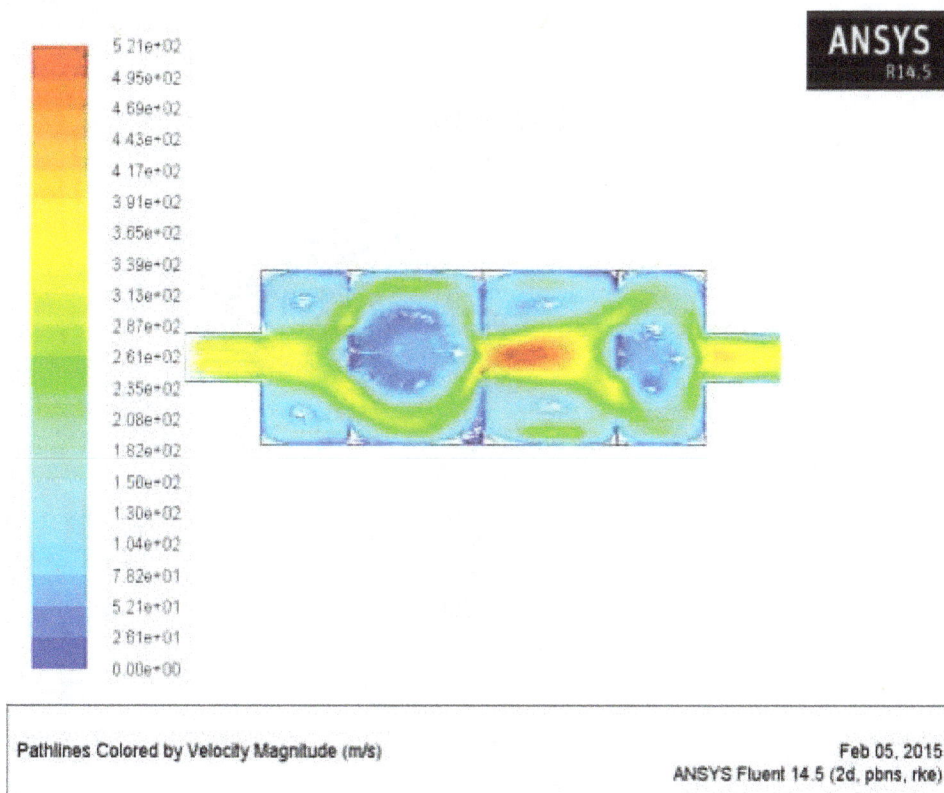

Fig. 8. Velocity field cut plot.

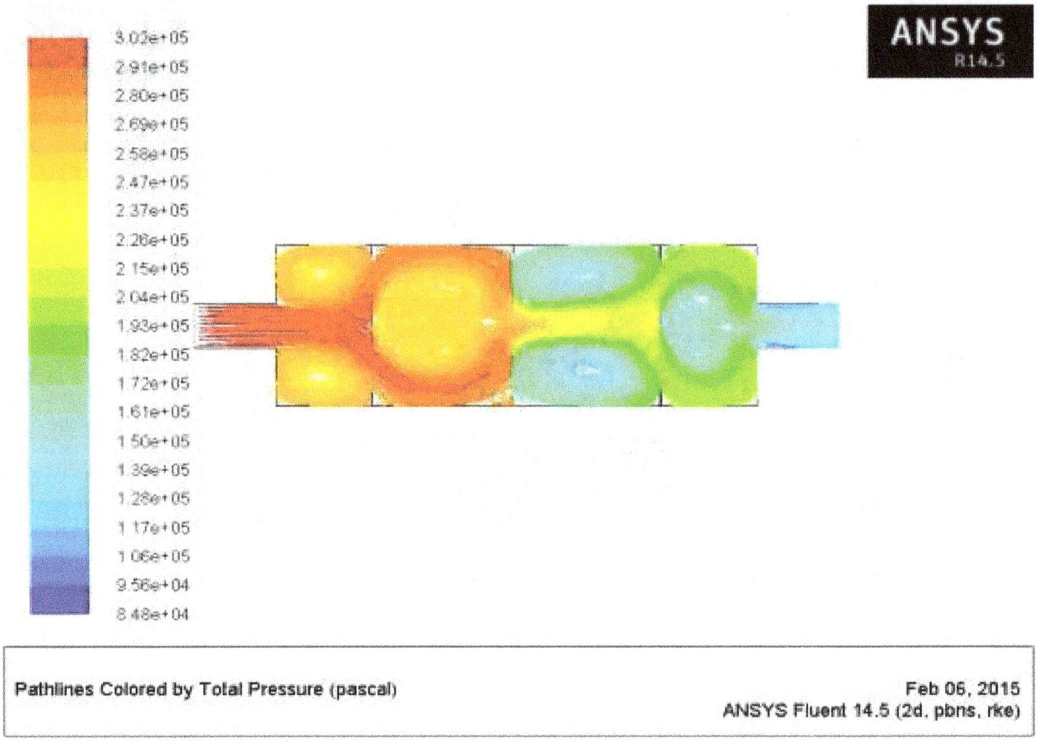

Fig. 9. *Pressure field cut plot.*

XY PLOTS

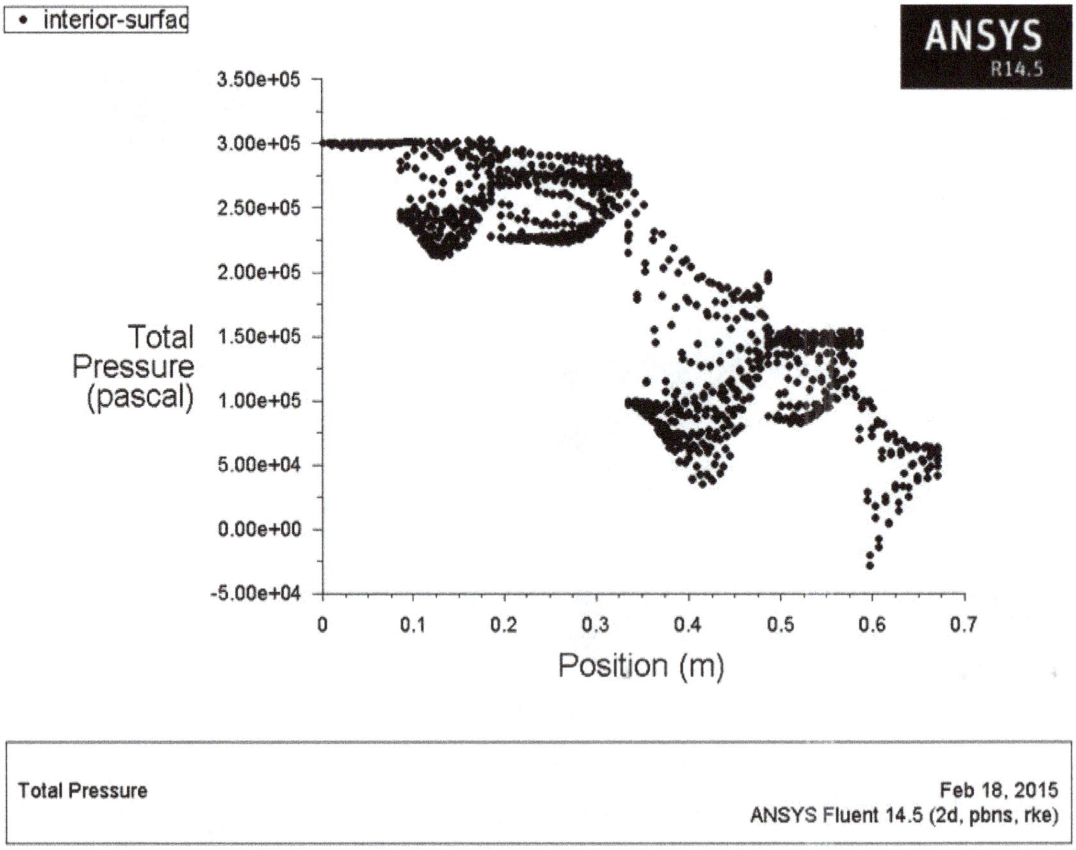

Fig. 10. *XY plot (Total pressure vs position).*

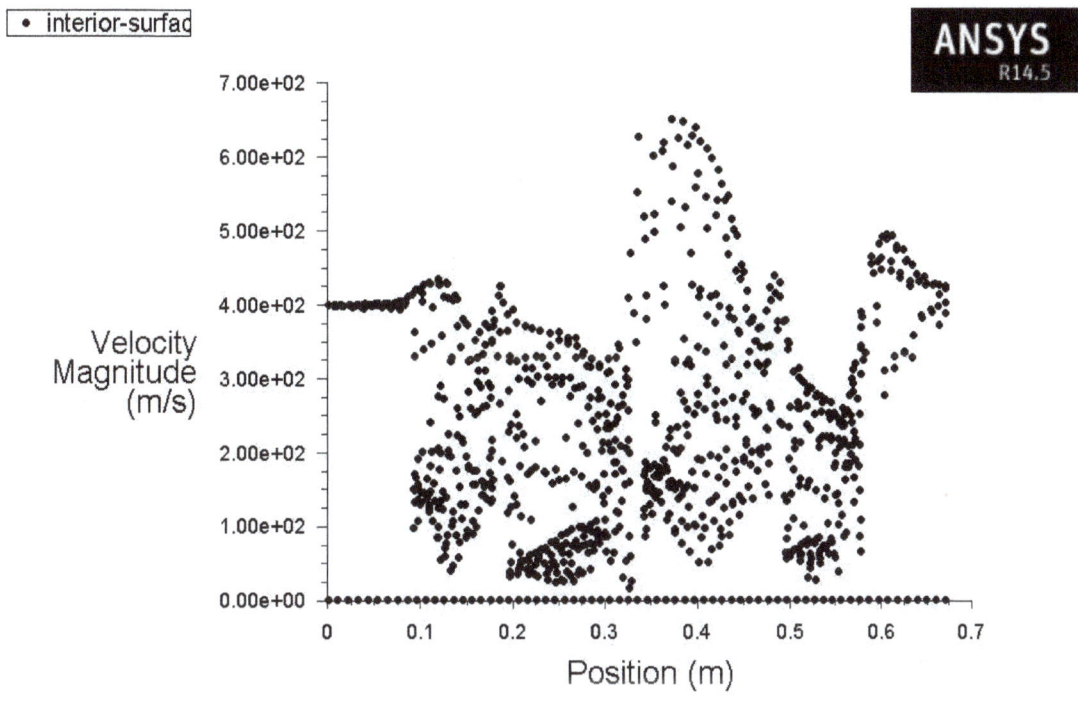

Fig. 11. *XY plot (Velocity magnitude vs position).*

6.3. Velocity Inlet Boundary Condition for Model 2

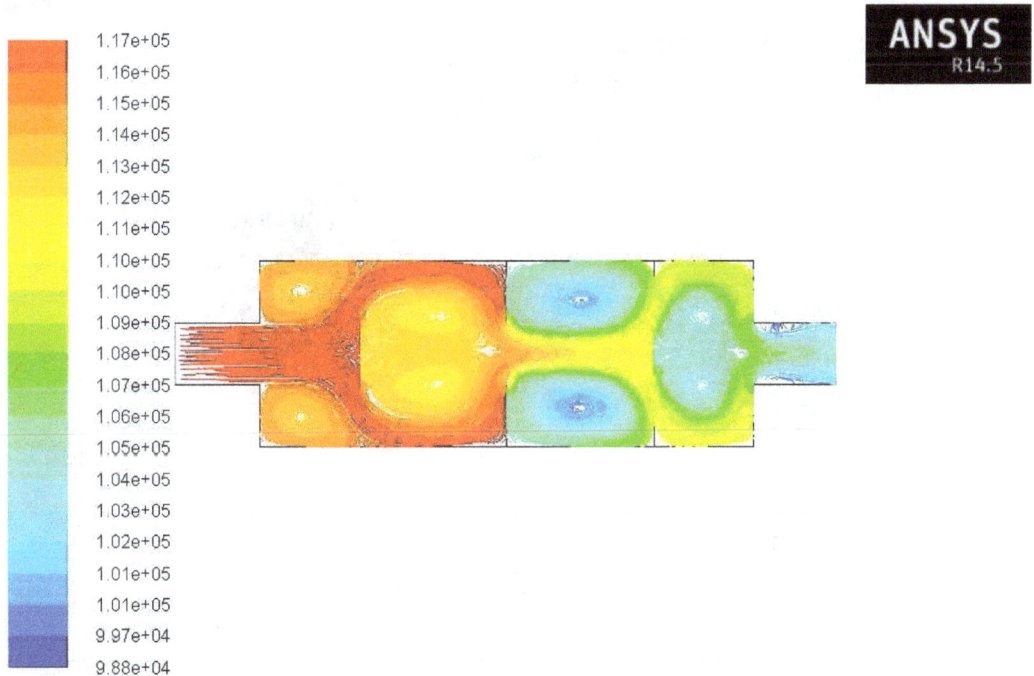

Fig. 12. *Pressure field cut plot.*

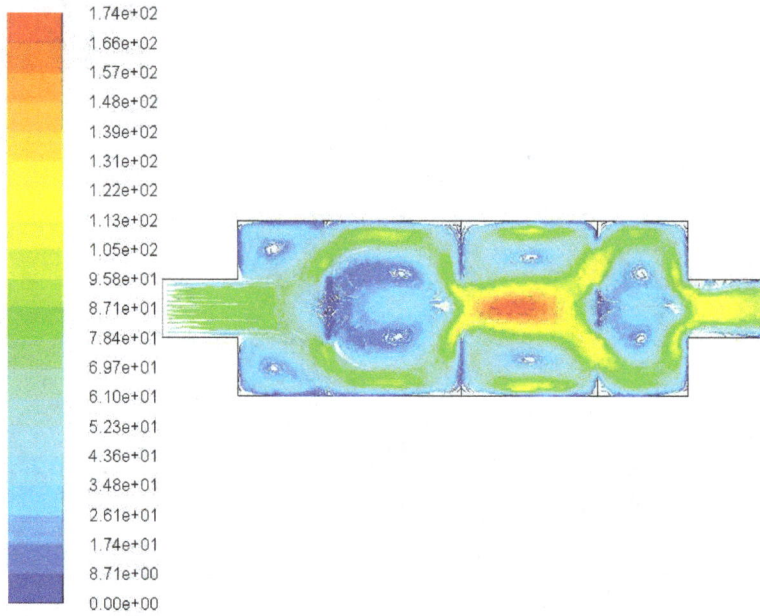

Pathlines Colored by Velocity Magnitude (m/s)

Feb 18, 2015
ANSYS Fluent 14.5 (2d, pbns, rke)

Fig. 13. *Velocity field cut plot.*

XY PLOTS

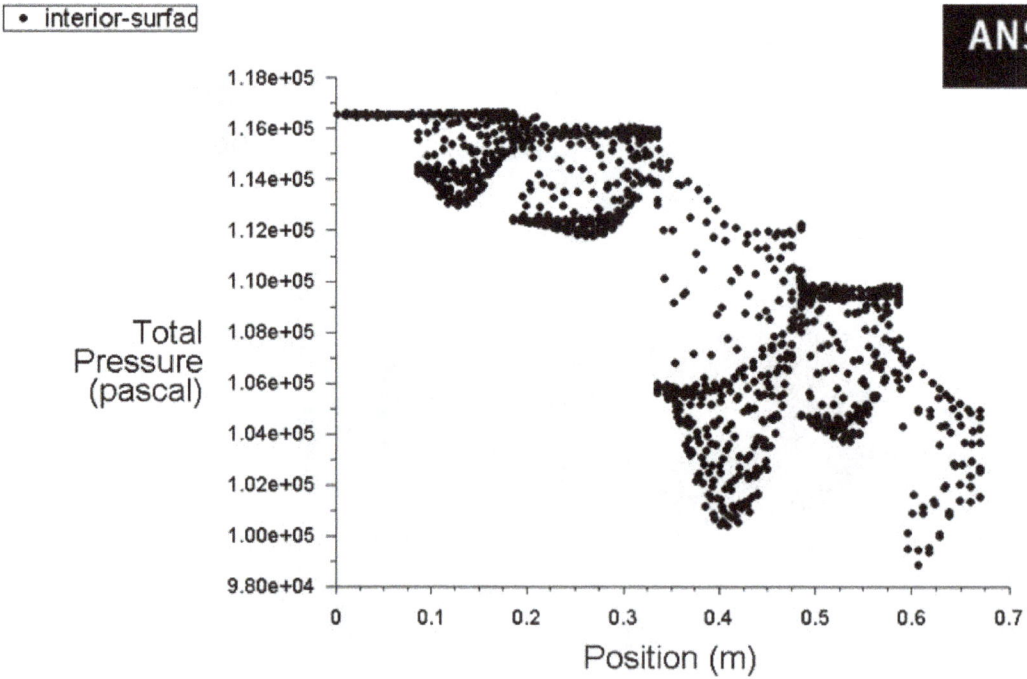

Total Pressure

Feb 18, 2015
ANSYS Fluent 14.5 (2d, pbns, rke)

Fig. 14. *XY plot (Total pressure vs position).*

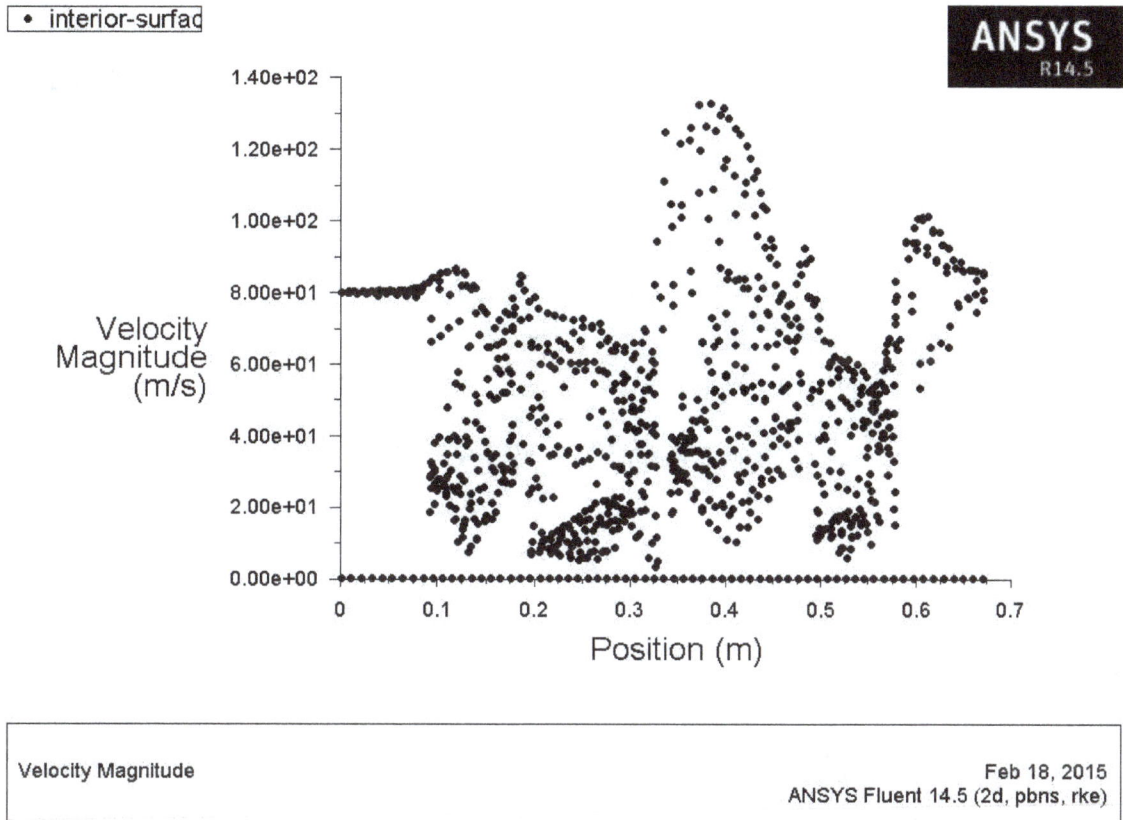

Fig. 15. *XY plot (Velocity magnitude vs position).*

6.4. Pressure Inlet Boundary Condition for Model 2

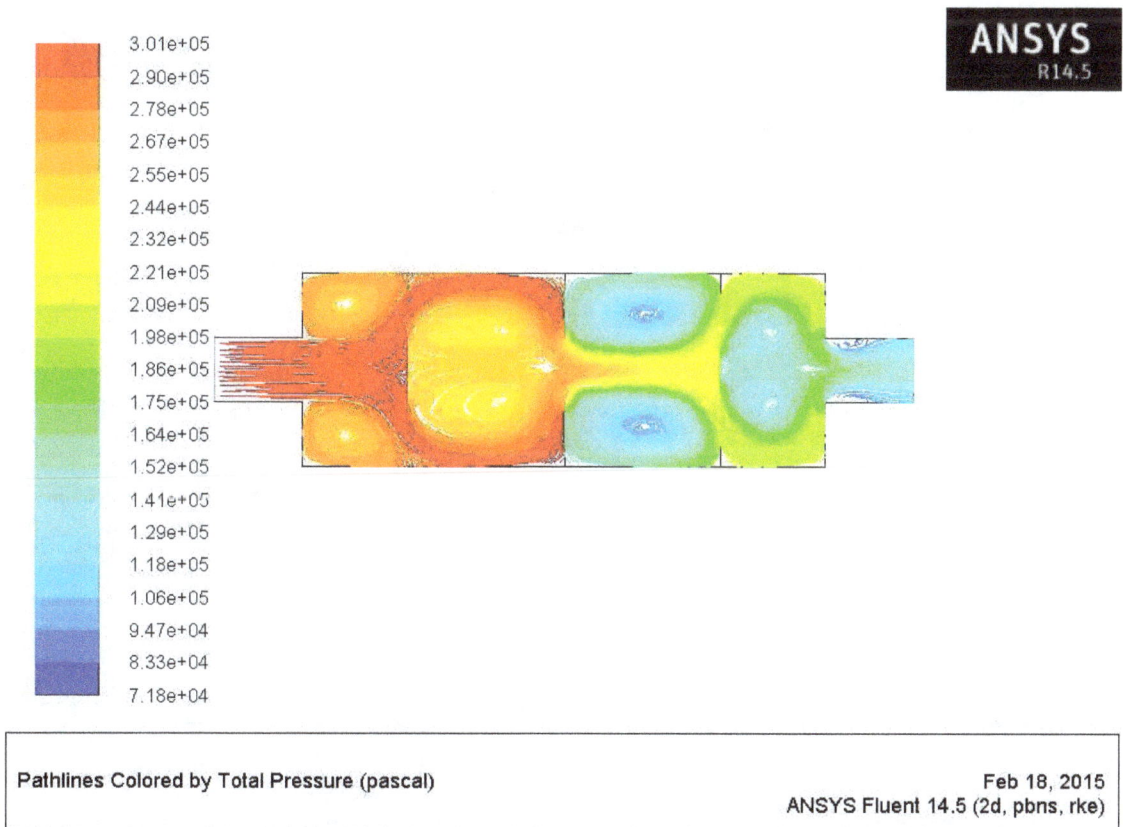

Fig. 16. *Pressure field cut plot.*

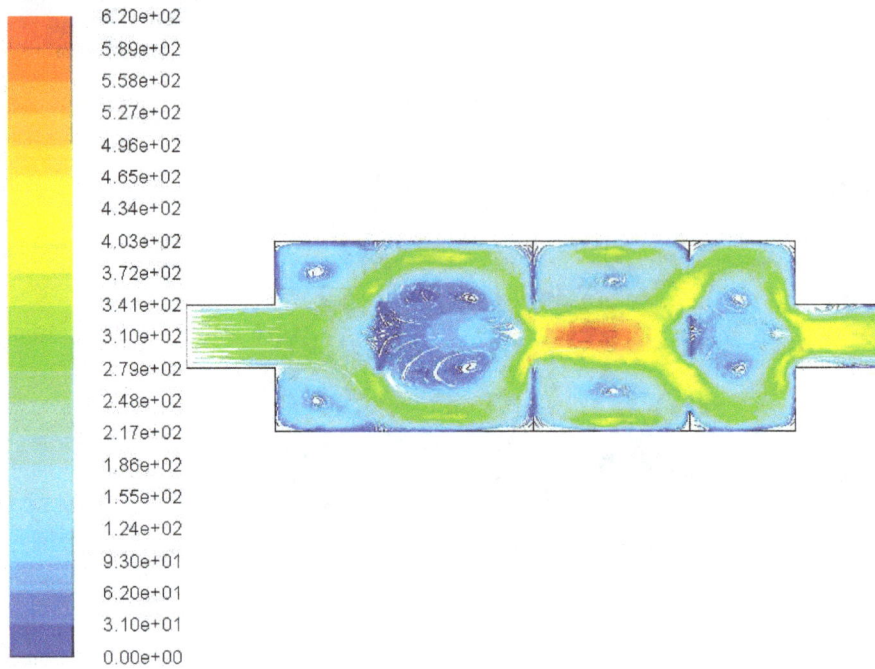

Pathlines Colored by Velocity Magnitude (m/s)

Feb 18, 2015
ANSYS Fluent 14.5 (2d, pbns, rke)

Fig. 17. Velocity field cut plot.

XY PLOTS

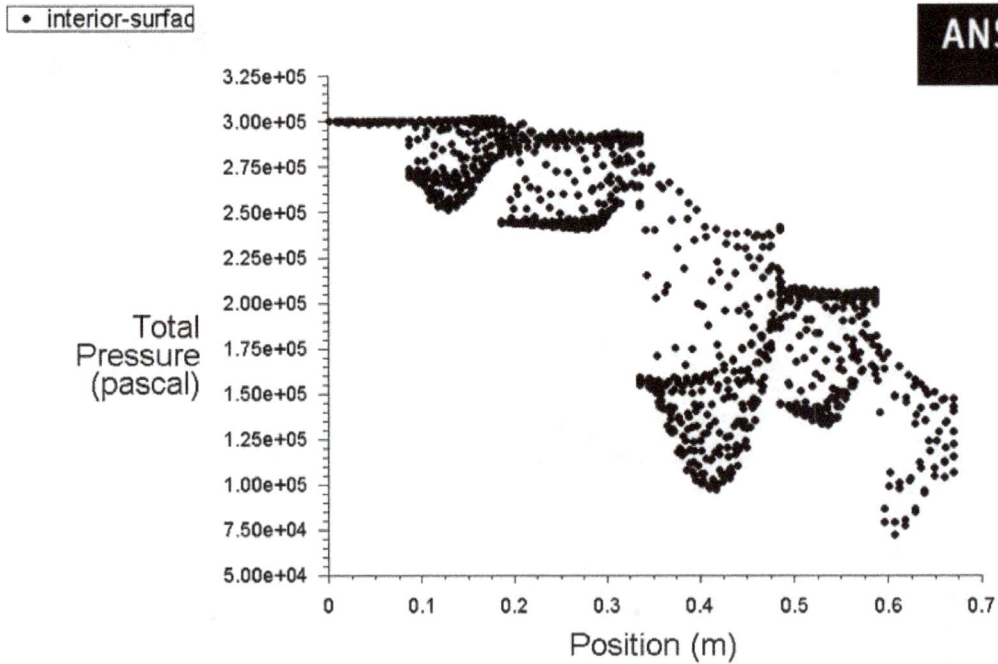

Total Pressure

Feb 18, 2015
ANSYS Fluent 14.5 (2d, pbns, rke)

Fig. 18. XY plot (Total pressure vs position).

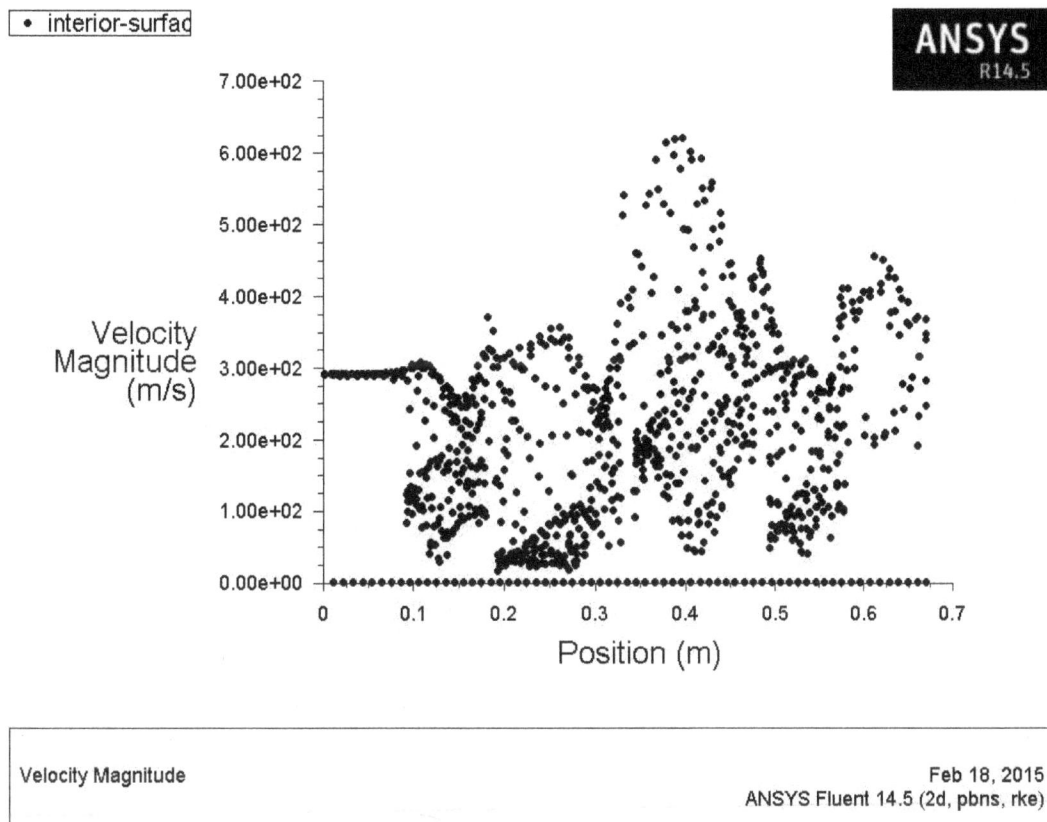

Fig. 19. XY plot (Velocity magnitude vs position).

7. Conclusion

In this work two different models of a muffler have been designed for the engine output of an LCV diesel engine and the flow has been simulated using ANSYS FLUENT. The flow characteristics obtained through the simulation were promising.

On comparing the results and performances of the two models, we observe that though both the models have same similar design parameters, the second model was more effective in reducing the exhaust pressure than the second one because of its internal baffle arrangement.

1. Maximum velocity in model 1 for velocity inlet boundary condition is 133m/s
2. Maximum velocity in model 2 for velocity inlet boundary condition is 176m/s
3. Exhaust pressure reduction in model 1 is 53.82%
4. Exhaust pressure reduction in model 2 is 57.14%

The reduction in pressure of exhaust in model 1 is 53.82 % whereas the reduction in exhaust pressure in model 2 is 57.14 %. Hence we conclude that model 2 is more efficient in reducing the exhaust pressure when compared to model 1.

Future Scope

This work deals with the flow analysis of an automobile muffler which has a circular shell. Further experimentation can be done by incorporating an elliptical shell and subsequently the length of the muffler can be reduced with similar performance or even an enhanced behaviour.

Along with the changes in the shape of the shell and the length of the muffler, the experimentation can be further implemented with changes in the following:

1. Boundary condition
2. Implemented solver
3. Can get in porosity (catalytic converter)

References

[1]   Erdem Özdemir, Rifat Yılmaz, Zeynep Parlar, Şengül Ar, AN ANALYSIS OF GEOMETRIC PARAMETERS' EFFECTS ON FLOW CHARACTERISTIC OF A REACTIVE MUFFLER, 17th International Research/Expert Conference "Trends in the Development of Machinery and Associated Technology" TMT 2013, Istanbul, Turkey, 10-11 September 2013.

[2]   Sileshi Kore, Abudlkadir Aman and Eddesa Direbsa, PERFORMANCE EVALUATION OF A REACTIVE MUFFLER USING CFD, Journal of EEA, Vol. 28, 2011.

[3]   D. Tutunea, M.X. Calbureanu and M. Lungu, The computational fluid dynamics (CFD) study of fluid dynamics performances of a resistance muffler, INTERNATIONAL JOURNAL OF MECHANICS Issue 4, Volume 7, 2013.

[4]   Prof. Amar Pandhare, Ayush Lal, Pratik Vanarse, Nikhil Jadhav, Kaushik Yemul, CFD Analysis of Flow through Muffler to Select Optimum Muffler Model for Ci Engine, International Journal of Latest Trends in Engineering and Technology (IJLTET),volume 4, issue 1, May 2014.

[5] Shital Shah, Saisankaranarayana K, Kalyankumar S. Hatti, Prof. D.G. Thombare, A Practical Approach towards Muffler Design, Development and Prototype Validation, SAE Paper 2010.

[6] Mr. Jigar H. Chaudhri, Prof. Bharat S. Patel, Prof. Satis A. Shah, Muffler Design for Automotive Exhaust Noise Attenuation - A Review, International Journal of Engineering Research and Applications, ISSN : 2248-9622, Vol. 4, Issue 1(Version 2), January 2014, pp.220-223

[7] Yunshi Yao, Shaodong Wei, Jinpeng Zhao, Shibin Chen, Zhongxu Feng, Jinxi Yue, Experiment and CFD Analysis of Reactive Muffler, Research Journal of Applied Sciences, Vol. 6 , (17): 3282-3288, September 2013.

[8] Rahul D. Nazirkar, S.R.Meshram, Amol D. Namdas, Suraj U. Navagire, Sumit S. Devarshi, DESIGN & OPTIMIZATION OF EXHAUST MUFFLER & DESIGN VALIDATION, Proceedings of 10th IRF International Conference,

[9] A.K.M. Mohiuddin, Ataur Rahman and Yazid Bin Gazali, SIMULATION AND EXPERIMENTAL INVESTIGATION OF MUFFLER PERFORMANCE, International Journal of Mechanical and Materials Engineering (IJMME), vol 2, 2007.

[10] M. Rajasekhar Reddy, Dr K. Madhava Reddy, Design And Optimization Of Exhaust Muffler In Automobiles, International Journal Of Engineering Research and Applications, ISSN: 2248-9622, Vol. 2, Issue 5, September-October 2012, pp.395-398.

[11] Vinod Sherekar and P. R. Dhamangaonkar "Design Principles for an Automotive Muffler" Volume 9, Number 4 (2014).

[12] Mrs.Varsha Chitale-Patil, Mr.Amol R.Patil, Prof.R.B.Patil "Design Analysis and Performance Evaluation of Reactive Silencer by SYSNOISE "Volume 3, India April 2014.

[13] Sunil, Dr Suresh P.M, Experimental Modal Analysis of Automotive Exhaust Muffler Using Fem and FFT Analyser, International Journal of Recent Development in Engineering and Technology(IJRDET).

[14] Sudarshan Dilip Pangavhane,Amol Bhimrao Ubale,Vikram A Tandon,Dilip R Pangavhane "Experimental and CFD Analysis of a Perforated Inner Pipe Muffler for the Prediction of Backpressure" Vol 5 No 5 Oct-Nov 2013.

[15] Mr. K.S. Tanpure, Dr. S.S. Kore "VIBRATIONAL ANALYSIS OF GENSET SILENCER USING FEA & FFT ANALYZER" Tanpure et al., 1(1): June, 2014.

[16] J.Kingston Barnabas, R.Ayyappan , M.R.Devaraj " DESIGN AND FABRICATION OF EXHAUST SILENCER FOR CONSTRUCTION EQUIPMENT " 2013 .

[17] OVIDIU VASILE, KOLUMBAN VLADIMIR "Reactive Silencer Modeling by Transfer Matrix Method and Experimental Study "Romania, June 24-26, 2008.

[18] Nitin S Chavan, Dr. S. B. Wadkar "Design and Performance Measurement of Compressor Exhaust Silencer By CFD" volume 2 September 2013.

[19] Shahid Nadeem, K.S.Shashishekar" DESIGN OPTIMIZATION OF INDUSTRIAL STEAM VENT SILENCER FOR NOISE REDUCTION CFD SIMULATION "volume 4, july-august 2013.

[20] Nicolae ENESCU, Ioan MAGHETI, Craita Daniela CARP-CIOCARDIA "Acoustical Silencers (Mufflers)" vol VII issue 1/2010.

8

Weld Quality Assurance Practices in the Metal Welding Industries in Ghana

Akpakpavi Michael

Mechanical Engineering Department, Accra Polytechnic, Accra, Ghana

Email address:

micakpakpavi@yahoo.com

Abstract: This study investigates the weld quality assurance practices in the metal welding industries in Ghana. The data for the study were collected using questionnaire, interview and personal discussions. One hundred and twenty informal sector welders, 80 formal sector welders, and 25 welding hardware shops/dealers were contained in the analyzed sample. The results of the analysis revealed that majority of the welding industries in the country lack weld quality assurance standards, weld quality testing equipments and materials as well as personnel to help assure weld quality practices in the metal welding industries. Also, equipments and materials for welding inspection and testing are not readily available in the welding hardware shops in the country. Moreover, even though most of the welders in the country have considerable years of working experience, they lack the skills of ensuring quality in their welding activities due to low levels of technical training. To instill professionalism into the welding industry, the suppliers should import the weld quality testing equipments into the country to be made available to the welders. The welders and the suppliers should take advantage of training programs offered by the training institutions in the country to enable them upgrade and update their knowledge and skills in weld quality assurance practices. The Government, the banks and other corporate organizations should help establish industries solely to produce more of the weld quality testing equipments locally in the country. This indeed, will help create more jobs and reduce the unemployment rates in the country.

Keywords: Ghana, Weld, Quality, Assurance, Practices, Industries

1. Introduction

1.1. Background

Welding activities in the Ghanaian welding and fabrication industry are classified under two sections: the formal and the informal sector. The formal welding sector comprises of companies operating under the metal production and manufacturing industries in Ghana and is classified as medium and large enterprises (MLEs). However, the informal welding sector consists of enterprises which are not recognized to be operating under the metal production and manufacturing industries in Ghana and are classified as micro and small scale enterprises (MSEs) [1].

The philosophy that often guides the fabrication of welded assemblies and structures is to "assure weld quality". Weld quality assurance is the use of technological methods and actions to test or assure the quality of welds, and secondarily to confirm the presence, location and coverage of welds. In manufacturing, welds are used to join two or more metal surfaces. Because these connections may encounter loads and fatigue during product lifetime, there is a chance they may fail if not created to proper specification [2]. It has been the goal of every metal welding industry globally to internalize, adopt and apply weld quality assurance practices in their welding activities. This will not only allow them to produce metal products that meet appearance and functional requirements, but also will enable them to remain competitive as welding firms in the worldwide market place [3].Welding as a manufacturing process has been practiced in almost all metal processing industries in Ghana in both the pre and post independence era. Unfortunately, however, quality assurance practices have not been injected and internalized by the metal welding industries particularly the micro and small welding firms that undertake most of the welding activities in the country. They lack the requisite knowledge in weld quality testing, and have not been able to achieve professionalism in their welding practices. A visit to most of the welding firms in both the formal and informal sectors across the country reveals that weld quality assurance practices are virtually absent in these enterprises. In fact, majority of them do not have basic weld quality tools and equipments as well as personnel to undertake weld quality tests to assure full compliance with standards.

The objective of this paper therefore, is to investigate weld quality assurance practices in the metal welding industries generally in the country. It is also to help educate and sensitize the welding industries in the country about the various weld quality testing methods and the need for the welding firms to start employing them in their welding practices.

1.2. The Informal and Formal Welding Industries in Ghana

The informal welding sector in Ghana comprises the micro and small scale enterprises (MSEs). The level of professionalism in this sector is low but welders in this sector are experienced as a result of constant practice in the welding trade. The main activities performed by this group include fabrication of plate and sheet metals, manufacturing of metal products, and maintenance and repairs of metallic products. Typically, some of their products include: cement block making machine, wagon and chassis (trucks), coal pot, fluid storage tanks, iron gates, car seat frames, billboards, burglary protection shield as well as agro processing products. The informal sector welding industry essentially employs basic hand tools in their manufacturing processes including hammer, chisel, grinding tool, and drilling tool. The fabricator uses his energy colloquially termed as "man-power" in cutting these materials. Again, the welding processes used in this sector largely include shielded metal arc welding (SMAW) and sometimes oxyacetylene welding (OAW). Furthermore, in this sector, welding operations are mostly carried out in structures along the roadside and sometimes in miniature workshops and under trees. Indeed, most of the welding jobs are done on the floor and this process normally involves that welders have to squat during the welding process since welding benches are rarely used. Moreover, close to about 65% of welding jobs in this country are carry out by welders in this sector. Customers prefer to go to them for reasons of proximity, accessibility and relatively low works charges.

However, the formal sector of the welding industry in Ghana comprises of companies operating as medium and large enterprises (MLEs) registered in Ghana's trade register with clear-cut business objectives. The formal welding sector among other things designs and manufactures such products as palm kernel crackers, corn mill, block making machine, vegetable processing machine, metal tanks, containers, gas and oil tanks, pipeline construction and installation and reconstruction of mining equipment. Most of the firms in this sector employ such tools as machining centre, materials handling and storage centre, welding machines (SMAW), welding generator, lathe machine, bending machine, drilling machine, forklifts, cranes, oxyacetylene equipments as well as the basic hand tools in their metal processing activities.

2. Methods

2.1. Study Area Description

The study was carried out in five major industrial areas in

Ghana namely: Tamale, Kumasi, Takoradi, Accra and Tema metropolis. The study was confined to these areas because a large majority of the metal welding and fabrication industries as well as the renowned welding and fabrication hardware suppliers/dealers in the country are located in these areas. The researcher classified the metal welding industries into two categories. Category A consists of metal welding industries in the formal sector, registered in the Ghana's trade register as MLEs. Category B represents metal welding industries in the informal sector generally known as MSEs.

A visit was also made to recognized tertiary institutions in the country such as the Kwame Nkrumah University of Science and Technology, University of Mines and all the ten Polytechnics in the country to find out if they run higher diplomas and degree programs exclusively in metal welding and practice.

2.2. Data Collection Techniques

The study used multi approach techniques in data collection. It involved observation, personal discussions as well as questionnaire administration to workers from each category of metal welding firms and metal welding hardware suppliers/dealers identified in the country. This approach was used because of its complementary effect of strengths and weakness associated with each method. This approach was chosen to increase the validity of the study by enriching the scope, depth, and knowledge derived from the data.

2.2.1. Observations and Discussions

Observations and discussions were made by the researcher by visiting the two categories of the metal welding sectors as well as the welding hardware suppliers/dealers in the country. During the trip, information about welding quality assurance practices by the welding firms as well as weld quality testing equipment and tools available in the welding hardware shops in the country were observed and discussed. The observations made were essentially recorded on an observational chart.

2.2.2. Survey

The researcher administered pre-tested questionnaires to some metal welders outside the selected regions. Identified mistakes were corrected and question rephrased to avoid ambiguity. The questionnaires were administered to respondents from the two categories of welding sectors as well as welding hardware and equipments suppliers/dealers identified. The questionnaire comprised both the open and close ended questions. Questions were centered on metal weld quality assurance practices utilized by the firms, availability of adequate weld quality testing equipment and tools in the country, as well as capability of the welding firms to undertake weld quality testing procedures.

2.2.3. Sampling

In all, 225 metal welding workers and welding tools and equipments suppliers/dealers who responded to the questionnaire make up the sample size. 120 respondents were workers from the informal welding sector, 80 respondents were from the formal welding sector, while 25 constitutes the welding hardware suppliers/dealers in the country. The welders from the two welding sectors were randomly selected, whiles the welding shop owners, supervisors, foremen, and the owners of the welding hardware shops visited were purposively selected. The sampling size was strongly influenced by willingness of respondents to participate in the studies.

2.2.4. Data Analysis

Data extracted from the administered questionnaires and interview were analyzed using statistical tools such as pie charts and bar charts for percentile analysis [4].

3. Results

3.1. Level of Technical Training of Welders

With regard to level of technical training, about 15% of the informal sector welders who responded to the questionnaire had welding certificates (ordinary and intermediate certificates), whiles 85% responded having apprenticeship training. This finding is as pictured in fig. 1.

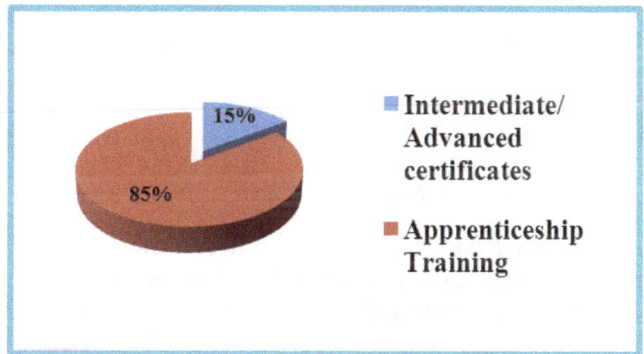

Fig. 1. *Training levels of informal sector welders.*

Moreover, the formal sector welders who responded to the questionnaire, about 11% had welding diplomas, 45% had intermediate welding certificates, 29% had advanced certificates whiles about 15% possessed training in the form of apprenticeship. This is illustrated in fig. 2.

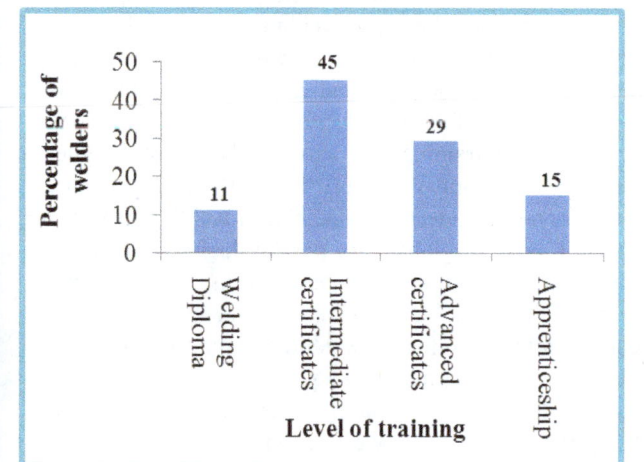

Fig. 2. *Training levels of formal sector welders.*

3.2. Working Experience of Welders in the Formal and Informal Sector

On the issue of number of years of working experience, 80% of the welders in both the formal and informal sector who responded to the questionnaire had more than 10 years working experience whiles the remainder 20% have had between 1-5 years working experience. This is depicted in fig. 3.

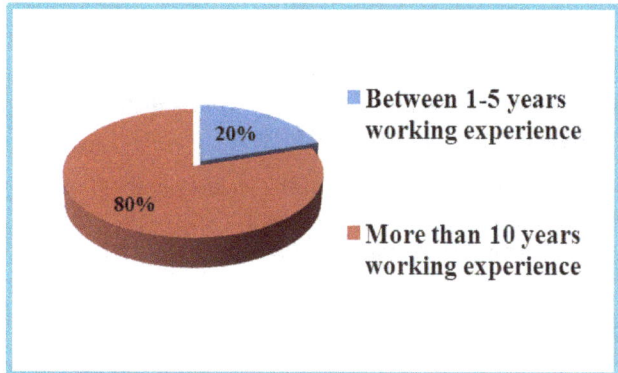

Fig. 3. *Working experience of welders in formal and informal sector.*

3.3. Technical Training Areas of Welding Equipments and Materials Suppliers/Dealers

The welding hardware shops who responded to the questionnaire on their technical training areas 12% obtained training in welding and other technical subject areas, 21% trained in marketing, 15% had trained in purchasing and supply, 24% had accounting and booking keeping training whiles the remaining 28% had some form of entrepreneurial training/education. This finding is as pictured in fig. 4.

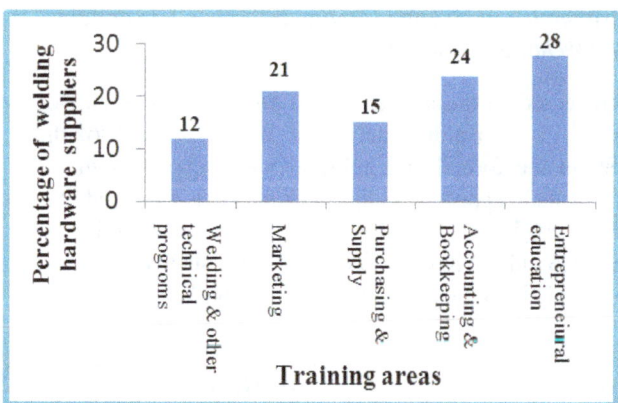

Fig. 4. *Technical training areas of welding hardware suppliers/dealers.*

3.4. Availability of Weld Quality Testing Manuals/Standards in the Welding Industries

On the issue of the availability of weld quality testing manuals/standards in the welding industries in the country, about 21% of the welders in both the formal and informal sector had weld quality testing manuals, whiles 79% did not have weld quality testing manuals/standards. This is shown in fig. 5.

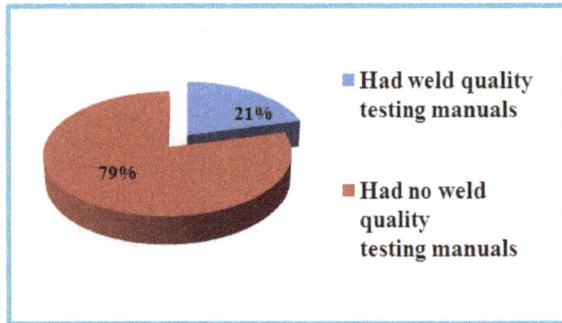

Fig. 5. Availability of weld quality testing standards in the welding industries.

3.5. Availability of Weld Quality Testing Tools, Equipments & Materials in the Welding Industries

On the above issue, about 12% of the welders in the formal and informal sector in the country together had weld quality testing equipment, tools and materials in-house for weld quality testing activities, whiles 88% had no weld quality testing equipments. This is illustrated in fig. 6.

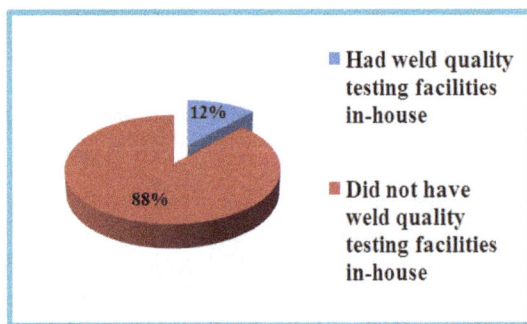

Fig. 6. Availability of weld quality testing equipments and materials in the welding industries.

3.6. Availability of Welding Quality Assurance Personnel in the Welding Industries

With regard to the availability of weld quality assurance personnel in the welding industries, only about 7% of the welders in the formal and informal welding sectors in the country who responded to the questionnaire indicated that they had certified weld quality assurance personnel in-house, whiles 93% did not have any certified weld quality assurance personnel. This finding is as illustrated in fig. 7.

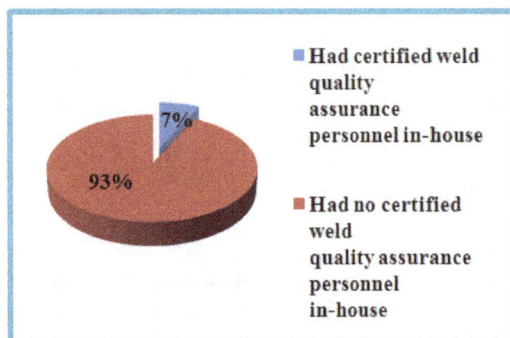

Fig. 7. Availability of certified weld quality assurance personnel in the welding industries.

3.7. Availability of Weld Quality Testing Tools, Equipments & Materials in the Welding Hardware Shops

On the above issue, only about 5% of the welding suppliers/dealers who responded to the questionnaire said they had stocked and supplied weld quality testing tools, equipments and materials to the welders in the country. The remainder 95% had no stock of weld quality testing tools, equipments and materials. This finding is as depicted in fig.8.

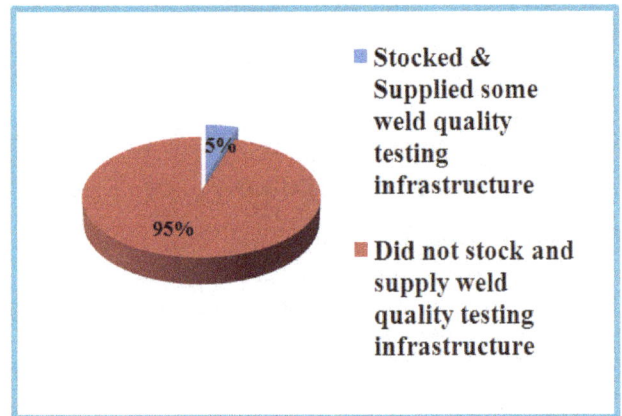

Fig. 8. Availability of weld quality infrastructure in the welding hard-ware shops.

4. Discussion

4.1. Technical Training Levels of Welders

Research available indicates that artisans with lower education and training are constrained by lack of adequate knowledge to enable them understand welding theory on their own. The research further shows that, higher level of education has a higher positive impact on weld products quality and this works to the effect that products made by artisans with primary education levels are of lower quality than those products made by their counterparts with secondary education levels [5].

Indeed, until recently, welding itself was a skill that craft people could learn without a real understanding of the science behind it. The scientific and engineering principles behind welding must replace the art of welding for it to achieve its potential as a preferred state-of-art manufacturing process [6]. To this end, there has been a paradigm shift in the training of welders in both the formal and informal training situations. Currently, welding training programs worldwide are designed to train welders not only to acquire the art of welding but also the scientific and engineering principles of welding. It is also expected of welders globally to possess high mathematics, reading, writing, oral and computer skills. In developed countries such as the USA, China, Germany and so on, students in formal or informal welding programs learn the art of heating and shaping metals. Required classes may include advanced mathematics, metallurgy, blueprint reading, welding symbols, pipe layout and a welding practicum [7]. Methods and techniques taught in welding classes include arc welding, soldering, brazing, casting and bronzing. Hands-on training often includes oxyacetylene welding and cutting, shielded metal arc welding, gas tungsten arc welding and gas metal arc welding. This is done to ensure that the products of their welding programs build successful careers in welding and remain competitive in the globalized market place. Certainly, welding education programs in these developed countries may culminate in a Welding Certificate of Achievement, Associate of Science in Welding or Bachelor of Science in Welding Engineering [8].

As identified in section 1.2, a chunk of metal welding activities in this country are carried out by welders in the informal sector due to reasons of proximity to customers and low works charges. Unfortunately, however, as indicated in figure 1, a very large number of these informal sector welders went through the apprenticeship form of training with few of them acquiring secondary and basic school educational levels. Majority of them are school dropouts. As a result of low educational levels, the welders in this sector can hardly read engineering drawings, drafts and other documents containing welding symbols. They are also completely unaware of the concepts of weld quality. Observations reveal that welders in the informal sector do not take time to properly design and prepare their welds and weld joints, select appropriate welding electrodes and other welding consumables, choose appropriate welding methods and above all, do not have the capacity to test and inspect their weld quality levels after welding operations. Indeed, even basic weld quality testing equipments and other welding logistical supports are virtually not available in the informal welding establishments.

In the formal sector, as indicated in fig. 2, most of the welders possess welding certificates (intermediate and advanced) qualifications essentially from the technical institutes in this country. From observations, the welders in the formal sector, by virtue of their training seem to understand weld quality concepts. Few of them try to achieve quality by taking the appropriate steps to ensuring weld quality before, during and after metal welding processes. They also have the capacity to read engineering drawings and drafts containing essentially welding symbols for proper weld design and execution.

Nonetheless, the level of weld quality check practices which is prerequisite for maintaining high quality standards in the welding industry in this country is rather low. Indeed, to help raise the quality levels of the metal welding carry out by the welders in both the formal and informal sectors, it is necessary that training institutions such as the universities, polytechnics, the technical training centers, Intermediate Technology Transfer Unit and so on organize in-service training for the welders in the country to help them upgrade and update their knowledge and competencies in weld quality assurance practices.

Moreover, a visit to the country's Science and Technology Universities as well as the ten Polytechnics which serves as training institutions in this country shows that none of these institutions runs a higher diploma or degree programs exclusively in welding and fabrication. They also do not have enough infrastructure and logistics on metal welding and

essentially on metal weld quality inspection and testing as is the case in most universities and polytechnics in advanced countries. It is therefore, expedient that the government, the banks, as well the corporate organizations in the country help in providing adequate welding logistics in these institutions. As plans are advanced to convert the country's polytechnics into technical universities, the infrastructure, programs and curriculum of the polytechnics should be expanded and upgraded to enable some of them run degree programs exclusively in welding and fabrication. This will help train welding specialists and weld quality assurance experts locally in the country. It will also go a long way to guarantee professionalism in the country's welding sector.

4.2. Working Experience of Welders in the Formal and Informal Sector

Welding as a profession and as one of the main joining methods employ in industry has evolved to the extent that computer simulations are being employed globally to help detect stress levels and other defects before the commencement of the main welding process. Again, highly precise fabrication is being demanded, and so microstructure and distortion control are very important. It is significant to determine welding process conditions by considering combinations of the various aspects of "process mechanics," including the appearance of a bead, penetration shape through a cross-section, fabrication efficiency, microstructure, weld defects during and after welding, environment, thermal distortion, residual stress, and so on [9-10].

Majority of the welders in both the formal and informal sector have more than ten years working experience as identified in section 3.2. Nonetheless, despite their long number of years of experience in the welding sector, observations reveals that, the welders, particularly those in the informal sector and some few ones even in the formal sector lack knowledge in basic welding procedures and methods. Obviously, they seem to lack adequate knowledge in basic metal welding concepts such as: acceptable levels of metal residual stresses, weld distortions, heat-affected zone (HAZ) properties, and welding defects including cracks, distortion, gas inclusions (porosity), non-metallic inclusions, lack of fusion, incomplete penetration, lamellar tearing, and undercutting.

Surely, in order to ensure a display of professionalism in the welding industry, it is imperative that the welders in the country take advantage of in-service training programs offered by technical training institutions such as Danest Weld located in Takoradi, Gratis foundation, ITTU and so on to upgrade their professional competency levels particularly in weld quality inspection and tests procedures. The welding sector, essentially those in the formal sector should endeavor to introduce computer weld simulations in their welding activities in order to enable them infuse efficiency, effectiveness and competiveness in their welding activities and in the welding sector in the country at large.

4.3. Technical Training Areas of Welding Equipments and Materials Suppliers/Dealers

Welding sales representatives serve a very important function in the welding industry. They are the conduit for information concerning new technologies that can improve the productivity, reliability, and more importantly, quality of welding processes. In the USA, Germany, China, India and elsewhere, knowledgeable welding sales representatives offer manufacturers /customers' invaluable technical support and guidance in the appropriate application of welding technologies. They also give technical advice on weld quality to their customers [11]. Currently, developments in ICT has even make it possible for customers in the welding sector worldwide to buy products, check specifications and compare prices over the internet and many frown on the idea of having to waste time on technically untrained people who are little more than a walking brochure. In most developed countries and elsewhere it is expected of welding supply industry representatives/dealers to be trained to the highest level and must be able to give accurate information with regard to the correct welding procedures to adopt, the most appropriate equipment for the application, give crucial information on the latest Health & Safety requirements as well as maintain a high standard of professional and ethical conduct [12-13].

However, as identified in figure four of this paper, only a small number of the welding equipments/suppliers in the country have certificates in welding and other required technical areas. A very large number of the suppliers in the country rather trained in non technical areas. An informal discussion with the dealers shows that a majority of them knows virtually nothing about welding processes, welding methods as well as quality issues in welding. The dealers/suppliers also lack the ability to give expert advice to welders on appropriate welding equipments, materials, tools, welding consumables etc. This indeed, is largely due to the suppliers/dealers lack of training in the appropriate technical areas such as welding, metallurgy and mechanical engineering which is rather the case in other jurisdictions.

Moreover, the situation whereby a hefty number of the welding equipment suppliers/dealers in the country do not possess the requisite technical qualifications does not portend well for the welding industry in Ghana. It is actually expected of the suppliers not only to furnish their equipments to the welding industry, but also be in a position to offer connoisseur information on welding materials, equipments, welding methods, welding standards, weld quality information and so on. They are also expected to have knowledge about the welding supplies required by the welding sector and therefore make them available in the hardware shops. Hence, it is expected of the welding hardware suppliers in the country to take advantage of training programs offered by the technical training institutions in the country as well as abroad to enable them enhance their technical competencies. This will go a long way to sanitize the welding sector and inject professional and quality in the welding industry in the country.

4.4. Availability of Weld Quality Assurance Standards/Policy in the Welding Industries

To provide a well-defined basis for planning welding operations and to ensure a system for quality control during welding, it is important that organizations involve in welding issue the relevant welding procedure specifications, ensure that welders are qualified and appropriately trained and tested to do the work safely and precisely. Today, the metalworking industry worldwide has instituted weld quality assurance standards to guide welders, weld inspectors, engineers, managers, and property owners in proper welding technique, design of welds, how to judge the quality of weld, Welding Procedure Specification, how to judge the skill of the person performing the weld, and how to ensure the quality of a welding job [14-15-16]. However, as indicated in figure five of this report, a very large number of the welding industries in the country do not have welding quality policy nor are certified under any welding quality assurance standards. The few companies that have weld quality policy are essentially in the formal welding sector and they operate under standards such as the ISO 9001 and ISO 3834 welding quality standard.

Additionally, those welding industries in the formal and informal welder sectors which do not have any welding quality policy nor are certified under any welding quality assurance standard perform welding operations based on standards provided to them by their customers. Hence, in the absence of quality welding standards, the welding firms carry out welding operations under no quality standard. However, the few welding industries in the country who utilize quality standards in their welding activities do obtain the standards from third party certification bodies including: American Bureau of Shipping (ABS), Bureau Veritas (BV), British Standards (BS), American Society of Mechanical Engineers (ASME), as well as some domestic certification bodies such as Sonic Control Engineering, and Probe Engineering.

Undoubtedly, this situation whereby a very large majority of the welding firms in the country do not work to any standard specification is rather unfortunate. This indeed means that professionalism and adherence to acceptable quality levels are virtually absent in the welding industry in the country. To help address this situation, it is crucial that the Ghana Standard Boards, the Garage Association of Ghana, the Ministry of Trade and Industry, as well as the relevant regulatory bodies in the country ensure that the actors in the welding sector adopt and adapt the use of quality assurance standards in their welding activities. This will not only bring about sanity in the welding sector but will also raise the efficiency and competency levels of the welders in this country. The oil and gas industry in the country will also be more than willing to assign contracts to the welding firms in the country.

4.5. Availability of Weld Quality Testing Equipments and Materials in the Welding Industries

Many distinct factors influence the strength of welds and the material around them, including the welding method, the amount and concentration of energy input, the weldability of the base material, filler material, and flux material, the design of the joint, and the interactions between all these factors. As a result of this, welds are normally tested to specifications to help detect, control and correct defects in weld joints prior to usage. To test the quality of a weld, either destructive or nondestructive testing methods are commonly used to verify that welds are free of defects, have acceptable levels of residual stresses and distortion, and have acceptable heat-affected zone (HAZ) properties. Typical welding defects include cracks, distortion, gas inclusions (porosity), non-metallic inclusions, lack of fusion, incomplete penetration, lamellar tearing, and undercutting. The common non-destructive methods are: Visual Inspection-this employs testing equipments such as rulers, fillets, fillet weld gauges, squares, magnifying glasses, and reference weld samples; Dye Penetrant Test-this employs fluorescent penetrating liquids and developers; Magnetic Particle Test-this employs iron particles or fluorescent, ultraviolet light; Radiographic Test-this employs X-ray or gamma ray, film processing and viewing equipment, penetrameters; Ultrasonic Test-this employs ultrasonic units, probes, and reference comparison patterns. A few examples of destructive testing include macro etch testing, fillet-weld break tests, transverse tension tests, guided bend tests, acid etch testing, back bend testing, tensile strength break testing, nick break testing, and free bend testing [17].

Moreover, most of the afore-mentioned welding testing equipments are available in the shop-floor and laboratories of the metal welding industries in most developed as well as developing countries such as India and Malaysia [17-18-19]. Unfortunately, however, as depicted in figure six of this report, large majority of the metal welders in the informal and formal sector in the country have no or inadequate weld quality testing equipments and materials at their shop-floor or laboratories. Discussion with the welders shows that they are not able to procure these equipments due to inadequacy of capital investments and financial difficulties. The situation is such that the few formal sector welders who particularly manufacture oil and gas equipments undertake weld quality assurance activities. This they essentially do by relying on outside weld quality testing firms both locally and abroad upon request by their customers. In the informal sector, apart from not having the technical knowledge and expertise to carry out weld quality testing activities, they also do not have enough capital to procure the needed weld quality testing equipments. The consequence of this is that a very large number of the welding industries in the country do not have weld quality testing equipment and materials in place for weld quality testing purposes, which is undesirable.

Nonetheless, to be able to compete favorably in the globalized market place, and to inject efficiency and effectiveness in their welding activities, it is crucial that the welding establishments in the country procure the relevant weld quality testing equipment and utilize the latter to enhance their weld quality levels. The informal welding sector which have been characterized by low capital investments should endeavor to acquire visual weld quality testing equipments such as: Rulers, Magnifying Glass, Squares, Dye Penetrants, Pocket Telescopic Inspection Mirror, Oval Inspection Mirror, Welding Profile Gauge, Digital Pit Depth Gauge, Fillet Weld Gauge /Calculator, Pre Inspection Pocket Fillet Weld Gauge, Taper Gauge, Dial/Pit Depth Gauge, reference weld samples. These are relatively easier to use and are also cost effective. Additionally, the formal sector welders with relatively large capital investments can obtain the afore-mentioned simple weld quality testing equipments in addition to the more advanced equipments such as: Magnetic Particle Test Sets, Radiographic Test Sets, Ultrasonic Test Equipments, WeldPrint analyzers, computer software simulation packages on weld quality test etc.

Certainly, as indicated in section 3.1 of this report, more than 60 percent of the welders in the country are in the informal sector and had their training through the apprenticeship system of training. Hence, by procuring these basic testing equipments the trainees/learners in the apprenticeship metal welding training system will become aware of the basic concepts in weld quality testing and will also be pre-disposed to the use of weld quality testing equipments. The trainees will then be willing to internalize and practice them in their welding activities after their training programs. This will go a long way to bring about sanity and professional in the welding sector generally in Ghana.

4.6. Availability of Welding Quality Testing Personnel in the Welding Industries

To do high-performance welding, it is important for a company to have a quality weld inspection program in place. Welding inspection requires knowledge of weld drawings, symbols, joint design, procedures, code and standard requirements, and inspection and testing techniques. In order to do so, a company must have a welding inspector who is formally qualified and capable of performing a number of different testing methods or have the necessary knowledge and experience to conduct weld quality inspection. It is not practical for a person who is not well-versed in the necessary procedures to perform this task [19]. Welding enterprises in most developed and developing countries around the world tend to employ qualified welding inspectors who have the responsibility to be personally present where the welding operation is conducted. These on-site welding inspectors ensures that good quality welds are achieved, specifically by monitoring items such as: welders' certificates, weld joint preparation, condition of electrodes, welding technique, welding current, welding machines and tools, slag cleaning, incomplete welds and other similar tasks [20-21].

Unfortunately, a very large majority of the welding enterprises in both the formal and informal sectors have not been able to mobilize welding team which essential includes certified weld quality assurance personnel as illustrated in fig.7. Certainly, weld quality demands that quality be built into a product right from the onset of the weld, but not after welding operations. The absence of certified weld quality assurance personnel in the welding industries in the country means that weld quality is checked in these firms only after welding operations but not before and during welding activities. In the informal welding sector, weld quality testing is hardly done, hence, a significant proportion of them do not employ the services of weld quality assurance experts on-site. However, in the formal welding sector some few of them do undertake weld quality testing by employing the services of external testing officers. Additionally, an informal interaction with most of the formal sector welding firms reveals that they hardly employ permanent weld quality assurance staff.

Obviously, the situation whereby majority of the welding firms do not permanently engage the services of weld quality assurance experts is rather not a healthy situation. Definitely, the engagement of permanent weld quality assurance staff may increase labor cost of welding; nevertheless, their presence in the welding industries will rather enable the latter to be noted for producing quality weld products and reduce their cost of fabrications. The welding firms will also stand a chance of gaining both local and international recognition and contracts.

4.7. Availability of Weld Quality Testing Tools, Equipments & Materials in the Welding Hardware Shops

Fig. 8 shows that about 95% of the welding hardware shops in the country have not been able to stock weld quality testing tools, equipments and materials for sale to welders. A visit to the renowned welding hardware shops in the country and in the study locations reveals that most of them deal in assorted welding supplies including: welding machines, welding electrodes and other welding consumables, welding shields, helmets, gloves, hacksaw blades, grinding stones etc. However, interactions with the suppliers/dealers show that they hardly stock weld quality testing equipments for sale to the welders. The situation is such that even simple visual weld inspection and testing equipments such as Rulers, Magnifying Glass, Squares, Dye Penetrants, Pocket Telescopic Inspection Mirror, Oval Inspection Mirror, Welding Profile Gauge, Depth Gauge, fillet gauges, magnetic particle units, radiography equipments, reference weld samples etc. are not readily available in these welding hardware shops to enable the welders in the country to procure them. Again, the welding hardware shop owners claim that the welders have not been demanding the testing equipments; hence, they see no reason for stocking them for sale. That will lead to them incurring loss on those items. Moreover, observations show that, currently there is no manufacturing firm in the country that is into the manufacture and supply of any weld quality testing equipments which is rather the case in most developed as well as developing countries such as India and Malaysia [17].

Indeed, the situation whereby weld quality testing equipments and materials are virtually non available in the welding hardware shops in the country is undesirable. This is because a very large majority of the informal welders as well as a good number of the formal sector welders in the country have relegated weld quality inspection and testing practices in-house to the background due to their inability to obtain the needed weld quality testing and inspection logistics from the local welding hardware shops available in the country. To improve this situation, there should be effective interaction between the welders and the welding suppliers/dealers in the country. This will enable the latter to be aware of the weld quality equipments and materials needed by welders so they can import them into the country to be procured by the welders. Again, the government, the banks, as well as corporate organizations should help to establish industries solely to produce weld quality equipments and other similar equipments locally to make them available and accessible in the country.

5. Conclusion

In this paper, weld quality assurance practices in the metal welding industries in Ghana have been investigated. The study shows that weld quality assurance practices in the welding industries in the country are rather low and abysmal. This situation may well be attributed to the non-availability of welding standards, weld quality assurance personnel, weld quality assessment equipments and materials in the welding industries as well as low technical training levels of the welders in the country. Again, the welding equipments suppliers/dealers in the country are also not in a position to offer useful advice on weld quality assurance practices to the welders due to the suppliers' inability to train in relevant technical areas. Moreover, simple visual weld inspection and testing equipments such as Rulers, Magnifying Glass, Squares, Dye Penetrants, Pocket Telescopic Inspection Mirrors, Oval Inspection Mirrors, Welding Profile Gauges, Depth Gauges, reference weld samples, radiographic equipments, magnetic particle sets etc. are hard to come by in most of the welding hardware shops available in the country. For these reasons and many others, most of the welding industries in the country have relegated weld quality assurance practices to the background. To help reverse this trend and to infuse professionalism, efficiency and effectiveness in the welding sector in the country, it is absolutely imperative that the welders as well as the welding equipments suppliers/dealers take advantage of training programs offered by training institutions such as ITTU, GRATIS foundation, Danest West and so on to enable the welders and the suppliers to upgrade and update their technical competencies and technical training levels particularly in weld quality assurance practices. The welding industries in the country should do well to work according to acceptable welding standards. The welding hardware suppliers/dealers in the country must endeavor to import weld quality testing equipments into the country and make them available in their hardware shops to be procured by the welding industries in the country. Moreover, the government, the Banks as well as corporate organizations should help establish industries exclusively to produce more of the weld quality testing equipments and materials locally in order to make them available in the country. The curriculum and facilities on welding in the country's training institutions such as the universities and the polytechnics should be upgraded to enable them run higher diplomas and degree programs in welding technology. This will ensure that weld quality assurance specialists are trained locally to help instill professionalism in the welding industry in Ghana.

References

[1] S. E. Edusah, "The Informal Sector, Micro-Enterprises and Small-Scale Industries: The Conceptual Quan dary",2013.

[2] C. Hayes, 1998, "The ABC's of Nondestructive We Examination".

[3] Taylor and Francis, 2005, "Quality Assurance of Welded Construction", Elsevier Science Publishers Ltd, 2nd Edition, pp. 1-10.

[4] I.R. Levin, 1989, "Qualitative Approaches to Manage- ment". McGraw-Hill, Singapore.

[5] M.M.C Ondieki, E.T. Bisanda, and W.O. Ogola, "Impact of Education Level on Product Quality: Case Study of Arc Welding in Small Scale Metalworking Enterprises in Kenya",2013, pp. 001-008.

[6] American Welding Society, 2011, Vision for Welding Industry.

[7] Study.com, 2002-2015, Factsheet: Welding Education Requirements and Career [Online]. Available at http://study.com/welding_education.html [Accessed on 20th April, 2015]

[8] Wikipedia, the Free Encyclopedia, 2015, Factsheet: Welding [Online]. Available at http://en.wikipedia/wiki/welding. [Accessed on 25th April, 2015].

[9] A. S. Yasir, 2012, Factsheet: Study the Effect of Welding Joint Location on the Fatigue Strength and Fatigue Life for Steel Weldment [Online]. Available At http://www.asian-transactions.org/journals /vol02issue04/ate/ate-80212044.pdf. [Accessed on 18th March, 2015].

[10] Springer Link, 1999, Factsheet: Fatigue Behavior of Welding Joints [Online]. Available at http://link.springer.com/chapter/10.1007% 2F978-94-009-2277-8_24. [Accessed on 15th March, 2015]

[11] American Welding Society, 2015, Factsheet: Certified Welding

Sales Representative [Online]. Available at http://www.aws.org/certification/detail/ certified-welding-sales-representative. [Accessed on 5th April, 2015].

[12] Alcotech, 2015, Factsheet: Welder Qualifications [Online]. Available at http://www.alcotec.com/us/en/education /Training-and-Certification.cfm. [Accessed on 20th April, 2015].

[13] The Association of Welding Distribution, 2015, Factsheet: Careers in Welding [Online]. Available at http://www.awd.org.uk /careers.asp. [Accessed on 10th March, 2015].

[14] Wikipedia, 2015, Factsheet: Weld Quality Assurance [Online]. Available at http://www.awd.org.uk/careers.asp. [Accessed on 1st April, 2015].

[15] TWI, 2015, Factsheet: Welding Quality [On line]. Available at http://www.twi- global.com/capabilities/joining-technologies /welding-engineering/welding-quality/. [Accessed on 20th April, 2015].

[16] Inspecta, 2012, Factsheet: Welding Quality Assurance (ISO 3834) [Online]. Available at http://www.inspecta.com/en/Our-Services /Certification/Management-Systems/ISO-3834- Welding-Quality-Assurance/ [Accessed on 15th March, 2015]

[17] Caltech Engineering Services, 2015, Factsheet: Welding Inspection Instruments & Accessories [Online]. Available at http://www.caltechindia.com/Welding-Testing- Instruments-and-Accessories.htm [Accessed on 5th March, 2015].

[18] Alibaba.com, 2015, Factsheet: welding testing equip ment [Online]. Available at http://www.alibaba.com/showroom/welding- testing-equipment.html [Accessed on 5th April, 2015].

[19] Tony Anderson, 2007, Factsheet: The Fabricator.Com: Quality Inspections [Online]. Available at http://www.thefabricator.com/article/ testingmeasuring/quality-inspections. [Accessed on 1st May, 2015].

[20] Tony Anderson, 2002, Factsheet: Establishing Quality System for Welding [Online]. Available at http://www.thefabricator.com/article/ shopmanagement/establishing-quality- systems-for-welding. [Accessed on 14th March, 2015].

[21] Welding, 2012, Factsheet: The Importance of Visual Welding Inspection [Online]. Available at http://weldingdesign.com/safety-regulatory MalikVisualInspection. [Accessed on 5th March, 2015].

9

An Overview on Voltage Stability Indices as Indicators of Voltage Stability for Networks with Distributed Generations Penetration

Haruna Musa

Department of Electrical Engineering, Bayero University, Kano, Nigeria

Email address:

harunamusa2@yahoo.co.uk, hmusa.ele@buk.edu.ng

Abstract: The increase in Distributed Generation (DG) penetration in distribution network can be used as a means of addressing the increasing load demand without the upgrade of transmission lines. However, this increasing demand despite the supplementary supply from DG sources can cause the system to operate at its maximum capacity or at point of voltage collapse. It therefore, becomes necessary to determine the maximum capacity limit of the system before voltage collapse occurs due to instability. This paper reviews the concept of voltage stability index (VSI) as an indicator of a weak bus that is closed to its maximum allowable limit or the most critical line to voltage collapse in a network. The review also evaluates various voltage stability indices including those originally developed for transmission systems. Subsequently, their adequacies are quantitatively compared and from the comparison, it is observed that the existing voltage stability indices would be inadequate for assessing the most sensitive bus or line on the verge of voltage collapse for modern distribution systems.

Keywords: Distributed Generation, Penetration, Voltage Stability, Voltage Stability Index, Voltage Collapse

1. Introduction

The impact of DG on distribution network is positive as it improves voltage stability of a network apart from other improvement, such as power loss reduction, voltage profile improvement and power quality improvement provided it is well sited. The improvement in the voltage profiles at load terminals vary depending on their relative locations with respect to the switching point. In general, analysis of voltage performance shows that DG can support and improve the voltage profiles at load terminals. This can extend the stability margin of dynamic loads, which can result in loss of stability with voltage dips. One of the main factors restricting increase of load served by distribution system is the voltage stability which many distribution companies consider as one of the planning objectives especially with increased penetration of DG in the existing networks [1]. The increase is due to the fact that DG has brought new changes to traditional power system and their connection to distribution networks has improved the voltage stability of the network [2].

In voltage stability assessment one of the most important considerations is to know the distance to maximum loadability point from the current operating point. One of the

methods for such assessment is by use of Voltage Stability Index (VSI) which many researchers have attempted using in order to find the distance between initial loading points to the maximum loadbility point [3-6]. This index gives a measure of how far the present operating point is to voltage collapse point which is expressed in term of real or/and reactive power as well as the line parameters.

A number of methods for voltage stability analysis have been tried by many researchers such as P-V and Q-V analysis, Modal Analysis, voltage instability proximity indicator, multiple load flow solutions based indices, Line stability index, Line stability Factor, Reduced Jacobian Determinant, Minimum Singular Value of Power Flow Jacobian, and other voltage indices methods [7]. All these methods are used for the determination of the distance to maximum loadability point from the current operating point.

2. Voltage Stability

Voltage stability is defined as the ability of a system to maintain voltage at all nodes within the acceptable limits when subjected to disturbance [8]. Normally a voltage stable power system is capable of maintaining the post-fault voltages near the pre-fault value. However, if the system is unable to maintain the voltage within acceptable limits, the system can suffer from voltage collapse [9]. In general for the purpose of simulation these methods are categorized into two categories namely the static and dynamic. The dynamic voltage instability is the major cause of interruptions due to the non-linear nature of load involved unlike the static voltage instability that uses linear load and employs assumption that the system is operating in steady state [10].

2.1. Static Voltage Stability with DG

Static voltage stability analysis can be conducted by assuming the system is operating in the steady state and the load is a linear load. The static load models are expressed interms of active and reactive powers which are functions of the bus voltages.

The level of penetration and the type of DG always determine the nature of impact it will have on the power system. At the same time these two parameters always determines the voltage collapse limit. Networks with high induction generation DG penetration have the tendency of weakening the system voltage stability much more than DG with high penetration of synchronous generation. The reason is that induction machines cannot generate and control reactive power instead they consume reactive power from the system. Also the number of central generations with voltage control ability is lower when part of the demand is supplied by DG units and this can results in less reactive power supply from the central generation [11].

For the purpose of supplying extra power needed by the load during heavy load demand which if not supplied can results in instability that can lead to voltage collapse, DGs can be embedded in the distribution system to supply the extra load demand as proposed in [12] for the purpose of maintaining system stability.

2.2. Dynamic Voltage Stability with DG

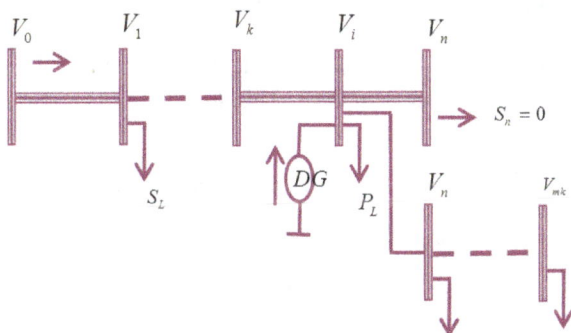

Fig 1.0. *Radial Distribution feeder with DG.*

The rapid increase of generators especially wind turbine in to the power system can have big influence on the power quality and voltage stability of entire network. The optimal placement and sizing of DG in radial distribution system as proposed for capacitor is employed with n bus, laterals and sub-laterals as shown in Fig 1.0 for radial distribution system [13].

The intermittent nature of wind and the characteristic of wind generators can cause serious problem to the power system as well [14]. Voltage instability problem can be more prominent if induction motors are the main load, since a decrease in its terminal voltage will increase the reactive power consumption greatly even though the active power consumption will slightly decrease. For the purpose of avoiding unintentional islanding during fault condition, the wind turbines are immediately disconnected at the same time other sources are made to increase generation so as to cover for the load around these turbines.

In situations where large numbers of turbines are connected with grid during short-term fault, voltage stability problem can occur due to high reactive power consumption of the generators. This is one of the main technical challenges with large wind generation penetration which usually leads to transient instability that can results in switch off of a large number of the wind generators.

The induction generators are the most popular generators for wind turbines and these generators absorbs reactive power during normal operation which can create low voltage issue in the power system. Furthermore the power flow pattern and system dynamics characteristics changes with high penetration of wind generators into the grid. This therefore calls for system dynamic simulations to ensure the stability of the power system.

3. Evaluation of Voltage Stability by Using Voltage Stability Indices (VSI)

Even though voltage stability is a dynamic problem, static indexes still plays a very important role in voltage stability analysis and helps operators to know how close the current operation point is to static stability limit. Voltage stability index (VSI) was introduced as far back as early 80`s together with other computational methods for evaluating voltage stability in networks which has resulted in opening of new perspective for predicting voltage collapse [15]. The VSI formula is derived from simple current flow equation based on figure 2.0 that indicates the sending and receiving buses as 1 and 2 respectively.

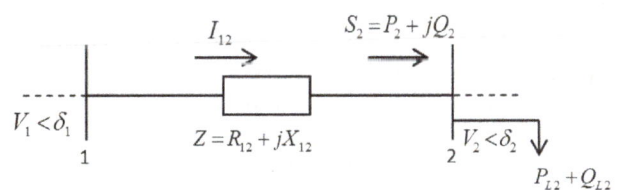

Fig. 2.0. *Simple 2 bus system for a network.*

The VSI has been identified as one of the useful index that could facilitate early prediction of instability in any network. Despite the useful perfections of the existing method, urgent attention is still required in the area for the introduction of newer formulations in order to fully exploit the advantages offered by these methods. The advances in formulations are actually needed for enhancement of capability for detecting instability and early warning signal to system operators. Many formulations were derived to represent the VSI as an indicator of voltage collapse. Among the most popular equations is Fast Voltage Stability Index (FVSI) which was proposed by [15]. In their equation, the authors assumed the power angle between the buses in the network to be small and hence equated to zero. Thus, the formula for FVSI becomes simpler and easier to analyze.

In addition to that, the analysis on VSI was also employed in other power system analysis by using VSI as a constraint as proposed by [16] where the voltage stability indices were included in the optimal power flow algorithm as a constraint during the load curtailment procedure. By implementing this idea, the total capacity of load curtailment was found to increase as voltage stability margin increases for the system. However, the proposed algorithm is restricted to large system with large loads if better solutions are required. Besides that, [17] have used the VSI value as an indicator during optimization of reactive power in a network. In this study, the reactive power was found to be an important factor in the network stabilization by means of VSI indicator. The reactive power injection was minimized during the optimization. In another development, the concept of lower voltage limit in the analysis of VSI was introduced by [18] in order to improve the analysis of VSI to become much faster than traditional VSI. The authors used the angle parameter as a VSI indicator to determine how far the system is from the collapse point.

Moreover, the analysis on VSI is not only to the transmission or to the grid network as been discussed earlier but was also extended to distribution network. Use of VSI was proposed by [19-20], on distribution network and the only difference with analyses on transmission network is on those line parameters that cannot be simplified due to big difference in their values as load changes. For transmission analysis, the authors assumed that the difference in power angle between two buses is small and as such the assumption cannot be applied to distribution systems. In some cases some researchers have assumed that the sending voltage is very close to receiving voltage. All these assumptions or simplification cannot be used for the analysis due to the high R/X ratio associated with the of distribution systems. Furthermore, with the existence of DG in the distribution network the VSI value will differ from what it used to be.

The static indexes that are currently used by system planners and operators are four and these indexes are all kept less than the critical value for system stable operation. All the four indexes are based on power flow concept [21-22]. The four indexes are fast voltage stability index (FVSI), line stability index (LMM), on line stability index (LVSI) and line stability factor (LQP).

The voltage stability index can be evaluated at all nodes for radial distribution as presented by [23]. The equations for the indices were formulated for solution in load flow problem for radial distribution system.

3.1. Fast Voltage Stability Index (FVSI)

This VSI was proposed by [24] based on simple two bus system having sending and receiving bus. The sending bus is assumed to be the reference bus $(\delta_1 = 0 \text{ and } \delta_2 = -\delta)$ and employing this assumption to the index in equation (1) simplifies the expression;

$$\frac{4Z^2Q_{12}X_{12}}{V_1^2(R|_{12}Sin\delta - X_{12}cos\delta)^2} \leq 1 \qquad (1)$$

The fast voltage stability index (FVSI) is the simplified equation is expressed as;

$$FVSI = \frac{4Z_{12}^2Q_{12}}{V_1^2X} \qquad (2)$$

From this expression, the *FVSI* is directly proportional to reactive power while active power is indirectly related through the bus voltage.

3.2. Line Stability Index (L$_{mn}$)

This index was formulated by [25] and was also based on power flow through a transmission line using π model representation for a two bus system. The power at the receiving end is expressed as;

$$S_2 = \frac{V_1V_2}{Z_{12}} < (\theta - \delta_1 + \delta_2) - \frac{V_2^2}{Z_{12}} < \theta \qquad (3)$$

$$Q_{12} = \frac{V_1V_2}{Z_{12}}Sin(\theta - \delta_1 + \delta_2) - \frac{V_2^2}{Z_{12}}Sin\,\theta \qquad (4)$$

$$P_{12} = \frac{V_1V_2}{Z_{12}}Cos(\theta - \delta_1 + \delta_2) - \frac{V_2^2}{Z_{12}}Cos\,\theta \qquad (5)$$

where θ is the angle of the line impedance δ is the angle difference between the supply voltage and the receiving end voltage.

Putting $\delta_1 - \delta_2 = \delta$ into (4) and solving for V_2;

$$V_2 = \frac{V_1Sin(\theta - \delta) \pm \sqrt{[V_1\,sin(\theta - \delta)]^2 - 4Z_{12}Q_{12}Sin\theta}}{2Sin\theta} \qquad (6)$$

For real value of V_2, the discriminate of equation (6) must be greater than or equal to zero. This is a stability criteria which must be satisfied as;

$$[V_1Sin(\theta - \delta)]^2 - 4Z_{12}Q_{12}Sin\theta \geq 0 \qquad (7)$$

Since $X_{12} = Z_{12}Sin\theta$ the line index is expressed as;

$$L_{mn} = \frac{4X_{12}Q_{12}}{[V_1\,sin(\theta - \delta)]^2} \qquad (8)$$

The line index is also directly related to the reactive power and indirectly related to the active power through the voltage.

3.3. Online Stability Index (LVSI)

The online stability index expresses the relationship between line active power and the bus voltage of a line as proposed by [26-27]. By using the discriminant of voltage quadratic equation which must be greater or equal to zero, the receiving end bus voltage V_2 can be derived from equation (5) as;

$$V_2 = \frac{V_1 Cos(\theta - \delta) \pm \sqrt{[V_1 Cos(\theta - \delta)]^2 - 4ZP_2 Cos\theta}}{2Cos\theta} \quad (9)$$

The discriminant of equation (9) is;

$$[V_1 Cos(\theta - \delta)]^2 - 4Z_2 P_2 Cos\theta \geq 0 \quad (10)$$

For $R = Z_2 Cos\theta$

$$\frac{4R_{12}P_2}{[V_1 Cos(\theta - \delta)]^2} \leq 1.00 \quad (11)$$

Therefore, LSVI is defined as;

$$LVSI_{12} = \frac{4R_{12}P_2}{[V_1 Cos(\theta - \delta)]^2} \quad (12)$$

The LVSI relationship is a direct relationship with active power of the receiving end bus and is indirectly related with reactive power through the sending end voltage unlike the case for FVSI and L_{mn}.

3.4. Line Stability Factor (LQP)

Using the same concept of load flow studies [28] has derived the LQP relationship such that the active and reactive power on receiving bus can be expressed as;

$$P_2 = \left[(V_1 cos\delta - V_1) \frac{R_{12}}{R_{12}^2 + X_{12}^2} - V_1 sin\delta \frac{X}{R_{12}^2 + X_{12}^2} \right] V_2 \quad (13)$$

$$Q_2 = \left[(V_1 cos\delta - V_2) \frac{X}{R_{12}^2 + X_{12}^2} + V_1 sin \frac{R}{R_{12}^2 + X_{12}^2} \right] V_2 \quad (14)$$

The assumption adopted is that the line is a lossless line with $\frac{R}{X} \ll 1$ and equations (13) and (14) can then be expressed as;

$$P_2 = \frac{V_1 V_2 sin\delta}{X} \quad (15)$$

$$Q_2 = \frac{V_1 V_2 cos\delta - V_2^2}{X} \quad (16$$

Then employing the trigonometry identity giving by;

$$Sin^2\delta + Cos^2\delta = 1 \quad (17)$$

Equations (15) and (16) can be expressed as;

$$\left(\frac{XP_2}{V_1 V_2} \right)^2 + \left(\frac{XQ_2 + V_2^2}{V_1 V_2} \right)^2 = 1 \quad (18)$$

Thus

$$V_2^4 + (2XQ_2 - V_1^2)V_2^2 + X^2 Q_2^2 + P^2 X^2 = 0 \quad (19)$$

Equation (18) is quadratic equation and for real solution the discriminant must be greater or equal to zero.

$$(2XQ_2 - V_1^2)^2 - 4(X^2 Q_2^2 + P_2^2 X^2) \geq 0 \quad (20)$$

For a lossless line $P_1 = -P_2$, then;

$$4\left(\frac{X}{V_1^2} \right)\left(\frac{P_1^2 X}{V_1^2} + Q_2 \right) \geq 1 \quad (21)$$

The LQP is expressed as;

$$LQP_{ij} = 4\left(\frac{X}{V_1^2} \right)\left(Q_2 - \frac{P_1^2 X}{V_1^2} \right) \quad (22)$$

The relationship here is that both active and reactive powers are directly related.

4. Comparison of Indexes

The comparison of the four indexes which are all based on a line connecting between two buses is to establish their critical values based on variables involved and the necessary assumption to be made (if any) as shown in Table 1. Evaluation of these figures will determine the closeness of the line to its transmission limit and at this point any small increase in load can lead to voltage collapse of the entire system for a static load model.

Table 1. Comparison of the Indexes

Index	Formulation	Relative variables	Assumption	Critical value
FVSI simplified	$\frac{4Z^2 Q}{V_1^2 X}$	Q V Z X	$\delta = 0$	1.0
FVSI	$\frac{4Z^2 Q_j X}{(V_i)^2 (Rsin\delta - Xcos\delta)^2}$	Q V δ Z X R	No	1.0
L_{mn}	$\frac{4X_{12}Q_{12}}{[V_1 sin(\theta - \delta)]^2}$	Q V δ R X	No	1.0
LVSI	$\frac{4R_{12}P_2}{[V_1 cos(\theta - \delta)]^2}$	P V δ R Q	No	1.0
LQP	$4\left(\frac{X}{V_i^2}\right)\left(Qi - \frac{P_i^2 X}{V_i^2}\right)$	P Q V X	$^R/_X \ll 1$	1.0

For the simplified FVSI where the formulation is based on the assumption that the voltage angle difference is zero, the

implication is that such assumption is only applicable to lightly loaded lines and cannot be applicable to heavily loaded lines since the difference is large. The case of FVSI and L_{mn} is however different in the sense that the two are equivalent and are having direct relations with the receiving end reactive power and indirect relationship with the bus voltage of the sending bus. They are therefore more sensitive to reactive power changes. In the case of LVSI the relationship is direct with active power on receiving end while the bus voltage on sending bus LVSI is more sensitive to δ than FVSI and L_{mn} because Cos (θ - δ) changes much faster than (θ - δ) by around 90^0 and when θ - δ approaches 90^0, Cos (θ - δ) approaches zero as Cos (θ - δ) is in the denominator. There may be a dramatic increase in LVSI due to this influence. It implies that a healthy line may be identified as a critical line by LVSI (bottleneck). From all the indexes only LQP have direct relationship with both active and reactive power on the receiving end. Hence, LQP may perform well for either reactive power change or active power change, and is likely going to be better than FVSI and L_{mn}.

5. Conclusions

An overview on voltage stability indices for DG penetrated networks was presented in this paper. The penetration has great improvement on power quality supplied to consumers but has impact on voltage stability of the entire network. This paper consequently, reviews the techniques for the determination of maximum capacity limit for a system before voltage collapse occurs using voltage stability index. The paper also discusses the adequacies of the indices for assessing the most sensitive bus or line on the verge of voltage collapse for modern distribution systems. Out of the indices the line stability factor (LQP) is the only index that directly relates active and reactive power of the receiving end bus and is the most likely index that can perform better for modern distribution systems.

Acknowledgements

The author H. Musa acknowledges with gratitude the financial support offered by Bayero University Kano Nigeria and the provision of suitable research facilities.

References

[1] Prada, R.B.; Souza, L.J.; "Voltage stability and thermal limit: constraints on the maximum loading of electrical energy distribution feeders", IEE Proceedings-Generation, Transmission and Distribution, Volume 145, Issue 5, Sept. 1998 Page(s) : 573 – 577

[2] N. G. A. Hemdan and M. Kurrat, "Efficient integration of distributed generation for meeting the increased load demand," Int. J. Electr. Power Energy Syst., vol. 33, no. 9, pp. 1572–1583, Nov. 2011.

[3] D. Devaraj and J.P. Roselyn, On-line voltage stability assessment using radial basis function network model with reduced input features, International Journal of Electrical Power & Energy Systems, vol. 33, no. 9, Nov 2011, pp. 1550-1555.

[4] M.S. Kumar and P. Renuga, Application of Bacterial Foraging Algorithm for Optimal Location of FACTS Devices with Multi-Objective Functions, International Review of Electrical Engineering-IREE, vol. 6, no. 4, pp. 1905-1915, July2011.

[5] V. Jayasankar, N. Kamaraj, and N. Vanaja, "Estimation of voltage stability index for power system employing artificial neural network technique and TCSC placement," Neurocomputing, vol. 73, no. 16-18, pp. 3005-3011, Oct.2010.

[6] A.K. Sinha and D. Hazarika, A comparative study of voltage stability indices in a power system, International Journal of Electrical Power & Energy Systems, vol. 22, 2000,pp. 598-596

[7] "http://www.dispersedgeneration.com/," [accessed 5-2-2015].

[8] Kundur P, Paserba J, Ajjarapu V, Anderson G, Bose A, Canizares C, et al. Definition and classification of power system stability. IEEE Trans Power Syst 2004;19(3):1387–401

[9] Kundur P. Power system stability and control. New York, US: McGraw-Hill; 1994

[10] Global Wind Energy Council (GWEC) and Greenpeace International, "Global Wind Energy Outlook 2008," http://www.gwec.net/ fileadmin/ images/ Logos/ Corporate/ GWEO_A4_2008_lowres.pdf, Aug. 2008

[11] Thong V.V., Driesen J., Belmans R. (2005). Power quality and voltage stability of distribution system with distributed energy resources. Int J Distrib Energy Resour.; 1(3): pp. 227–40.

[12] Thyagarajan, K., Davari, A., Feliachi, (2005). A. Load sharing control in distributed generation system," System Theory, 2005. SSST' 05. Proceedings of the Thirty-Seventh Southeastern Symposium on , vol., no., pp. 424- 428, 20-22.

[13] Baran, M.E. and Wu, F.F (1989). Optimal Capacitor Placement on Radial Distribution System", IEEE Trans. Power Delivery, 4(1): 725-734

[14] Jauch, C., Matevosyan, J. M., Ackermann, T., and Bolik, S. M. (2005). International Comparison of Requirements for Connection of Wind Turbines to Power Systems. Wind energy, 8 (3): pp. 295-306

[15] Ismail, M. and Rahman, T. K. (2005). Estimation of maximum loadability in power systems by using fast voltage stability index (FVSI). Journal of Power and Engineering Systems, vol: 25,pp. 181-189 DOI: 10.2316/Journal.203.2005.3.203-3392

[16] Huang, G.M., Nair, N.-K.C. (2002). Voltage stability constrained load curtailment procedure to evaluate power system reliability measures. Proceedings of the IEEE Power Engineering Society Transmission and Distribution Conference, 2, , pp. 761-765

[17] Devaraj D. and Roselyn, J.P (2011). On-line voltage stability assessment using radial basis function network model with reduced input features. International Journal of Electrical Power & Energy Systems, Vol. 33, no. 9, pp. 1550-1555.

[18] Kataoka, Y., Watanabe, M., Sakaeda, S. and Iwamoto, S. (2010). Voltage Stability Preventive Control Using VMPI Sensitivities. Electrical Engineering in Japan, vol. 173, no. 4, pp. 28-37

[19] Hamada, M.M., Wahab, M.A.A. and Hemdan, N.G.A (2010). Simple and efficient method for steady-state voltage stability assessment of radial distribution systems. Electric Power Systems Research, Vol. 80, no. 2, pp. 152-160

[20] Hamouda N. and Zehar, K (2011). Stability-index based method for optimal Var planning in distribution feeders. Energy Conversion and Management, Vol. 52, no. 5, pp. 2072-2080.

[21] H Musa, B Usman, SS Adamu (2013) Improvement of voltage stability index using distributed generation for Northern Nigeria subtransmission region. Computing, Electrical and Electronics Engineering (ICCEEE), 2013 ...

[22] Zhang X, Wong C. K. (2011). Comparison of Voltage Stability Indexes Considering Dynamic Load. IEEE Electrical Power and Energy Conference.

[23] Chakravorty, M. and Das, D (2001). Voltage stability analysis of radial distribution network. International Journal of Electrical Power & Energy Systems, Vol. 23, pp. 129-135

[24] Musirin, I.; Abdul Rahman, T.K. (2002). Novel fast voltage stability index (FVSI) for voltage stability analysis in power transmission system. Research and Development, 2002. SCOReD. Student Conference 2002; vol., no.265- 268 DOI:

[25] Moghavvemi, M., Omar, F.M. (1998). Technique for Contingency Monitoring and Voltage Collapse Prediction" IEEE Proceeding on Generation, Transmission and Distribution, Vol. 145, N6, pp. 634-640

[26] Moghavvemi, M. Faruque, M.O. (2001). Technique for assessment of voltage stability in ill-conditioned radial distribution network. IEEE Power Engineering Review pp. 58-60.

[27] H Musa, S.S. Adamu (2013) Enhanced PSO based multi-objective distributed generation placement and sizing for power loss reduction and voltage stability index improvement Energytech, 2013 IEEE, PP 1-6

[28] Mohamed, A., Jasmon,G.B., Yusoff,S.(1989). A Static Voltage Collapse Indicator using Line Stability Factors. Journal of Industrial Technology, Vo1.7, NI, pp. 73-85,

10

Used Oil Storage and Disposal Practices in Automobile Repair Garages in Ghana

Akpakpavi Michael

Mechanical Engineering Department, Accra Polytechnic, Accra, Ghana

Email address:

micakpakpavi@yahoo.com

Abstract: This study investigates the used oil storage and disposal practices in automobile repair garages in Ghana. The data for the study were collected using questionnaire, observations and personal discussions. One hundred and fifty informal sector garages and 100 formal sector garages made up the analyzed sample size. The results of the analysis reveal that a very large majority of the auto repair garages in the country lack used oil storage and disposal standards. Also, the garages do not organize training programs to educate and sensitize the mechanics about the health and environmental hazards of used oils. Again, proper used oil storage and disposal practices are virtually absent in the auto repair garages, coupled with lack of licensed used oil collection centers, transporters and recyclers in the country. This therefore, has caused used oil generators in the country to engage in an undesirable used oil disposal practices which pollute the environment excessively. To help address this issue, the Environmental Protection Agency (EPA)-Ghana's by-laws on used oil handling and disposal practices must be vigorously enforced. The EPA must also intensify its visits to the garages to educate and sensitize them on proper used oil storage and disposal practices. The Government, the Banks, philanthropies and wealthy individuals in the country should help establish used oil collection centers in the country to enable the mechanics sell off their generated used oils for recycling. Government must assistant the endowed garages in the country to enable them develop the capability to recycle their generated used oils. This will help create more jobs and enhance the revenue base of government as a result of cut downs on the importations of virgin motor oils into the country.

Keywords: Used Oil, Storage, Disposal, Automobile Repair Garages, Ghana

1. Introduction

1.1. Background

Automotive repair garages play a critical role in the automotive industry in every economy since they help repair and maintain the vehicles on the roads. Typical vehicle maintenance activities include oil and filter changes, battery replacement, light metal machining et cetera. Potential wastes generated as a result of vehicle maintenance and repair activities are: used oils, spent fluids, spent batteries, asbestos brake pads and linings, metal machining wastes, spent organic solvents, and tires. These wastes have the potential to be released to the environment if not handled properly, stored in secure areas with secondary containment, and/or protected from exposure to weather. If released to the environment, the impact of these releases can be contamination of surface waters, ground water and soils, as well as toxic releases to the air [1-2]. Moreover, used motor oil contains numerous toxic substances, including polycyclic aromatic hydrocarbons, which are known to cause cancer. In addition, tiny pieces of metal from engine wear and tear, such as lead, zinc and arsenic, make their way into lubricants, further contributing to the polluting potential of used motor oil. Because used motor oil is heavy and sticky, and contains an extensive concentrated cocktail of toxic compounds, it can build up and persist in the environment for years [3-5]. Indeed, a paper presented by Nwachukwu et al (2012) shows the environmental impact assessment of used engine oil (table 1).

Table 1. Environmental impact assessment of local uses of changed engine oil.

Local uses of used oil	Application	Environmental effect
Road construction	On the ground	Soil pollution
Rust prevention	On a metal device	Stains on contact
Old engines emergency lubrication	Automobiles, Generators	Air pollution,
Wood preservation	Timber, roofing, fencing	Land pollution
Mixed with grease for gear oil	Gear box lubrication	Spills, soil pollution
Burning, Boilers, furnaces	Burners, bakery, incinerators	Off-gas, air pollution
For pest, weed, and dust control	Garden, workshops	Soil pollution
Ball joint oil and nuts loosing oil	Ball and socket joints, nuts	Stains on contact
Block and Balustrade mold lubricant	Block, balustrade molds	Spills, land pollution
Dust and tick control	Land, floor	Land pollution

Source: Nwachukwu et al (2012)

To help address the problem of used oil pollutions in the environment, laws, regulations and standards on proper storage and disposal of used oils are enacted and enforced in most developed and developing countries to ensure that vehicle repair garages in these countries properly store and reasonably dispose off their generated waste oils.

Unfortunately, the disposal of waste oil into gutters, water drains, open vacant plots, farms and so on is a common practice in Ghana especially by motor mechanics. Conspicuously, large quantities of used oil generated in the garages in the country tend to be dumped around the cities in drains which ultimately contaminate and pollute water bodies including rivers, lagoons, streams, etc. Examples of water bodies polluted by the dumping of used lubricating oils, greases, etc. from automobile service garages abound and these include the Odaw River and Korle Lagoon in Accra, Fosu Lagoon in Cape Coast, and so on [6]. Undoubtedly, this indiscriminate dumping of waste oils in the water bodies in the country could largely be attributed to the mechanics gross lack of knowledge regarding the physical and environmental hazards of the used oils.

This study therefore, investigates motor oil storage and disposal practices in automobile repair garages in Ghana. It is the believe of the researcher that this paper would help educate and sensitize the automobile repair garages in the country about the hazards associated with used oils and their proper methods of storage and disposal approaches to avoid polluting the environment unjustifiably.

1.2. The Automobile Repair Garages in Ghana

In Ghana, the most common automobile repair garages are the local garages where a group of automotive mechanics come together to offer automotive maintenance and repair services. They operate under the informal sector with usually one master owning the shop. The garages are normally built on a piece of land hired from a landlord. Again, in a classical local automotive repair garage not less than three master

mechanics are available with each master having specialized area of auto repairs as automotive electrical specialist, mechanic specialist, vehicle body works specialist, brake binding specialist et cetera. Moreover, each master normally has apprentice trainees who study under them.

The second categories of vehicle repair garages are those that operate under the formal sector and they include: automobile repair workshops that are owned by one person but employs automotive mechanics specialized in different areas. These garages engage in the repair and maintenance of vehicles and are few in number in the country. Workshops owned by international automobile companies also operate under the formal sector and are opened to the general public for general vehicle repair and maintenance. Typical examples include Silver Star Ltd, Toyota Ghana Ltd, Rana Motors, Mechanical Lloyed, PHC Motors etc. Again, in Ghana, there are automotive repair garages in government organizations or institutions such as Ministry of Food and Agriculture, Ministry of Health, Ghana Water Company Ltd, Ghana Audit Service etc. In these organizations, the vehicles of the institutions are repaired and maintained. Moreover, most of these workshops are not opened to public vehicle owners for vehicle servicing activities. Besides, all these garages operate under the formal sector.

2. Methods

2.1. Study Area Description

The study was carried out in four main industrial areas in Ghana, where particularly local automobile repair garages were dominant including: Kumasi, Takoradi, Accra and Tema. The survey was confined to these places because most industrial activities are concentrated there [7]. Certainly, this finding is still the same now. Also, the four centers mentioned are all urban areas. Hence, it was taught that the views of the respondents from these centers would adequately represent the whole population.

The researcher classified the automobile repair garages into two categories. Category A represents vehicle repair garages in the informal sector generally known as micro and small scale enterprises (MSEs). These informal micro and small scale garages are numerous and they are dotted all over the country. Hence, a very large number of the automotive garages in the country are in the informal sector. They are patronized by many vehicle owners in the country because of their low works charges and proximity. Category B consists of automobile repair garages in the formal sector, registered in the Ghana's trade register as small and medium scale enterprises (SMEs).

2.2. Data Collection Techniques

The study used multi approach techniques in data collection. It involved observation, personal discussions and questionnaire administration to workers from each category of automobile repair garages identified. This approach was chosen to increase the validity of the study by enriching the scope, depth and knowledge derived from the data.

2.2.1. Observations

Observations were made by the researcher by visiting the two categories of the automobile repair garages identified three times on different occasions informally. During the trip, used oil storage and disposal practices by each category of garage was observed, noted and where necessary recorded.

2.2.2. Sampling

In all, 250 automobile repair workers who responded to the questionnaire make up the sample size. One hundred and fifty respondents were from the informal micro and small scale garages, whiles one hundred respondents were from the formal small and medium scale garages in Ghana. All the mechanics from both formal and informal sectors were randomly selected. However, the garage owners, workshop supervisors and transport officers in both the formal and inform sectors were purposively selected and they were also included in the total sample.

2.3. Data Analysis

Data extracted from the administered questionnaires and interview were analyzed using statistical tools such as pie charts, tables and bar charts for percentile analysis [8].

3. Results

3.1. Educational Levels and Working Experience of Mechanics in the Automotive Repair Garages

With regard to educational levels of mechanics in the informal auto repair garages in the country, about 5% had education up to tertiary levels, 21% had secondary/technical levels of education, 30% have had basic education levels whiles close to 44% of the mechanics in the informal sector had apprenticeship form of education. This finding is as illustrated in fig. 1.

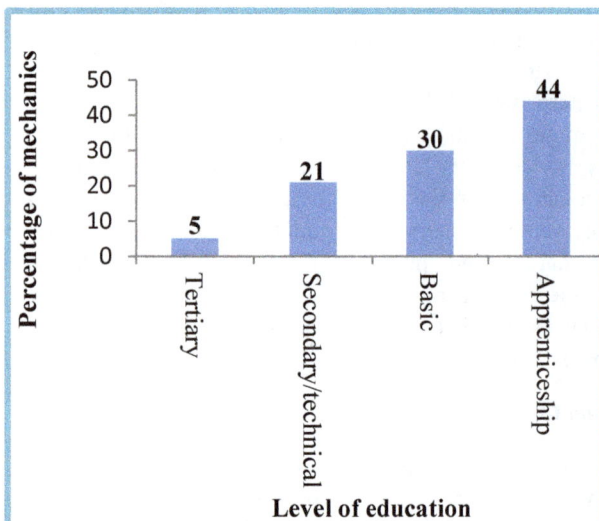

Fig. 1. Educational levels of mechanics in the informal sector garages.

Moreover, the mechanics in the formal sector automobile repair garages who responded to the questionnaire on their educational levels indicated that about 55% had tertiary levels of education, 26% had secondary/technical levels of education, 11% had basic education levels whiles the remaining 8% had apprenticeship form of education. This is pictured in fig. 2

On the issue of working experience of mechanics in the country, about 70% of the mechanics in both the formal and informal auto repair garages in the country had more than ten years working experience whiles the remainder 30% have had between 1-6 years working experience. This is pictured in fig. 3

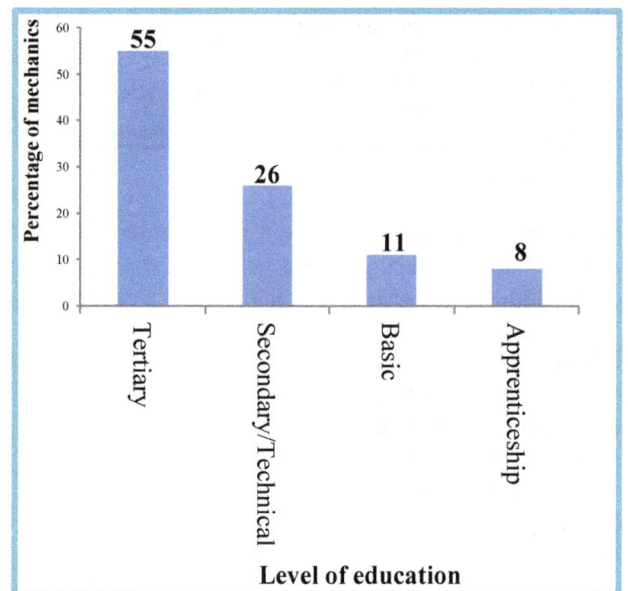

Fig. 2. Educational levels of mechanics in the formal sector garages.

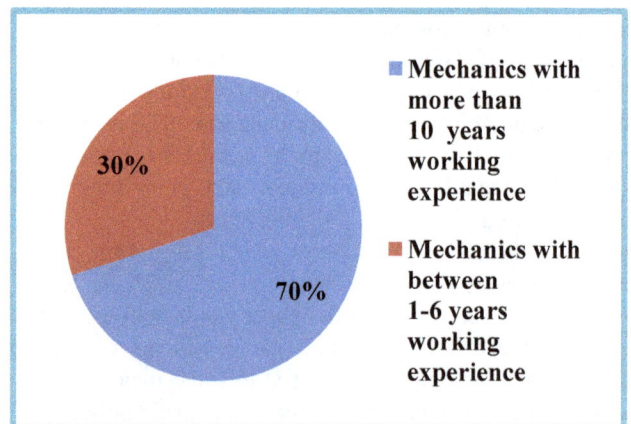

Fig. 3. Working experience of mechanics in the garages in the country.

3.2. Availability of Used Oil Management Standards in Automobile Repair Garages

Used oil disposal standards play a critical role in ensuring that vehicle repair garages properly handle and dispose of their used oils in such a manner to minimize oil leakages into the environment. However, on the issue of availability of used oil disposal standards in the garages in Ghana, about 5% of the vehicle repair garages in the formal and informal sectors had adopted some standards; whiles 95% had no waste oil disposal standards. This is shown in fig. 4.

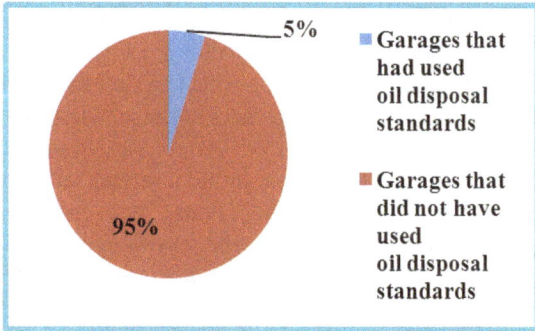

Fig. 4. *Availability of Used oil management standards in the vehicle repair Garages*

3.3. Education and Training on Hazards of Used Oil for Auto Mechanics

On the issue of education and training to help raise the awareness of the auto mechanics on the health and environmental hazards of used oil, about 5% of the mechanics in the informal micro and small scale sector had had it, whiles the remainder 95% were completely oblivious of the environmental hazards of the used oils. This is illustrated in fig. 5.

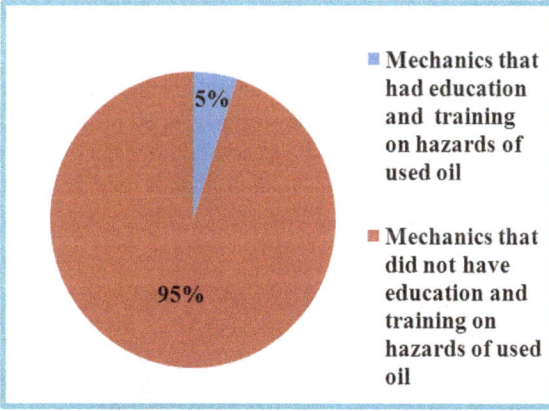

Fig. 5. *Training on used oil hazards for informal sector mechanics.*

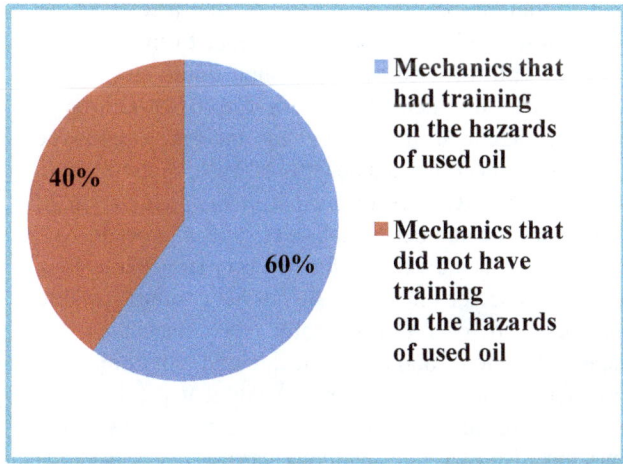

Fig. 6. *Training on used oil hazards for formal sector mechanics.*

Moreover, on the same issue of education and training on health and environmental hazards of used oil, about 60% of the mechanics in the formal sector had it, whilst the remainder 40% did not have it. This finding is as presented in fig.6.

3.4. Problems of Storage of Used Oils and Other Materials by Garages

On the above issue, about 85% of the garages in the formal and informal sector did have no problems with storage of their generated used oils for subsequent disposal whiles 15% did have storage problems of their generated used oils. This finding is depicted in fig. 7

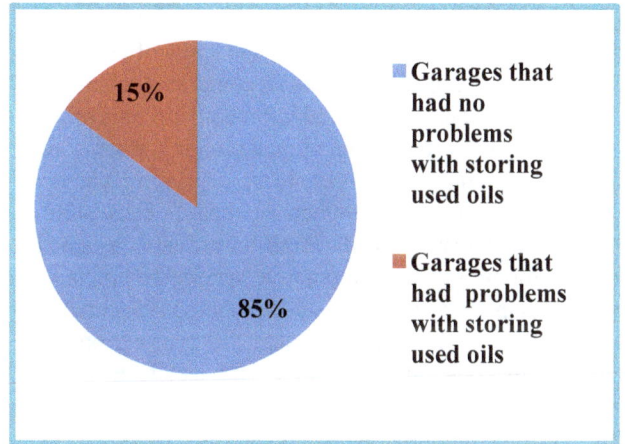

Fig. 7. *Problems of storage of used oils by garages.*

3.5. Disposal of Used Oil by Garages

With regard to the responses given by the garages on their used oil disposal approaches, 60% of the garages sold their generated used oil, 25% of the garages poured their generated used oil into the environment, 10% gave away the used oil for free whiles the remaining 5% of the vehicle repair automobile garages gave away their generated used oil to their clients. This finding is illustrated in fig. 8.

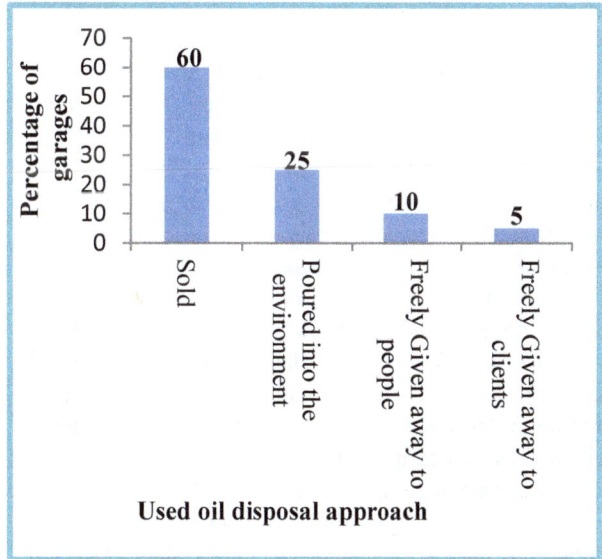

Fig. 8. *Used oil disposal methods by garages.*

4. Discussion

4.1. Educational Levels and Working Experience of Mechanics in the Auto Repair Garages

Automobile vehicle repair garages generate lots of wastes. The major environmental wastes generated by the automotive manufacturing industry include: machine lubricants and coolants; aqueous and solvent cleaning systems; paint; and scrap metals and plastics. Hazardous cleaning chemicals are very common and are likely to require special waste management arrangements [9]. Surely, auto-mechanics requires relatively higher levels of formal and informal education to enable them appreciate issues of good housekeeping including the use of preventive maintenance in an effort to reduce the number of leaks and spills of used oils that occur. Certainly, in most developed countries such as the USA, Germany, Japan, China etc. a high school diploma or the equivalent is typically the minimum requirement for someone to work as an automotive service technician or mechanic. High school courses in automotive repair, electronics, computers, mathematics, and English provide a good background for prospective automotive service technicians. Therefore, completing a vocational or other postsecondary training program is considered as just the preparation for entry-level positions in automotive service technology. Moreover, in developed countries as well as developing countries such as Malaysia, India and currently Ghana, some service technicians can earn an associate's degree. Courses usually include basic mathematics, computer, electronics, and automotive repair. Some programs add classes in customer service, English, as well as other necessary skills in automotive wastes generation and recycling [10].

As identified in fig. 1 of this work, a large majority (74%) of the auto mechanics, in the informal micro and small scale auto repair sectors in the country possesses formal education up to basic levels. Really, as a result of their low educational levels, the mechanics in the informal sector tend to have general apathy towards the environment despite their appreciable number of years of working experience (fig.3). Undeniably, visits to most of these garages shows that automotive repair wastes in the form of liquids, gases and solids are indiscriminately and instinctively dumped by the mechanics in their auto servicing workshop environments.

The situation whereby most of the mechanics in the informal sector tend to have gross disregard for the environment due to their low levels of formal education call for serious concern. This is because most of the auto repair activities carried out in this country is undertaken by the garages in the informal micro and small scale sector as observed by the Director in charge of Environmental Education at the EPA-Ghana, Salu (2013), who made a presentation on the theme, "The status of waste management in small garages". According to him, a research conducted by the EPA on four small garages in Accra and Tema shows that most of the small-scale garages in the country operated in residential areas, particularly in densely populated areas, and there is the need for the operators such as the auto mechanics

to understand the responsibilities of automobile servicing and repairs and take the necessary actions to improve the environmental management of their operations. He further indicated that, the Environmental Protection Agency (EPA) had, for a long time, been concentrating its monitoring activities on large garages, a situation which has left the micro and small scale informal sector garages to have a field day [11].

Frankly, to enable the mechanics in the informal sector to identify with basic environmental issues and therefore develop positive attitudes towards the environment, it is expedient that the mechanics and their masters in this sector take advantage of the education offered by the non-Formal Education Division of the Ministry of Education, which is concerned with adult literacy to upgrade and update their formal education and skills. Again, the auto mechanics can also enroll on the part time Higher National Diploma (HND) programs run in some polytechnics in the country such as Accra Polytechnics to enable them acquire higher certificates, diplomas and degrees particularly in automotive technology. Ultimately, an implementation of the model of education as proposed by Donkor (2006), will serve a useful purpose of helping to educate mechanics who essentially had their education through the apprenticeship system of education. Indeed, this model of apprenticeship envisages apprenticeship training to take place at workshops (on-site) and through distant education (off-site), via radio and television broadcasts, audio-tapes and video-cassettes [12].

Surely, the afore-mentioned measures will culminate in the proper understanding of the environment by the mechanics and will consequently lead to their proper handling and safe disposal of their generated solid, gas and liquid wastes. There will also be significant drop in both indoor and outdoor pollution levels caused by the garages in the country leading to improved health of the people.

4.2. Education and Training on Hazards of Used Oil for Auto Mechanics

Used oils can have detrimental effects if spilled or released into the environment, and therefore are considered as hazardous substances. To this end, most garages globally tend to organize training programs to educate auto mechanics on the good housekeeping approaches to adopt to manage and minimize the spill of used oil in the shop floors or garages. Good housekeeping methods include improved employee training, management initiatives to increase employee awareness of the need for and benefits of waste minimization, and requiring increased use of preventive maintenance in an effort to reduce the number of leaks and spills that occur.

Unfortunately, as indicated in fig. 5, a very large percentage of the informal sector mechanics do not have training on the hazards of used oil. Consequently, the mechanics are completely oblivious about the health and environmental hazards of used oil. Observations reveals that the informal sector auto mechanics are not able to identify, recover and appropriately dispose of all hazardous wastes such as used oils, oil filters, oil contaminated rags, etc. Again, the mechanics are also ignorant about basic environmental issues such as global warming, ozone depletion, air and water pollution, worker safety and chemical risks, hazardous waste labeling, waste minimization methods, recycling waste and use of recycled products, spill prevention and other related issues.

Really, visits to most of these informal sector small scale garages show that spills from motors, differentials and transmissions, brake and power steering units as well spills from oil filters are allowed to drop freely the shop floor of the garages in enormous proportions; a situation which is objectionable as far as the environment is concerned. Furthermore, broken gallons of used oils as well as oil filters and rags soaked in used oils are voluminously speckled around the service compounds of these informal sector garages exposing the mechanics ignorance about the health and environment hazards associated with these materials. In fact, the mechanics carry out their activities of blanket dumping of hazardous materials with impunity and without any regard for environmental laws or standards.

Again, discussions with the garages in the informal sector reveals that the Environmental Protection Agency-Ghana hardly visit the micro and small scale garages in the country to educate the mechanics about the health and environmental hazards of used oil and other related issues. This definitely is regrettable. The EPA-Ghana should regularly visit the garages in the informal sector which are strewed all over the country and educate them on environmental issues. This will enable the automotive repair garages to shun the haphazard dumping of their generated wastes such as used oils in their work service compounds. Other training centers in the country such as Gratis Foundation, Intermediate Technology Transfer Units (ITTU), Technical training institutions etc. available in the country should also help to educate and alert the informal sector auto mechanics on environmental issues in relation to their operations.

Moreover as indicated in fig. 6, a good number of the automotive garages in the formal sector frequently organize training programs to educate the auto-mechanics on relevant issues of auto repair. Again, visits to the garages in the formal sector which includes garages belonging to the international automotive companies such as Silver Star Ltd, Toyota Ghana Ltd, Rana Motors, Mechanical Lloyed, PHC Motors etc. clearly shows that they organize in-service training programs to educate and train their mechanics and staff on how to repair and maintain essentially their imported vehicles. Indeed, as a result of education and training, the mechanics in these garages endeavor to avert used oil spills and improper handling and disposal of oil filters and rags containing used oils. Generally, these formal sector garages rather have positive attitudes towards the environment and make an effort to achieve pollution free environmental operations.

Nonetheless, an informal discussion with the mechanics in the formal sector indicates that they still need more education on the health and environmental hazards of used oil. Again, discussions with the formal sector garages reveals that the Environmental Protection Agency-Ghana, pays visits to these garages at least twice in a year to ensure that these garages

operate their workshops in such a manner so as to avoid polluting the environment with prejudice. The EPA should do more to educate the mechanics about the hazards of used oil.

4.3. Availability of Used Oil Storage and Disposal Standards in Automobile Repair Garages

Used oil contains toxic substances that can kill plants and animals, spoil water supplies and even cause cancer. When used motor oil is spread on the land, soil productivity is reduced and surface or ground supplies may be contaminated. One quart of used oil will foul the taste of 250,000 gallons of water. Oil in surface waters severely disrupts the life-support capacity of the water. Oil encourages the growth of organisms that deplete the dissolved oxygen supply available to aquatic life, such as fish. Without oxygen, they die. Oily films on the surface block sunlight, affecting the growth of plants and the direct entry of airborne oxygen. The toxic substances in used oil kill the small organisms that support the rest of the food chain. Used motor oil must be handled carefully because it contains carcinogenic and other toxic substances. Among the more serious health hazards is the large amount of lead found in much of the used oil. Leads and other dangerous compounds accumulate through the combustion process in the engine of a vehicle. Used motor oil must be processed to remove toxic contaminants before it can be safely reused. Mensah-Brown (2013) presented a comparative analysis of Society of Automotive Engineers (SAE) new oil, used oils and re-refined oil (table 2). The study clearly indicates the relatively large amounts of lead and other metals contained in used oils [6].

Table 2. Characteristics of used and re-refined motor oils.

	Sample		
Test	New SAE 40 Oil	Used Oil	Re-Refined Oil
Water Content (% V/V)	0.0	0.1	0.0
Fuel dilution (%V/V)	0.0	2.9	0.0
Pour point, °C	-20	-20	-20
n-pentane Insolubles (weight %)	0.0	1.69	0.0
Benzene Insolubles (weight %)	0.0	0.911	0.0
Neutralization Number (mgKOH/g)	0.40	1.645	0.06
Metal Contents (ppm)			
Lead	0.0	2813	1.86
Calcium	543.2	606.5	9.56
Copper	0.0	21.25	2.17
Iron	0.0	89.5	0.0
Potassium	0.0	19.0	0.0
Nickel	0.0	2.0	0.0

Source: Mensah-Brown (2013)

Contamination by the toxic substances in used oil can be a serious health hazard to humans in different ways such as: people may consume the contaminated water directly; toxic substances in used oil become concentrated in water, plants and animals and poison people who may eventually eat them; and direct skin contact over an extended period of time has been identified as a potential cause of cancer.

As a result of the numerous environmental and health challenges associated with used oils, automotive repair garages which generate or handle used oil are expected to follow certain good housekeeping practices. These mandatory practices, called "management standards," are developed by the EPA for businesses that handle used oil. The management standards are common sense, good business practices designed to ensure the safe handling of used oil, to maximize recycling, and to minimize disposal. The standards apply to all used oil handlers, regardless of the amount of the oil they handle.

Fundamentally, the EPA's requirements are meant to be followed by all types of businesses that handle used oil including: used oil generators, collection centers and aggregation points, transporters, re-refiners and processers, used oil burners and marketers. Expressly, the EPA's standards on used oil managements principally includes: used oil storage management practices, used oil leaks and spills, record keeping on used oils, mixing of used oils and hazardous wastes, effective management of used oil filters as well as standards on effective cleanup of used oil practices [13-14].

Unfortunately in Ghana, as shown in fig. 4, a very large percentage (95%) of the garages in both the formal and informal sector have not obtained any used oil management standards to guide them to safely and properly re-claim, store and dispose their used oil to minimize environmental pollution. The non-availability of used oil management standards in the vehicle repair garages has resulted in a situation whereby the mechanics engage in gross inappropriate handling and disposal of used oils to the detriment of the environment and human health. In fact, as a result of low formal educational levels of the mechanics in the informal sector coupled with lack of training, the mechanics in the micro and small scale garages in the country just allow waste oils from sump, gallons, oil filters, saw dust, rags etc. to tire out and dispense profoundly on their shop floors. Undeniably, some of the mechanics deliberately discharge unacceptable volumes of used oils on their service compounds with the intention of checking soil erosion (fig.9).

Not surprisingly, most of the garages in the informal sector as well some from the formal sector have their auto repair servicing compounds utterly covered with thick coat of used oils. Indeed, the mechanics do not find the need to take urgent steps to clean up spill of used oils. Used oil filters are indiscriminately discarded allowing used oil to drain heavily on the ground to cause environmental pollution. It is not uncommon to find situations like spills-wet patches not cleaned up, absorbent materials not available, floor surface slippery when wet and other similar situations persisting in the informal automotive repair garages in the country with neither the employer nor the mechanics taking responsibility for the on-site health and safety of the whole workforce in the garage. Most garages do not use drip pans and the occurrence of fluids leaking from automobiles and parts placed on the floor are common. Figure (9a and b) shows used oil drained unacceptably on the floor in a small scale garage in Accra.

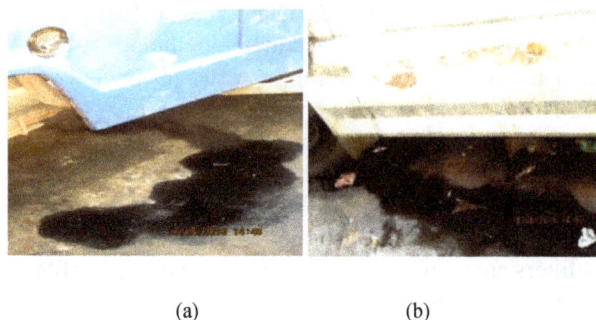

(a)   (b)

Fig. 9. Used oil drained on the floor from vehicles in a small garage in Accra

Nonetheless, by using drip pans, shop floors will remain cleaner and hence less frequent cleaning of the floors would be required. Added benefits would be a reduction in the use of rags and adsorbent to clean the floors, and a safer work environment.

Meanwhile, used oil filters are difficult to completely drain of oil, even after draining, may contain oil trapped inside the filter. Used filters are considered to be source of liquid waste and cannot be placed in a bin or skip for disposal. Filters need to be managed by Hydraulic filter crushers which separate the dirty oil from the filter. Best practice management of oil filter wastes involves the use of a mechanical crusher to remove and recover most of the oil they contain. The free oil removed from the filter must be contained, managed and stored separately (for collection into waste oil storage drums) from the filters. Drained and mechanically crushed oil filters (not containing free liquids) are then classified as solid waste. Some metal recyclers may accept the crushed filters that can then be passed to scrap metal dealers with a much reduced chance of environmental pollution [15-16-17]. Fig. 10 shows typical hydraulic oil filter crushers.

Again, Sawdust and rags are commonly used to absorb a spill of used oils and other hazardous substances in almost all the formal and informal garages in the country. Saw dust used to clean waste oil might be classified as hazardous waste and be subjected to hazardous waste regulations. Undeniably, most garages currently in the country dispose of the used saw dust and rags in the trash. To reduce the environmental effects associated with disposal of dirty rags which might be classified as hazardous, the automobile repair garages in the country should seriously consider the use of leasing arrangements in which a laundry service picks up the dirty rags, cleans them, and returns them to the shop.

Furthermore, management initiatives to increase employee awareness of the need for and benefits of used oil minimization, the increased use of preventive maintenance in an effort to reduce the number of leaks and spills that occur in the garages in the country will help reduce the environmental pollution caused by used oil drains. Additional ways to reduce or minimize waste oil spills include: improve used oil inventory control practices by the garages, use of first-in, first-out (FIFO) policy, minimization of storage quantities, increase storage area inspections as well as employment of spill containment techniques by the automobile repair garages in the country.

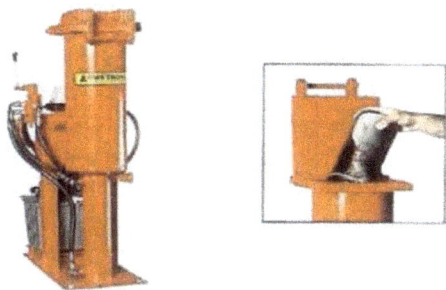

Fig. 10. Hydraulic filter crushers.

4.4. Used Oil Storage in Automobile Repair Garages

EPA's used oil storage requirements demands that, all used oil tanks and containers be properly labeled, kept in good condition, be prevented from rust, leak, or deterioration. In addition, structural defects need to be fixed immediately. Certainly, under the EPA's regulations, used oil is never supposed to be stored in anything other than tanks and storage containers. Also, storage of used oil in lagoons, pits, or surface impoundments that are not permitted under EPA regulations and requirements is prohibited [13, 18]. As shown in fig. 7, about 85% of the auto repair garages in the formal and informal sector in the country claims they do not have problems of storing their generated used oils.

However, visits to the informal auto repair garages shows that tanks and gallons which are not properly covered, labeled and not in good conditions are rather utilized by the informal sector mechanics to store used oils particularly outdoor due to lack of storage rooms. Again, these micro and small sector garages lack storage spaces. They only possess small wooden structures in which they store most of their tools and equipments for safe keeping and protection. Certainly, these methods of storing used oils in bare gallons and containers out-door by the mechanics causes water and other contaminants to get into the stored used oils and contaminate it.

Moreover, if used oil is mixed with hazardous waste, it probably will have to be managed as a harmful waste. Hazardous waste disposal is a lengthy, costly, and strict regulatory process. Again, Certified center managers will not be in a position to accept used motor oils that have been contaminated with other fluids such as antifreeze, solvents, gasoline, or water. The only way to be sure that used oil does not become contaminated with hazardous waste is to store it separately from all solvents and chemicals and not to mix it with anything.

As a result of lack of proper used oil storage facilities in the micro and small scale informal auto repair garages, the mechanics ends up discarding their generated used oils into containers in household garbage bins which end up in landfills. Some mechanics also engage in inappropriate use of the used oil including pouring on weeds, spraying on roads as dust suppressant, cleaning tools and protecting timber posts and fences from termites. Again, significant volumes of used oils are disposed directly into the environment, with large majority of the garages in the country pouring more on the ground. This is because most of the work areas in the garages are bare soils that could be used for disposal. Used oil utilization in road construction and in marking play grounds result in release of the used oil components into the environment as identified by Nwachukwa et al (2012). These practices are harmful because the used oil can then enter the soil and leach through to contaminate ground water.

Clearly, this shows that there is a need for information about waste oils, especially for the mechanics in the informal micro and small garages in the country. Training and making available information on handling and storing hazardous materials to these mechanics would significantly reduce their poor storage practices of used oils which will result in reduction in environmental pollution. To help educate and sensitize the mechanics in the informal sector who incidentally undertake most of the vehicle repair activities in the country, the EPA must extend and intensify its education and training programs in these garages. The technical training institutions available in the country including the ITTU, Gratis Foundation, and so on must help to train and educate the mechanics particularly on the environmental impacts of used oils.

On the other hand, the formal sector medium scale auto repair garages, including the international automobile repair garages in the country rather take measures to appropriately store their used oils in well sealed gallons. They tend to be environmental conscious and therefore take steps to reduce used oil spillage in their automotive servicing garages. Indeed, the environmental consciousness exhibited by some of the mechanics in the formal sector could largely be attributed to the mechanics relatively high educational levels, regular training and information being made available to these mechanics. In addition, regular visits by the EPA- Ghana to these formal sector garages in the country also enable the latter to be proactive in ensuring safe environment practices.

Ultimately, the EPA must vigorously enforce its by-laws on the proper handling, storage and disposal of waste oil by the garages in the country. This will enable the mechanics in the country to develop a sense of responsibility towards the environment and do well to avert its pollution disproportionately.

4.5. Disposal of Used Oil by Garages

Currently in Ghana, none of the automobile repair garages is able to recycle its generated used oil for re-use. There are also no accredited used oil collection centers available in the country. Additionally, the transporters or buyers of used oil in the country are also not certified and accredited. Subsequently, as shown in figure 8, a large majority (60%) of the garages in both the formal and informal sector in the country do sell their generated used oil essentially to unlicensed used oil boys, popularly called 'gwagwa boys' who are particularly illiterates or school drop-outs and therefore are not informed about the environmental hazards of used oil. From observations, these boys hop from one garage to another during the early hours of the day and acquire the used oils from the various garages. A discussion with these 'gwagwa boys' reveals that they normal purchase one gallon of the used oil (that is 5 Liters) for paltry

₵2 (currently, equivalent to $0.56) from the garages. Again, discussions with the 'gwagwa boys' also indicate that they do sell the used oil they have procured from the garages to timber merchants for spray on timber species for protection against termites' attack. Furthermore, the boys also sell off some of the used oils to corn millers, road constructors, who mix the used oils with coal-tar for road constructions, as well as other unlicensed used oil dealers.

Moreover, the Korle Lagoon in Accra has become one of the most polluted water bodies on earth. It is the principal outlet through which all major drainage channels in the city empty their wastes into the sea. Large amounts of untreated industrial waste emptied into surface drains has led to severe pollution in the lagoon and disrupted its natural ecology.

Undeniably, the banks of the Korle Lagoon in Accra, has become one of the major unlicensed used oil collection centers currently in the country, where the 'gwagwa boys' send their used oils for sale after procuring it from the garages. Visits to the Korle lagoon geographical area reveal that gallons of used oils are undeservedly lined up at the banks of this river (fig. 11a and b). Incongruously, these gallons of used oils are allowed to drain continuously into this lagoon on daily basis causing severe pollution of both the lagoon and its contiguous environments. In fact, this situation persists even now. The worrying aspect of it all is that the Korle lagoon has its principal outlet through which all major drainage channels in the city empty their wastes into the sea. This implies that intolerable quantities of used oils are allowed to drain into the sea and other water bodies in the country from the Korle lagoon.

Definitely, to help curb this situation and save the water bodies in the country, the EPA-Ghana, the Accra Metropolitan Assembly as well as other regulatory bodies in the country should as a matter of urgency visit the Korle Lagoon Geographical areas and educate these unlicensed used oil dealers on proper handling methods of used oils. Where necessary, these regulatory bodies should warn the 'gwagwa boys' to desist from using the banks of the Korle lagoon as used oil collection centers. This will help prevent further polluting the lagoon and its vital environs with used oils.

As depicted in table 2, used motor oil is insoluble, persistent, and can contain toxic chemicals and heavy metals. When disposed improperly, used oil can contaminate soil and water (table 1). Fortunately, used motor oil can be recycled and either re-refined into new oil, processed into fuel oils, or used as raw materials for the petroleum industry.

(a)   (b)

Fig. 11. Gallons of used oil at banks of Korle Lagoon in Accra

To help ensure proper storage and more particularly appropriate disposal of used oils, most developed countries such as the USA, China, Japan, Australia, and so on encourage the recycling of used motor oil by sitting and certifying used oil recycling collection centers in most of its cities and towns . This ensures that certified used oil collection centers take used motor oil from the public at a fee for recycling. Essentially, these countries also promote the establishment of tribal used oil collection and recycling centers where the tribal and/or community members send their generated used oils to a drop-off location run by the tribe [19].

Even so, currently in Ghana, there are no licensed used oil collection centers in any of the regions or communities in the country leading to improper handling and disposal of the used oils by the garages in particular. The government of Ghana through the public-private partnership initiatives should help establish licensed used oil collection centers in strategic locations in the country where all used oil producers could be encouraged to send their generated used oils for sale, instead of selling them to the unlicensed 'gwagwa boys' who lack the requisite knowledge in proper used oil handling methods. The banks, corporate organizations, worthy business men and women in the country should also help in this direction.

Globally, four major reasons are driving this momentum of refining used oil. Firstly, the quality of re-refined base oils is improving dramatically. There has been an overall upgrading in the refining processes and in the quality of used oils collected due to better collection practices. Secondarily, rising crude oil prices have driven up base oil prices, making re-refining economically attractive. Thirdly, recent studies by the US Environmental Protection Agency have also revealed that there are significant economic benefits of waste oil refining. Tests conducted have shown that reprocessing used oil consumes far less energy and resources than refining crude oil from the earth. For example, it takes only 4 litres of used oil to produce 2.5 litres of new engine oil. But when starting from scratch, that same 2.5 quarts would require 160 litres of unrefined crude oil to manufacture. Finally, environmental standards across all countries, developed and developing, are becoming stringent. Tests have shown that 4 litres of oil dumped into a river can contaminate 4 million liters of water and have disastrous effects on our planet's ecosystem [2].

In fact as a result of many benefits associated with recycling of used oils, many local governments in most developed countries provides funds and logistical supports to the vehicle repair garages to enable them recycle their generated used oils in-house. These supports includes; awarding grants, subsidies, loans, and other financial supports to municipalities, parishes, political subdivisions, and vehicle repair garages to enable them establish and provide continuous operation of collecting services and facilities for used oil [19].

Moreover, Statistics from the National Petroleum Authority of Ghana and the Tema Lube Oil Company Limited indicate that the total domestic consumption of lubricating oils as far back as 2013 was about 45 million liters which amounts to about US$1050 million at current exchange rate of US$1.00 = GHC4.20 (20/6/2015). This constitutes a huge chunk of the foreign exchange requirement of the country with a nominal Gross Domestic Product (GDP) of US$48.678 billion for 2013 [20].

The government of Ghana should as a matter of urgency help partner the local garages in Ghana to procure the necessary equipments to enable them recycle their generated used oils in-house. Definitely, the establishment of re-refining and/or reprocessing as a strategy for used oil management in Ghana would significantly reduce the amounts of this common hazardous material being disposed arbitrarily into the environment to pollute it. Also, government spending on the importation of virgin engine oils into the country will considerably trim down.

5. Conclusion

This paper investigates used oil storage and disposal practices in the automotive repair garages in Ghana. The study shows that used oil storage and disposal practices in the auto repair garages in the country are rather indecorous, inapt, futile and unprofessional leading to nauseating environmental pollution by the auto repair garages in the country. The study also reveals that large majority of the garages in both the formal and informal sector do not possess waste oil standards to guide them to properly reclaim, store and dispose off their generated waste oils. Again, most of the mechanics in the auto repair garages in the country are completely oblivious about the health and environmental hazards associated with used oils due to lack of training and education. Furthermore, none of the garages currently in the country recycles used oil for re-use. There are also no certified used oil collection centers in any part of the country. This has created a situation whereby the mechanics ends up engaging in improper handling and disposal of the used oils including discarding it into containers in household garbage bins, pouring the used oil on weeds, spraying it on roads as dust suppressant, cleaning tools and protecting timber posts and fences and so on. Moreover, significant volumes of used oils are disposed directly into the environment, with large majority of the garages in the country pouring more on the ground. To help reverse this trend and to enable the mechanics adopt and adapt proper methods of storing and disposing used oils without polluting the environment excessively in the country, the EPA's by-laws on used oils handling and disposal practices must be made stringent and vigorously enforced. The EPA must also intensify its visits to the garages and educate and sensitize the mechanics about the health and environmental effects of used oils. This will enable the mechanics in the country to develop a sense of responsibility towards the environment and do well not to allow their generated used oils to pollute it unjustifiably. The government must provide financial and other logistical supports to the auto repair garages in the country to enable them procure the relevant equipments to recycle their generated used oils. The Banks, corporate organizations, philanthropies as well as wealthy individuals in the country should help establish used oil collection centers in most of the communities in the country to enable the mechanics sell off

their generated used oils for recycling. This will prevent the mechanics from pouring the used oils into the environment as well as selling it to unauthorized dealers such as 'gwagwa boys' who virtually knows nothing about the health and environmental hazards of used oils. Indubitably, the recycling of used oils in the country will not only reduce the unwarrantable dumping of the used oils by the mechanics and other unlawful dealers but will also create more jobs and enhance the revenue base of the government.

References

[1] Waste Oil Study Report to Congress, Prepared by the US Environmental Protection Agency (EPA), Washington, USA, 1974

[2] W.A. Irwin ,1978, Used oil: Comparative Legislative Controls of Collection, Recycling, and Disposal, pp.703-705

[3] S. Jhanani, J. Kurian, Used oil Generation and Manage ment in Automotive Industries, Environmental Assessment of Used Oil, vol. 2, 2011, pp. 135-138

[4] B.O. Okonokhua, B. Ikhajiagbe, G.O. Anoliefo, and T.O. Emed, "The Effects of Spent Engine Oil on Soil Properties and Growth of Maize (Zea mays L.)" 2007, pp. 147 - 152

[5] M.A. Nwachukwu, J. Alinnor, and H. Feng, "Review and Assessment of Mechanic Village Potentials for Small Scale Used Engine Oil Recycling Business, 2012, pp.465-474

[6] H. Mensah-Brown, "Optimization of the Production of Lubricating Oil Using Response Surface Methodology", 2013, pp. 749-756

[7] J. Powell, "Survey of Engineering Manufacturing Industries in Ghana. Technology and Enterprise Development Project". A Report of DFID, London, UK. 1995.

[8] I.R. Levin, "Qualitative Approach to Management", McGraw-Hill, Singapore, 1989

[9] RCRA In Focus, "Vehicle Maintenance", 1999

[10] Study.com, "Automechanic Training Programs and Requirements", 2015

[11] ModernGhana, Daily Graphic/Ghana, "Waste Dis-posal by Garages Endangers Public Health", 2013

[12] Francis Donkor, Department of Technology Education, University of Education, Winneba "Enhancing Apprenticeship Training in Ghana Through Distance Learning",2006

[13] USEPA, "Managing Used Oil: Advice for Small Business", 2015

[14] USEPA, "Standards for the Management of Used Oil, Part 280, 2015

[15] EPA, "Guides to Pollution Prevention, the Automotive Repair Industry", 1991

[16] M. G. Elnour, H.A. Laz, "Clean Production in Autorepair Workshops", 2013, pp. 66-77

[17] D.V. Jacobs, "How to Design and build your AutoWorkshop, publish. motorbooks intern, USA, 1998

[18] Ohio EPA, , "The Regulation of Used Oil: An Overview for Ohio Businesses who generate Used Oil", 2015

[19] R. Arner, 2012, Used Oil Recycling in America

[20] The State of the Ghanaian Economy in 2013. Institute of Statistical, Social and Economic Research (ISSER), University of Ghana, Legon, 2014.

11

Study Shapes and Alignments of Rotational States in Some Rare-Earth Nuclei

N. A. Mansour, N. M. Eldebawi

Faculty of Science, Physics Department, Zagazig University, Zagazig, Egypt

Email address:

Eldebawi@yahoo.com (N. M. Eldebawi)

Abstract: The ground state rotational band members for the studied nuclei have been identified and their energies calculated to about ±0.3%. These energies provide good test of various models for rotational bands, and impressive agreement is observed with a theoretical calculations by the variable moment of inertia (VMI) model and cubic polynomial (CP) formula. In addition to giving an excellent fit to the excitation energy of the ground-band levels in all nuclei. Also reproduces a critical spin I_c beyond which the square rotational frequency ω^2 decreases although the moment of inertia $\Im$ keeps on increasing, thus resulting in back bending in $\Im$ -ω^2 plot for some of the studied nuclei. This feature is predicted to appear not only in the deformed nuclei but also at quite low spins in nearly spherical nuclei as well. Also the formula contains the low and high spin behavior in agreement with the available experimental data.

Keywords: Rotational Parameter, Moment of Inertia, Back-Bending

1. Introduction

The study of nuclear states has in recent years expanded to include still higher angular momenta and excitation energies [1-3] by an interplay between deformation and alignment effects. Theoretical calculations predict various shapes: some are prolate or triaxial, others (near closed shells) become oblate for a certain spin interval. But at the highest spins the nuclei have a tendency towards a large triaxial deformation and sometimes towards very large prolate deformations (super deformations). Directly involved in their shape changes [3] are aligned orbitals which come down to the Fermi level as the nucleus rotates more rapidly. At a certain frequency these orbitals become populated and cause large alignments. These some orbitals also come down in energy as the deformation increases and therefore constitute a driving force towards large deformations. Moments of inertia are sensitive to shapes and alignments. In the present paper, I shall focus on a moment of inertia, which is proportional to the height of angular momentum obtained in the de-excitation of compound nuclei; the moment of inertia is a smooth and single-valued function of the rotational frequency. The nuclides $^{154}Sm_{92}$ [6], $^{158}Gd_{94}$ [7,8], $^{162}Dy_{96}$ [9,10], $^{164,166}Er_{98}$ [11,12], $^{168}Yb_{100}$ [13], $^{174}Hf_{102}$ [14,15] $^{178}W_{104}$ [16] and $182Os_{106}$ [17,18,19,20,21,and 22] have been chosen for the present work as an illustration out of an extensive investigation concerning the behavior of the ground state rotational bands in doubly even nuclides. These nuclides have been chosen because they show a quite different behavior of the band structures near the top of the bands and any theoretical description will ultimately have to explain these differences. The nuclei studied exhibits a behavior similar to the one ^{168}Yb, i-e, the moment of inertia is a smooth and single-valued function of the rotational frequency [5].

2. Description of the Models and the Formalism in the Analysis

Our procedure from adjusting the excitation energies of partially level schemes for the studied nuclei is discussed in more general terms. In the absence of detailed microscopic calculations attempts have been made to describe the dependence of $\Im$ on I (or ω^2) phenomenological. The analysis is based on a variational expression for the energy (using VMI model).

$$E_I = [I(I+1)/2\Im] + V(\Im) \quad (1)$$

$$V(\Im) = c(\Im-\Im_0) \quad (2)$$

where c and $\mathfrak{I}_0$ are parameters characteristic for a given nucleus. $\mathfrak{I}(I)$ is determined by the requirement

$$\partial E / \partial \mathfrak{I}|_I = 0 \qquad (3)$$

were shown to describe with surprising accuracy the energies of yarest bands in even-even nuclei ($I\pi = 0^+$, 2^+, 4^+, 6^+, 8^+,....,up to I_c, the critical angular momentum at which a sudden increase of $\mathfrak{I}$ occurs [back bending behavior Ref. (23) has appeared in Fig. (6)]). The limit of validity of the equations 1, 2 and 3 is reached for $\mathfrak{I}_0 = -\infty$, giving $E_I \alpha$ [I (I + 1)] $^{1/2}$. Between $\mathfrak{I}(0) = 0$ and $\mathfrak{I}(0) = \infty$, the ground state moment of inertia is $\mathfrak{I}(0) = 0$. The extension of equations 1, 2 and 3 to $\mathfrak{I}(0) \to -\infty$ permitted the definition of the average moment of inertia $\mathfrak{I}_{02} = \frac{1}{2}[\mathfrak{I}(0) + \mathfrak{I}(2)]$, which laid the basis for a macroscopic description of the effective moment of inertia . As $\mathfrak{I}_0$ reaches larger and larger negative values, the nuclear resistance to cranking increases until, at $\mathfrak{I}_0 = -\infty$, the threshold energy $\frac{1}{2} c (\mathfrak{I}_0)^2$ diverges [24] . The coefficients c has been determined by a least-squares fit weighted by the inverse square of the measured energy, that is, minimize the relative errors. The components of the rotational frequency ω and the moment of inertia $\mathfrak{I}_I$ are deduced from the data by defining

$$\hbar\omega = E (I \to I\text{-}2) / I(I+1)^{1/2} - [(I\text{-}1)(I\text{-}2)]^{1/2} \qquad (4)$$

and

$$2\mathfrak{I}_I / \hbar^2 = (4I\text{-}2) / E (I \to I\text{-}2) \qquad (5)$$

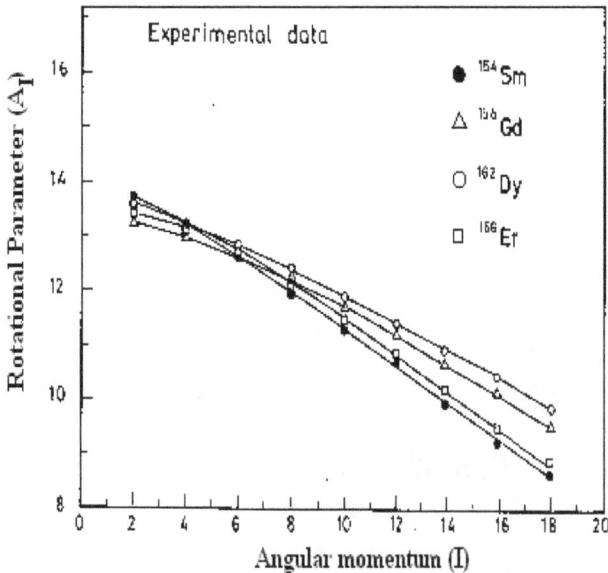

Fig. (2). *The calculated rotational constants plotted against angular momentum, I for the same nuclei in Fig. (1).*

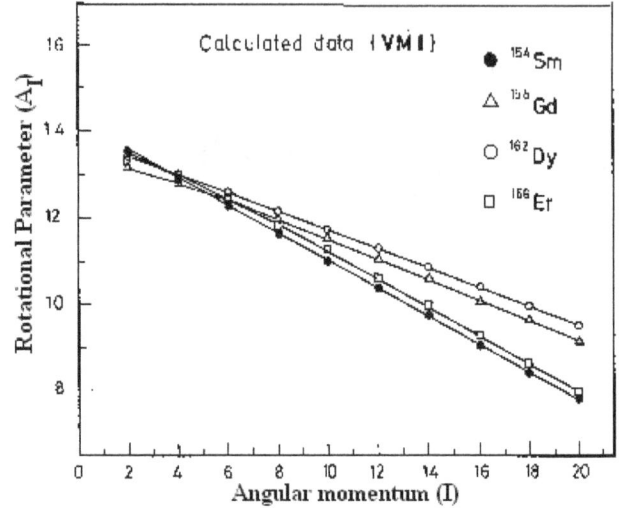

Fig. (3). *Experimental and calculated rotational constants plotted against angular momentum, I for $^{168}Yb_{98}$ nucleus.*

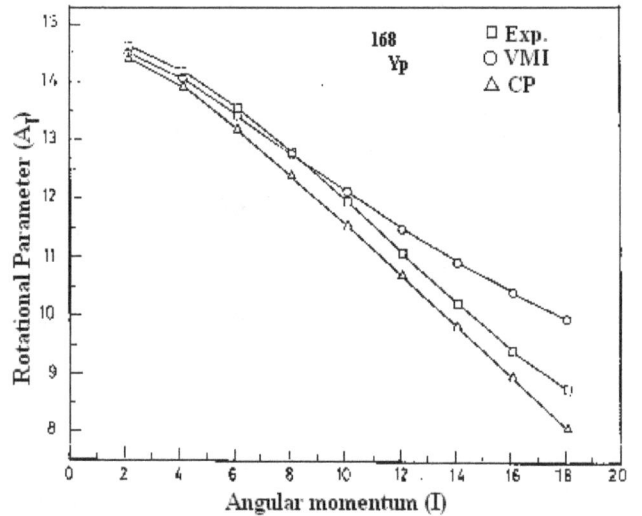

Fig. (1). *Experimental rotational constants (appropriate to the transitions) plotted against angular momentum, I for $^{154}Sm_{92}$, $^{158}Gd_{94}$, $^{162}Dy_{96}$, and $^{166}Er_{98}$ nuclei.*

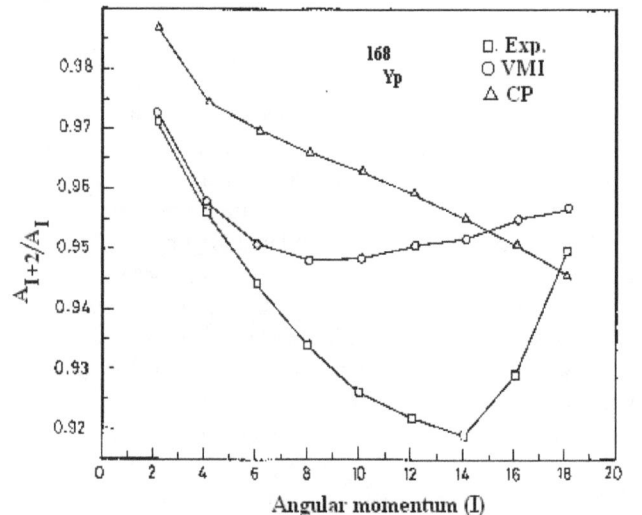

Fig. (4). *Ratio of successive experimental and calculated rotational constants, A_{I+2}/A_I, plotted against angular momentum, I for $^{168}Yb_{98}$, nucleus.*

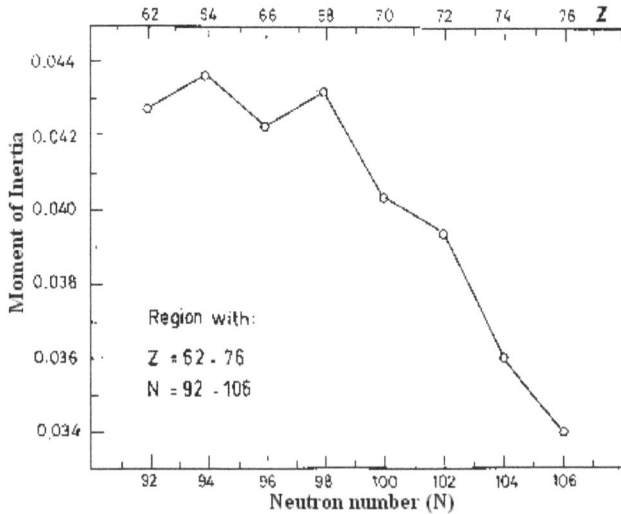

Fig. (5). *The inertial parameter for the ground state band plotted against neutron number in the region with N= 92-106 and Z=62-76.*

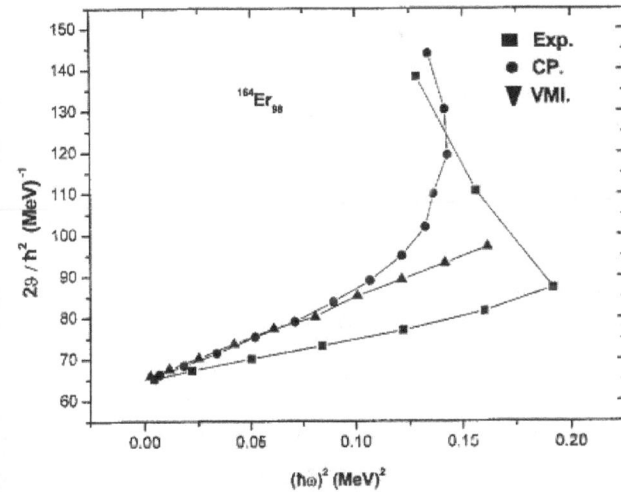

Fig. (6). *The observed moments of inertia $\Im$ as a function of the square of the angular velocity $(\hbar\omega)^2$ showing experimental and theoretical curves for ^{164}Er. The figure shows the predicted value of the critical spin at which back bending is expected from the calculations.*

From cubic polynomial (CP) formula the following expressions are derived by using a simple energy expression $E(I)=aI+bI^2+cI^3$ in equations 4 and 5 ,then

$$2\Im_I / \hbar^2 = (2I+1)[a+2bI+3cI^2]^{-1} \quad (6)$$

And

$$(\hbar\omega)^2 = [(I^2+I+1)/(I^2+I+¼)] \ [a+2bI+3cI^2]^2 \quad (7)$$

The calculations yield positive a and b and negative c for all the nuclei studied. For high spin values the first factor in equation (7) is practically unity, and ω^2 is seen to reach its maximum for the spin value $I_m = b / 3| c |$. Since the spins of interest are only even integers we designate the even integer next nearest to I_m as the critical spin I_c such that $\omega^2 (I_c) \le \omega^2 (I_c-2)$. Calculations show that I_c value ($I_c \approx 20\hbar$) agree reasonable well with those experimentally observed so far ($I_c \approx 16\hbar$) .

3. Theoretical Calculations and Discussion

In the present calculations for all studied nuclei with neutron number N=92-98 (from Sm to Er nuclei), the ground state of several neutron deficient Sm, Gd nuclei, known experimentally to be essentially spherical, are therefore generally somewhat deformed (often oblate in the calculations). One typical features of some of the light rare earth nuclei is the fact that the ground state distortion, whether oblate as in ^{148}Sm, ^{150}Gd or prolate as in ^{154}Sm$_{92}$,^{158}Gd$_{94}$,^{162}Dy$_{96}$,^{166}Er$_{98}$ and ^{168}Yb$_{98}$ appears to contract in deformation parameter β with increasing I. The reason for the diminishment in β can be directly understood from an inspection of a figure exhibiting neutron orbitals as functions of β [Ref.(1)]. For the heavy rare earth with N≥ 100 the opposite effect is noticeable. Thus for ^{174}Hf$_{102}$, ^{178}W$_{104}$, and ^{182}Os$_{106}$ the effect of stretching is very apparent.

For the present calculations only transitions whose association with the ground-state rotational band is highly probable. From the transition energies, E (I → I-2), we define the rotational constant, A_I as follows:

$$A_I = \hbar^2/2\Im_I = (E_I - E_{I-2}) / (4I-2) \quad (8)$$

where $\Im_I$ represents the moment of inertia appropriate to the transition. Figures 1 and 2 show the analysis of the experimental and calculated values of the rotational parameter A_I in terms of the angular momentum I for the nuclei studied. For comparison Fig. (2) included the result from a calculation of the variable moment of inertia VMI approach (which is one of the successful phenomenological models for even-even nuclei) in a detailed analysis of rotational bands in such nuclei. At low spin, the general features of this plot are well known :(1) a regular decrease in A_I with increasing spin, and (2) smaller slopes (more perfect rotors) associated with lower A_I values. The similarity in rotational properties of all these isotopes at higher spins is very pronounced.

The points in all cases are, or become with increasing spin, quite linear with a common limiting slope of 0.2 or0.3% decrease in A_I per state. It can be ruled out that at still higher spins the A_I values will diverge (Fig. 1, 2 and 3) again; however, the most tentative data at the highest spins rather suggests that they may converge to a single group. Thus from the lowest to the highest spins observed, the lighter rare-earth nuclei such as ^{154}Sm and ^{158}Gd seem to have rotational constant 3-6% lower than others in the region studied here (96-106 neutrons inclusive). A possible explanation is that this is due to reduction of the pairing correlation due to the energy gap in the Nilsson diagram [25]. This effect seems to be reproduced in Nilsson calculations of the moment of inertia based on the pairing model [25]. The very nearly identical behavior of the moments of inertia observed at high spin values for the nuclei studied suggests that a very general property of rotating nuclei must be involved. The average change in moment of inertia with spin observed in this study is about a factor of two, which indicates that an attempt to explain this should avoid using

perturbation theory.

Table (1). *Experimental and calculated rotational parameter values with the corresponding angular momentum in $^{154}Sm_{92}$, $^{158}Gd_{94}$, $^{162}Dy_{96}$ and $^{166}Er_{98}$ nuclei.*

Nucleus	^{154}Sm		^{158}Gd		^{162}Dy		^{166}Er	
I $\diagdown$ A_I	Exp.	Cal.	Exp.	Cal.	Exp.	Cal.	Exp.	Cal.
2	13.67	13.51	13.25	13.12	13.64	13.49	13.43	13.38
4	13.21	12.90	12.99	12.82	13.12	13.06	13.17	13.02
6	12.61	12.28	12.62	12.41	12.86	12.62	12.75	12.46
8	11.97	11.64	12.20	11.97	12.34	12.18	12.16	11.86
10	11.32	11.00	11.74	11.51	11.94	11.74	11.55	11.28
12	10.72	10.36	11.22	11.05	11.43	11.30	10.80	10.58
14	10.12	9.72	10.88	10.58	10.96	10.86	10.24	9.94
16	9.48	9.08	10.36	10.06	10.46	10.42	9.58	9.28
18	8.74	8.44	9.84	9.64	9.90	9.98	8.93	8.63

A comparison of the experimental data with the calculations by using the variable moment of inertia (VMI) and Cubic polynomial formula fit (CP) for the $^{168}Yb_{98}$ nuclei is shown in Fig. (4). this case $^{168}Yb_{98}$, represents one of the best fits of our nuclei. The ordinate here is the ratio A_{I+2}/A_I,

$$A_{I+2}/A_I = [E (I+2 \rightarrow I) / 4I+6]/ [E (I \rightarrow I-2) /4I-2] \quad (9)$$

which is related to the slope of Fig. (1), and is used primarily because it gives a plot which is very sensitive to the transition energies. A least-square fit has been used to calculate the energy spectra to the ground state bands. For all studied even – even nuclei, the model parameters a, b and c can be

determined from fitting the three energy levels E_2, E_4 and E_6. It can be seen that the quality of fit, with an error around $\pm 0.3\%$ between the calculated and experimental energy levels. The moment of inertia obtained for the studied nuclei are compared. The value of the inertial parameter tends to fluctuation until N= 98 and after that tend to decrease smoothly with neutron number as shown in fig. (5). Again there is a systematic similarity between the nuclei under investigation: although the interactions in these nuclei are stronger, both interactions initially decrease with neutron. These comparisons strongly suggestive of an underlying structural similarity of the intrinsic states in these nuclei.

Table (2). *Experimental and calculated ratios of successive rotational constants with the corresponding angular momentum in $^{168}Yb_{98}$ nucleus .*

I	AI (Exp)	AI (VMI)	AI (CP)	AI t 2 (Exp)	AI t 2 (VMI)	AIt2 (CP)	$\frac{A_{I+2}}{A_I}$ (Exp.)	$\frac{A_{I+2}}{A_I}$ (VMI)	$\frac{A_{I+2}}{A_I}$ (CP)
2	14.62	14.48	14.457	14.201	14.080	14.267	0.9713	0.97237	0.9869
4	14.20	14.08	13.966	13.579	13.484	13.610	0.95626	0.9577	0.9745
6	13.58	13.48	13.227	12.825	12.820	12.830	00.9444	0.95104	0.96998
8	12.82	12.82	12.417	11.980	12.160	12.000	0.93447	0.94852	0.9664
10	11.98	12.16	11.574	11.100	11.540	11.150	0.92654	0.94901	0.9633
12	11.10	11.54	10.720	10.240	10.980	10.290	0.92252	0.95147	0.95988
14	10.24	10.98	9.854	9.420	10.960	9.420	0.91992	0.95264	0.9559
16	9.43	10.46	8.983	8.770	10.000	8.550	00.9300	0.95603	0.95179
18	8.77	10.00	8.100	8.340	9.580	7.670	0.95079	0.9580	0.9469

4. Conclusion

The prolate and oblate configurations in N≅92-106 nuclei interact strongly perturbing states. Levels perturbed in their way must contain contributions from both prolate and oblate wave functions. This is due to level mixing and shape coexistence previously seen in much lighter rare earth nuclei such as ^{154}Sm and ^{158}Gd. Also, the ground-state rotational band members for several nuclei have been identified and provide a good test of various models for rotational bands. An impressive agreement is observed with a variable moment of inertia and cubic polynomial calculation, where the rotational

parameters are related to I, give equally good results.

References

[1] A. Bohr and B.R. Mottelson ; Phys. Scripta 10A (1974) 13.

[2] R. Bengtsson ; S.E. Larsson ; G. Leander ; P. Möller ; S.G. Nilsson ; S. Äberg and Z. Szymanski, Phys, Lett. 57B (1975) 310.

[3] K. Neergaard and V.V. Pashkevich ; Phys. Lett. 59B (1975) 218.

[4] G. Andersson ; S.E. Larsson ; G. Leander ; P. Möller ; S.G. Nilsson ; I. Ragnarsson and S. Äberg ; R. Bengtsson and J. Dudek ; B. Nerlo-Pomorska, K. Pomorski and Z. Szymanski ; Nucl. Phys. A268 (1976) 205.

[5] N.A. Mansour and A.M. Diab ; Indian J. Phys.77A(3),289 (2003) and references there in.

[6] William ; M.C. Latchie and W. Darcey and J.E. Kitching ; Nucl. Phys. A159 (1970) 615.

[7] G. Løvhøiden ; T.F. Thorsteinsen ; E. Andersen and M.F. Kiziltan ; D.G. Burkem ; Nucl. Phys. A494 (1989) 157.

[8] H.H. Pitz ; U.E.P. Berg ; R.D. Hell ; U.Kneissl and R. Stock ; C. Wesslborg and P. Von Brentana ; Nucl. Phys. A492 (1989) 411.

[9] L. Henden ; L. Bergholt ; M. Guttormsen ; J. Rekstad ; T.S. Tveter ; Nucl. Phys. A589 (1995) 249.

[10] J. Rekstad ; A. Atac ; M. Guttormsen ; T. Ramsdy ; J.B. Olsen and F. Ingebretsen ; T.F. Thorsteinsen ; G. Løvhøiden and T. Rødland ; Nucl. Phys. A470 (1987) 397.

[11] J.P. Lestone ; J.R. Leigh ; J.O. Newton ; D.J. Hinde ; J.X. Wei ; J.X. Chen ; S. Elfström ; M. Zielinska-Pfabe ; Nucl. Phys. A559 (1993) 277.

[12] C.A. Fields and K.H. Hickes and R.J. Peterson ; Nucl. Phys. A440 (1985) 301.

[13] J.C. Lisle; D. Clarkc ; R.Champman ; F. Khazaie ; and J.N. Mo ; H. Hübel ; W. Schmitz and K. Thiene ; J.D. Garrett ; G.B. Hagemann ; B. Herskind and K. Schiffer ; Nucl. Phys. A520 (1990) 451, and procedings of the conference on nuclear structure in the nineties Oak Ridge. Tennesse, April 23, (1990).

[14] P.M. Walker ; G.B. Hagamann; J. Pedersen; G. Sletten ; D. Howe; M.A. Riley ; B.M. Nyako; J.F. Sharpey-Schafer; J.C. Lisle; E. Paul; Daresbury Annual report (1983/84) 43.

[15] P. M. Walker ; G. Sletten ; N.L. Gjorup ; J. Borggreen ; B. Fabricius ; A. Holm ; J. Pederson ; M.A. Bentley ; D. Howe ; J.W. Reberts ; J.F. Sharpey-Schafer ; Daresbury Annual report (1989/90) 66.

[16] A. Krämer-Flecken ; T. Morek ; R.M. Lieder ; W. Gast ; G. Hebbinghaus ; H.M. Jäger ; W. Urban ; Annual report, Jülich GmbH Institute Für kernphysik (1988) 36.

[17] H. Schnare ; A. Krämer-Fleckem ; D. Balabonski ; W. Gast ; G. Hekbinghaus ; R.M. Lieder ; M.A. Bontley ; D. Howe ; A.R. Mokhtar ; J.D. Morrison ; J.F. Sharpey-Schafer ; P.M. Walker ; Annual report, Jülich GmbH Institute Für kernphysik (1988) 43.

[18] A. Ansari ; M.Oi ; N. Onishi ; T. Horibata ; Nucl. Phys. A654 (1999) 558.

[19] Takatoshi Horibata; M.Oi ; N. Onishi; A. Ansari ; Nucl.Phys. A646(1999) 277.

[20] Takatoshi Horibata ; N. Onishi ; Nucl. Phys. A596 (1996) 251.

[21] T. Kutsarova ; R.M. Lieder ; H. Schnare ; G. Hebbinghaus ; D. Balobanski ; W. Gast ; A. Krämer-Flecken ; M.A. Bentley ; P. Fallon ; D. Howe ; A.R. Mokhtar ; J.F. Sharpey-Schafer ; P. Walker ; P. Chowdhury ; B. Fabricius ; G. Sletten ; S. Frauendorf ; Nucl. Phys. A 587 (1995) 111.

[22] T. Kibédi ; G.D.Dracoulis; A. P. Byrne ; P.M. Davidson ; S. Kuyucak ; Nucl. Phys. A567 (1994) 183.

[23] N.A. Mansour and A.M. Diab ; Indian J. Phys.77A(4), 377 (2003).

[24] Goldhaber A.S and Scharff-Goldhaber G. ; Phys.Rev.C17,1171 (1978).

[25] S.G. Nilsson and O.Prior, Kgl. Danske Videnskab.Selskab; Mat.-Fys.Medd. 32, No.16 (1991).

12

Typology of School-Mosque in Ilkhani, Timurid, Safavid and Qajar Eras

Ladan Asadi[*], Hamid Majidi

Department of Architecture, Art and Architecture Faculty, Islamic Azad University, Mashhad, Iran

Email address:

Asadi0596@mshdiau.ac.ir (L. Asadi), ha.majidi86@gmail.com (H. Majidi)

Abstract: Mosques were the first place used for education in the first centuries of Islam. Although in later periods, independent schools were created, due to religious instructions performed in schools, mosques and schools rejoined in different ways, and the school - mosque appears in Islamic architecture. However, little attention has been paid to this type of architecture. This study aims to investigate the emergence of religious educational centers in different historical periods, i. e. Mosque-Schools. The main objective of the study is to analyze the typology of these mosque-schools. Using descriptive-analytical research method, as well as literature review and field studies, this article aims to investigate the innovations and changes made in the general plan of mosque-schools in Ilkhani, Timurid, Safavid and Qajar Eras. To achieve this, one school has been selected in each era. At the end, regarding the theoretical framework of the study, the general features of these mosque-schools within different eras have been presented and discussed.

Keywords: Mosque-School, Ilkhanid, Timurid, Safavid, Qajar, Architecture

1. Introduction

Islam has always emphasized greatly on learning and education. According to Muhammad, profit of Islam (PBUH), the scientist's pen is superior to the martyr's blood (Amoli 2012). Muslims has been always learning and gaining knowledge in accordance with their religious teachings; thus there is a direct and strong relationship between religion and knowledge in Islam, and a common place was created for both education and religion propagandizing; it can be said that in the first four centuries after the advent of Islam, the most important secondary use of the mosques was education (Halen Brand, 1994). In the mosque the classes were formed in circles and the prophet taught the religious teachings for all new Muslims; thus, the mosque was considered as a common environment for education and prayer (Kiani, 1998).

Another Hadith from Muhammad, the prophet of Islam, which represents mosque's association with education is as follows: "...everyone who enters the mosque in order to learn or teach the goodness is like a Mujahid working for God." (Halen Brand, 1994: 107).

With the advent of Islam in Iran, knowledge prospered,

research and training centers emerged; they were at their onset the continuation of those traditionally circles of debate and lessons used to be held in mosques. In mosques, each circle was identified by the name of the course it represented, such as jurisprudence circle, hadith circle and etc. (Soltanzadeh 1985, 92). Moghaddasi, the famous geographer in the fourth century AH enumerated the formation of 120 teaching circles at Cairo's Grand Masque (Halen Brand, 1994: 107). Eventually, due to large number of students, and some conflicts between religious and educational functions in mosques, and the sequential problems caused by them, an independent building was allocated to education and Islamic architecture found its way to school environments (Soltanzadeh 1985, 92). Regarding religious education, there is an inextricable relationship between religion and knowledge on the one hand, and schools and mosques on the other hand; and a structural connection has also been established between teaching and prayer environment.

2. Review of Literature

School has been studied by some researchers of Islamic architecture, as one of the elements of Islamic architecture. Halen Brand has studied school as well as other elements of

Islamic architecture, in all Islamic countries, and Pirnia has introduced it within the Iranian Islamic architecture; each one devoted part of their books to this type of architectural structure.

Soltanzadeh (2000), which examined the mosque-schools in Tehran regarding their configuration, he reached to a categorization. Zarrinchian (1998), in a study titled as "The Relationship between Mosque and School" examined the semantic features and functions in each environment.

Sheikholhokamaee (2002), also explained in his study about the dedication letters of Memar bashi mosque-school. Mollazadeh, (2002) in a book entitled "Schools and Religious Monuments" have mentioned all schools Iran like an encyclopedia. Kazemi (2011) in a paper titled as "The Recognition of the Association between Religious School and Mosalla in Yazd Mosall" has also studied the relationship between educational and prayer spaces in a case study.

3. Mosque-Schools in Iran

There are not many mosque-schools in Iran, but they are all significant for including simultaneous performance of educational and religious activities and allocating a separate part of corporal environment to each performance, their specific yet diverse environment, also manifested the ingenious creativity of the Iranian artists and architects in the space creation areas (idea, process, ornamentation) (Haj Seyed Javadi, 1999).

Important aspects of the school-mosques in Islamic cities, especially in Iran, are their special status in terms of location, navigation, and adjacency and etc. Sometimes the location has been a determining factor in the survival and prosperity or stagnation and destruction of the school over time after the death of its founder. For example, many schools built on residential neighborhoods were abandoned and gradually demolished after the death of the founders, while those schools built along bazaars and in the main squares of the cities, due to their specific urban status, were in a much better situation and have been the center of attention of authorities, shop keepers, and other people for years (Haj Seyed Javadi, 1999).

4. Types of Mosque-Schools Regarding Their Origin and Founders

Iranian Masque-Schools regarding their origin of foundation are categorized into two groups: 1) governmental schools and 2) public schools, each category had special characteristics considering foundation, purposes, the realm of authority, and the duration of prosperity.

4.1. Governmental Masque-Schools

Special importance and the specific socio-political role of religious schools always made them the center of attention of kings, ministers and dignitaries at different periods of time; each one according to their needs and power, to implement their political intentions, have built one or more mosques and predicted specific endowments for the administration and

funding. In these schools, there were specific regulations and requirements for instructors and students, and the teaching and the study permit were issued only by the permission of their founders (Haj Seyed Javadi, 1999).

4.2. Public Masque-Schools

Another category of religious schools was formed by the benefactors and scholars. Teaching and studying in these schools needed no specific requirements and permissions, just observing the rules in Islam and Sharia sufficed. These schools often appeared as public authority centers against those centers of state power. This group of religious schools were formed at the beginning and during the reign of the tyrant and bloodthirsty kings of Abbasid and were in the form of private and hidden circles of lessons, but gradually took the form of scientific-educational centers often built adjacent to religious centers and holy shrines of Imams (Haj Seyed Javadi, 1999).

5. Functional Typology of Schools

According to its name, the functional system of school must meet the needs of those who use that architectural building in order to learn and gain knowledge. The system is shaped according to the needs of the habitants in the building, and at school the needs are based on an educational and religious system. In Iranian architecture many spaces are multi-functional and a combination of multiple functions is obvious in most buildings.

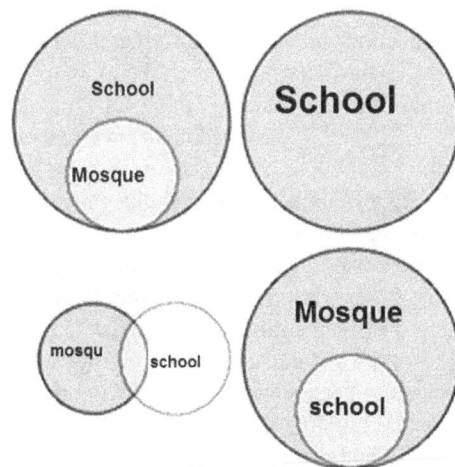

Figure 1. The functional system of school.

The functional system of school can be divided into several sections. School, mosque-school, school-mosque, Mosalla-school, and tomb-school.

5.1. School

The term school in functional Typology is used for a building which provides the needs of the students who live and study there. As an instance we can refer to those schools where prayer space is not public and is designed only for the use of their residents.

5.2. Mosque School

The term mosque-school in functional Typology is used for a building which meets all the religious needs of people and also is used by students for learning and studying. In such buildings, the priority is the mosque, i. e. the building is actually a mosque and is known as a mosque by people. In these buildings there are several rooms on the second floor or in the corners of the sub axes of Myansra where a few students live; these rooms may be the only common point between the building and a school. Hakim Mosque in Isfahan can be noted as an example of this type where a number of rooms exist on the ground and second floor.

In some school mosques the two functions have been merged so that giving the priority to one is difficult. These buildings, unlike former types of buildings, are known as mosque-school among people and the two mentioned functions have been merged in them with no functional interference of any function in the other. The functional organization in these mosque-schools has reached its consummation and these buildings can be considered as the pinnacle of architectural design of mosque-school. Aqa Bozorg mosque-school in Kashan, Seyed in Esfahan, Sheikh Abdul Hussein Tehrani and etc. can be referred to as some examples. The two functions of education and prayer, albeit functionally independent, are in different ways associated to each other.

5.3. School-Mosque

The term school-mosque in functional typology can be used for buildings whose functional priority is school where the combination of functions are organized so that it can provide the needs of the school residents and the only common functional feature of these buildings with mosque may just be part of their prayer room that was used in different periods for prayer purpose.

Many schools are also used as mosques; this may be due to the lack of mosques and adequate prayer spaces to meet the religious needs of the people in the context of the region, or in some cases in the dedication letter of the building, the religious function may be mentioned. In Salehieh Qazvin, the two functions of school and mosque work well together, but a separated entrance has been allocated to each, and the people who want to use the building for religious affairs, enter the building through a specific door.

5.4. Mosalla-School

Such a functional organization has been rarely seen in schools, and that is why we cannot categorize it as a single type. Here we mention one as an example.

Yazd Mosalla-School, is located in the new Moslla space, near Mir Chakhmaque square and still used by students of religious sciences.

5.5. Tomb-School

Due to various reasons, it is also difficult to categorize this kind as a separate type in schools functional system, but if we

want to use the term it can be said that, the term is used for schools which are associated with tombs in two ways. The first are those schools which after several years from their foundation became the tomb for their founders. In these cases a single room in the school is taken as the founder's tomb.

6. The Schools in Islamic Period

6.1. Ilkhani Era

Mongol raids destroyed many buildings in Iran. Most of the remained buildings have been constructed after the Mongol raids (Houshyari, 2013: 42). Baba Qasem or Imamieh School is one of the schools in Ilkhani era. It has a four- porch plan and is located in Isfahan. There is a domed house towards the qibla with an altar placed within. As seen a place for prayer is prominent in this Ilkhani School; however, this place is completely enclosed within the school area with no separate exit and is only for the use of the school residents. It can also be used as Madras. The domed house and altar can be seen in other Ilkhani schools which was typically aligned with qibla, like those in Ilkhini school of Shah Abolgasem Taraz, in Yazd (Houshyari, 2013: 42).

6.2. Timurid Period

This period can be considered as the golden age of Iranian schools (Halen Brand, 1994: 293). Ghiasiyeh Khargerd School has four domed houses built on its four corners of which the two on both sides of the entrance are larger and more distinguished. One of these two domed houses was used for prayer with a qibla not that much accurate (Godar 1989, 244). The tiled altar with its Mogharnas works in eastern domed house proved this; it has also been mentioned as a mosque (Blair and Bloom, 1995, 61). These two domed houses are directly linked to the outside, but the entrance porch is the only possible entrance; thus a kind of separation between the prayer and education space is visible in this school (Okin, 1987: 54).

6.3. Safavid (1501-1722 AD)

With the spread of Shiism, a new environment was created for growth and prosperity of art, architecture, and education. However no major change was made in the basis of education. Actually Safavid era was the period of evolution and perpetuate in educational environment in Iran. Meanwhile education system was better organized in comparison with the previous periods. The schools were administered by religious ulema, this brought a better integration in the school organization; thus, Safavid era was the most comprehensive and most integrated period in the history of education in Iran. With the improvement of the educational system, schools also benefited from this evolution. In Iran the Safavid schools have been identified as classic schools. Architecture of these schools like those in the past periods was according to that four porch pattern with no important innovation; there was however a more precise organization in plan and building components (Mahdavinezhad. 2013: 7).

6.4. *Qajar Era*

Despite its pretention to being religious and great communications with ulema, Qajar dynasty failed to give a religious tent to its rule. Religious schools in this period, compared to those in the Safavid period, were less prospered. With the expansion of economic and political relationships between Iran and Islamic countries, opportunities were made for Iranians to become more familiar with European culture and civilization. Establishment of Darolfonoon School and publication of Journal are of important events occurred in Qajar period. This school was the first cultural effort the government made toward establishing a school for educating the experts (Kiani, 2000: 131).

7. Baba Qasem (Imamieh) School - Ilkhani Era

Baba Qasem or Imamieh School is one of the Ilkhani schools with a four- porch plan in Isfahan. The school has a domed house towards the qibla within which an altar is located. So a place for prayer can be clearly seen in this Ilkhani School; but this place is completely within the school area with no separate exit and is allocated only to the school residents. It can also be used as Madras, the prayer space in ilkhani schools is prominent because they were going to be the tomb for their founders (Houshyari, 2013: 42).

Figure 2. The Ghiasiyeh school Khargerd (
http://tarikhnameh.persianblog.ir).

8. Ghiasiyeh Khargerd School in Khaaf, Timurid Period

Ghyasiyeh Khargerd School is located in Khaaf, a small village about 150 kilometers south of Mashhad. The building was built in the year 1444/848 on an order by Ghiyasuddin Bir Ahmad Khafi the Minister of Shahrukh Sultan of Timurid, and by the architects and the artists of that time, Ostad Qavammuddin and Ghiyasuddin Shirazi (Kiani, 1991).

There is a small space for prayer in Ghyasiyeh Khargerd School which is designed for the school residents and is not for public use, other spaces are organized to provide the needs of students (their educational and everyday needs of

life). In this school the main axis is not aligned with Qibla (Hassas, 2011: 6).

Figure 3. *Ghiasiyeh Khargerd School (source: the author).*

Figure 4. *Ghiasiyeh Khargerd School, Khaaf (http://4iranian.com).*

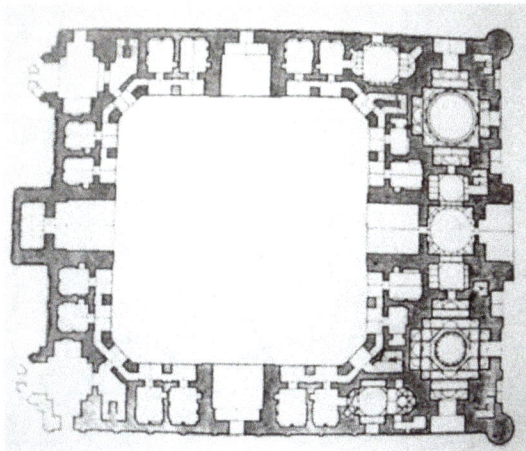

Figure 5. The Plan of Ghiasiyeh Khargerd School.
(http://www.safar-online.com).

This four-porch school has an entrance in south that leads to a square vestibule. There are also two domed shape spaces on both sides of the entrance porch which are decorated with ornamental plastering. The exterior facade of the school is beautifully designed that made it visible from distance, like a diamond, the exterior sides of the school entrance are encompassed by rectangular vaults. The praiseworthy façade of Ghiasiyeh Khargerd School is short and wide, an elegant gateway is situated deeply in the building, the symmetric walls on both sides of the entrance decorated with sharp headed vaults, are actually two rectangular frames which lead to short lateral towers. The building façade altogether induces a horizontal, sleeping position which was a quite new aspect in Timurid architectural style (Khazaee, 2009).

Ghiasiyeh Khargerd School has four domed houses built on its four corners of which the two on both sides of the entrance are larger and more distinguished. One of these two domed houses is for prayer with a qibla which is not that much accurate (Godar 1989, 244). The tiled altar with its Mogharnas works in eastern domed house is a proof of this; it has also been mentioned as a mosque (Blair and Bloom, 1995, 61). These two domed houses have direct exits to the outside, but the entrance porch is the only possible way to enter; thus a kind of separation between the prayer and education space is visible in this school (Okin, 1987: 54).

Figure 6. *Navab Mosque- School in Mashhad (before and after reconstruction) source: http://basarnews.ir.*

9. Navab School, Mashhad, Safavid Era, 1086 AH

Salehieh or Navab School is located on the northern part of Olia (Naderi) Street in Mashhad. Its architectural style is Isfahani. According to an inscription on its gateway, this school was built in the reign of Shah Soleiman, one of the Safavid kings, in 1086 AH. It has two floors and 84 rooms for student's residency which are still being used. The founder is Abu Saleh Razavi one of Mashhad nobilities and Sadat. He is

also known as Navab. That is why this building is known as Salehieh Navab or Navab (Bemanian, 2013: 22).

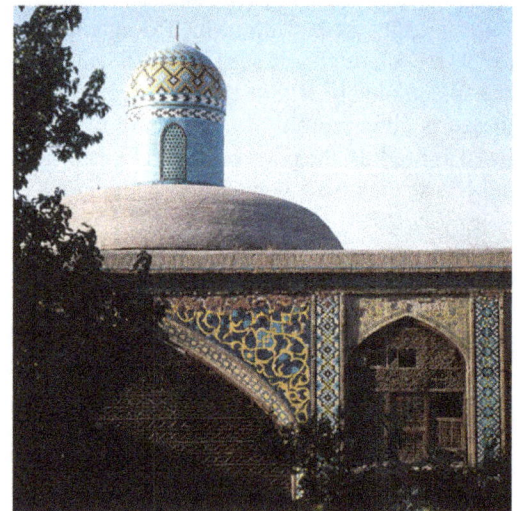

Figure 7. *Sardar Mosque - school, Qazvin (Source: http://seeiran.ir).*

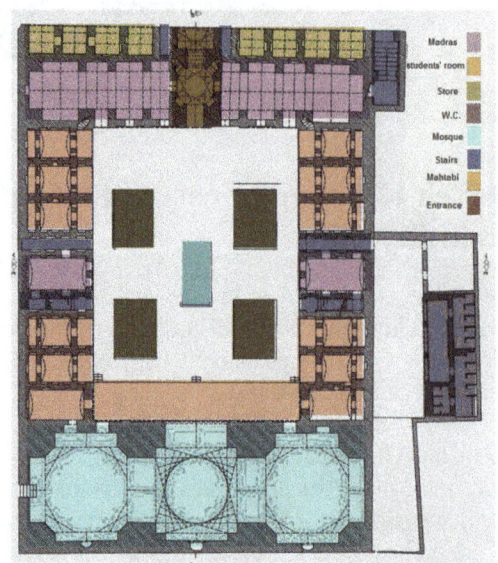

Figure 8. *The ground floor plan.*

10. Sardar Mosque - School, Qazvin, Qajar Era, 1231 AH

Sardar Mosque - school is located in Tabriz Street and Qmlaq district in Qazvin. Sardar School and Mosque used to be one of the most beautiful and privileged schools in the city. This building was built in 1231 AH by Hassan Khan and Hussain Khan Sardar rulers of Fath Ali Shah Qajar. The building plan is square-rectangular and the building is built in two floors. The entrance gate is located in the north and the middle of school and after a small vestibule leads to the courtyard. The school has a central courtyard and the plinth of the building is made of stone; the rest made of lathed bricks and decorated with colorful tiles. All around the school yard in four sides are decorated with inscriptions and lyrics of Mohtasham Kashani in nasta'liq in white color on purple tiles. There are 32 rooms in the western and eastern parts of the two earrings, connected to small porches; at the middle there is a small Madras or a large room and on its both sides are two narrow earrings which are the corridors of the second floor. There are also three rooms in the corridors leading to the earrings. Each room despite a separate back room, has a platform which is half a meter high and six square meters area (Mahdavinezhad, 2013: 11).

The school prayer space is located exactly on the main axis which is aligned with the Qibla and at the end of south side extended across the whole width of the building and is the largest space in the school. The Prayer room at these schools is designed for the use of more people than just the residents of the school and is more public (Hassas, 2014: 6).

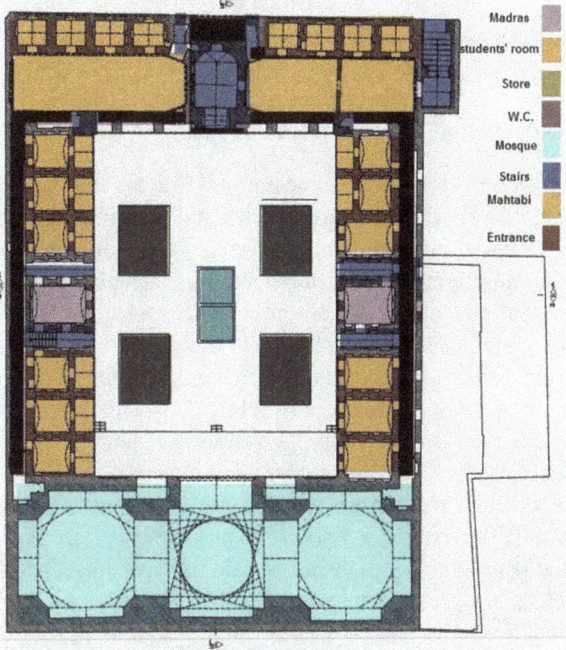

Figure 9. The first floor plan.

Table 1. Overview of the results.

Period	Ilkhanid	Timurid period	Safavid period	Qajar period
School Name	Baba Qasem School (Imamieh)	Ghiasiyeh School Khargerd	Nawab School Mashhad (Salehiya)	Sardar School of Qazvin
Year of construction	72AH / 703 AD	842 to 848 AH	1086 AH	1231 AH
Location	Isfahan	Khargerd Khaaf	Mashhad	Qazvin
Architect	Muhammad ibn Omar al-Sheikh	Qavam al-Din Shirazi & Qiath al-Din Shirazi	Abu Saleh Razavi of Mashhad nobility and Sadat	Hasan Khan and Hussain Khan Sardar of Fath-Ali Shah Qajar rulers
Geometry	An almost rectangular map	rectangular	rectangular	Square- rectangular
Pattern	Four-porch pattern	Four-porch pattern		Four-porch
Style and method		Azari style	Isfahani architectural method	
The number of floors	Two floors	Several rooms in two floors	Two floors	Several rooms in two floors for students residency
Function	The prayer space is completely encompassed within the school.	There are two mosque-like spaces on both sides of entrance porch.	Madras also known as prayer place was located on the northern porch on the ground floor, above which on the upper floor was the libarary.	for every two schools for training there was one domed house for prayer.
Area		An area with domed ceiling at the four corners.	Courtyard: 30*36 meters	With central courtyard
Entrance		The main entrance is a vestibule which leads to a spacious room from both left and right side.	The Entrance is located at the current location and has a beautiful gateway decorated with tilea and Mogharnas.	The main entrance is located on the north and in the middle.
Colors and Building materials		A coating of tiles and bricks in plaster mortar.	Brick, mud. The school floor is covered with bricks	The plinth of the building is made of stone; the rest made of lathed bricks.
Architectural elements	Including the entrance, vestibule, porch, courtyard, rooms on two floors, domed house, and altar.	A courtyard, four porches and rooms on two floors, the entrance vestibule.	Northern and southern symmetrical porches around which are rooms on two floors.	domed house, altar, Shabestan, vestibule
Decorations	It is beautifully Decorated with tile and parts of the porch and	In terms of ornamentation and tiling is among the	The exterior and interior view of the building are decorated with colorful	Decorated with colorful tiles. Inscriptions with lyrics

Period	Ilkhanid	Timurid period	Safavid period	Qajar period
	its vaults are covered with mosaic tiles in seven colors. The four porches are Also decorated with inscriptions in Mqly (masonry) and made the school so beautiful.	masterpieces of the 9th century AH.	mosaic tiles and inscriptions are very beautiful, And inside the porches and the outside coatings of walls are adorned with a combination of brick and tile.	of Mohtasham Kashani in nasta'liq in white on purple tiles

11. Discussion and Conclusion

Iranian art and architecture have enjoyed a great endurance throughout history. This art represents the way of thinking, the worldview, religious beliefs, and traditions of the people in the country. A glimpse into the development of architecture in Iran indicates that the architects of the country, whether from building a simple shelter or building the biggest and the most magnificent works of architecture, were not just after simple targets such as solving functional problems, but all human's physical and mental needs were significant for them. Iranian architects, considering climatic conditions and geography of this vast territory, have prospered and achieved innovations, and in each period created a describable masterpiece. Based on the studies done in this area, it can be said that mosque-school is an architectural space which is used for both religious and educational functions, and the spaces within the building are of almost the same significance. This kind of apace is known as mosque, school, or school-mosque. The investigated samples can be categorized into three types, the first type which is the most popular, are those mosque-schools with a single plan but separated functions. The second category includes those mosque-schools where the religious and educational functions are merged. Entering these buildings, we can see both religious spaces like Shabestan and educational spaces including Madras and the student's rooms altogether. The third category which is less popular than the previously mentioned categories, are those school-mosques which have separated spaces for each function. In this category the school and mosque spaces, without any interference, are related to each other through a common space.

Acknowledgements

This paper is taken from Hamid Majidi's master thesis entitled "The Campus of the Faculty of Quranic Sciences of Mashhad University" at Islamic Azad University of Mashhad and is guided by Professor Ladan Assadi.

References

[1]   Amoli, S. H. (2012). *Jame Al - Asrar va Manbao Al - Anvar*. Translated Muhammad Reza Jozi. Tehran. Hermes Publishing Co.

[2]   Beller SH, Bloom, J. M. (1995). *Islamic Art and Architecture*. Translated Ardeshir Araqi (2002). Tehran. Soroush.

[3]   Bemanian, M. R. Momeni, K. Soltanzadeh, H. (2013). *A Comparative Study of Architectural Design Features of Mosque-Schools in Qajar Era and Safavid Schools*. Armane Shahr Journal of Architecture and Urbanism, No. *11*, P 15-34.

[4]   Godar, A. (1987). *Iranian Works*. Translated Abolhasan Sarveghad Moghaddam. Islamic Research Foundation. Mashhad, Iran.

[5]   Haji SeyedJavadi, F. (1999). *Mosque Architecture*, Tehran Conference Proceedings of the Mosque, Past, Present, and Future. Tehran Art University.

[6]   Halen, B. R. (1994). *Islamic Architecture*. Translated Iraj Etesam. (2011). Tehran. Information Technology Organization of Tehran Municipality.

[7]   Hassas, N. (2014). *Spatial Elements of Schools and Their Use in Architecture*. The First International Conference of New Horizons in Architecture and Urbanism, Tehran, Iran.

[8]   Houshyari, M. M. (2013). *School Mosque Typology in Islamic Architecture (The Relationship Between Education and Prayer Space)*. Two- Quarterly of Iranian Architectural Studies. No. 3. P. 37-53.

[9]   Kiani, M. Y. (1998). *The history of Iran Architecture in Islamic Period*. Tehran. SAMT Publishing Co.

[10]   Kiani, M. Y. (1998). *The history of Iran Architecture in Islamic Period*. Tehran. 2nd Ed. SAMT Publishing Co.

[11]   Khazaee, M. (2009). *Structure and motifs of the Timurid schools in Khorasan*. Two - Quarterly of scientific - Research of Islamic Art Studies. No. 11.

[12]   Kiani, M. Y. (1998). *The history of Iran Architecture in Islamic Period*. Tehran. SAMT Publishing Co. Research and Development Center of Human Sciences, Tehran, Iran.

[13]   Mahdavi Nejad, MJ. Ghasempour Abadi, M. H. Mohammad Levi shabestary, A. (2013), *Typology of School Mosques in Qajar era*, Islamic Iranian City Studies, No. 11. P. 5-15.

[14]   Okin, B. (1987). *Timurid Architectur in Khorasan*. Translated Ali Akhshiri. Tehran. Islamic Research Foundation.

[15]   Soltanzadeh, H. (1987). *Schools were Established in Iran from Antiquity to Darolfonoon*. Tehran. Negah Publishing Co.

[16]   http://seeiran.i.

[17]   http://basarnews.ir.

[18]   http://www.safar-online.com.

13

Interdisciplinary Interation in Design – Relation Between Graphic and Interior Design

Rana Kutlu

Department of Interior Architecture and Environmental Design, Istanbul Kültür University, Istanbul, Turkey

Email address:

r.kutlu@iku.edu.tr

Abstract: Nowadays the concept of "identity" is worked on diligently as a marketing strategy and deemed as an important tool in concept works of design disciplines of different scales through physical space, which reaches out to consumers. In this study, through the literature review, impacts of identity on spatial design are discussed in terms of interdisciplinary communication and tools used for transferring the meaning, which transforms identity into image are listed. Besides, the impact of spatial design on users' perception is explained in connection with graphic design, architecture and interior architecture. Tools used for meaning transfer, which play a role in defining and developing design based on identity, are evaluated together with identity and image creating processes.

Keywords: Space, Graphic Design, Interdisciplinary Interaction, Relevance to Design Practice

1. Introduction

Design has continuously been evolving and developing under the influence of art movements and in the light of experience and technical and material-related developments throughout the history. Understanding and perception of design have resulted in changes in forms and their meanings (Bielefeld, Khouli 2010) The issue of defining branding-identity has come to prominence in a wide variety of fields from individual products to service industry as a result of increasing competition and globalization seen in design, as well, just like anywhere else. As expressed by Baudrillard (2002), the fact that the identity message given by the product has taken precedence over its value of use shows that nowadays its meaning is more important than the product itself. In this regard, designers have adopted a multi-disciplinary approach and started to work with not only engineers but also with teams comprising of representatives of various professions like artists, graphic designers, communication experts and to achieve a versatility in design thanks to interdisciplinary communication.

We observe a close link between architecture and art following the Industrial Revolution in De Stijl approach and in Bauhaus movement, where various and versatile art branches like sculpture, handicraft and painting can be found together. (Fielden, Geoffrey, 1963). Besides, nesting of

architectural design and art brings along conceptual integrity pursuant to spatial identity. Graphic design has played an important role in actualizing this conceptual integrity in terms of deepening meaning of the space, providing information transfer and creating an identity for the space. Especially in recent years graphic design has gone beyond advertising displays and billboards and become an influential discipline with regards to giving architecture and environment an identity through information transfer in environmental and spatial design.

2. Graphic Design and Historical Development

Any quality defining the entity and making it different from others of the same kind forms identity. Identity is the way an entity expresses itself. Image comes into being when identity is shaped in others' memory and perception (Dewey, J., 1997).

Graphic design is one of the most important disciplines and arts providing visual communication. It is also a visual tool, which provides the users with information and message esthetically and cutting extra corners. Designers use space to give a message in architecture. In line with this objective, architecture, which is discussed in different scales in physical environment designed for the user, will bear its fruits in cooperation with interior architecture, industrial design and graphic design.

Cave drawings dating back from ancient times to upper Paleolithic era and to 14000's B. C. can be deemed as the first graphic works of human beings, in which they use art as a way of communication. As a result of development of modern life following the Industrialization, invention of photography, establishment of printing press, products like graphic, catalogue or posters were first started to be designed by painters. For this reason, pictorial aspects were put first rather than typographic features. Thanks to printing techniques developed in time, graphic design has become a separate discipline.

Graphic design has been set apart from painting art towards the end of the 18th century thanks through book designs. It continued to develop rapidly especially pursuant to innovations brought by the Bauhaus School through reactive approaches to art and design. Later on, efforts to give space an identity through image and perception based design have gained momentum in order to create attraction, popularity, awareness and privilege even in building typologies having different functions than retail industry under the influence of semiotics which can be found in conceptual infrastructure of Post-Modern culture. In this period, graphic expressions that communicate with symbols in contrast to simplicity of the modern architecture have become more important in both architecture and environmental design.

Designers like Venturi, who have adopted Post Modernism and symbolic architecture, consider architecture to be the main system reaching out to the society. It is possible to increase diversity by combining different functions to strengthen the transfer of message in spaces under design and thereby to achieve an identity transfer. To this end, brand-identity-image works, which count as subjects of graphic design, have gained significance in architectural design. Spatial meaning can be transferred by symbol, photo, collage, illustration, pictogram and typography works.

Visual perception is of crucial importance in graphic design, which transfers meaning via materials and technology developed in recent years in construction industry. Uses and options of graphic design are diversified beyond the ordinary thanks to digital technology and interactive surfaces, which are also motion/user/environment sensitive, smart glass technologies with different levels of transparency, LED that gives wide range of opportunities especially in illumination and last but not least OLED technologies which are still developing.

3. The Relation Between Visual Perception – Spatial Design – Graphic Design

Architecture consists not only of eyesight since the visual difference which can be perceived by the user can be achieved by design, but it also involves other senses. It is also a tool that links us with space and time. (Pallasmaa, 2005), (Anderson, J., 2011). Architects often employ design methods to help them find more creative forms. These

methods make it possible to break free of the traditional canon of forms and established paradigms (Jormakka, 2008) The desire for designing and creating space is a way of communication for human beings using objects, materials, colors and forms rather than verbal tools(Brooker, Stone, 2010). Message carried by the space can be transferred to users via an interdisciplinary cooperation using tangible, abstract, audio, visual aspects (Pallasmaa, 2009). Interdisciplinary relation in transfer of identity related meaning in design can be seen in Figure 1.

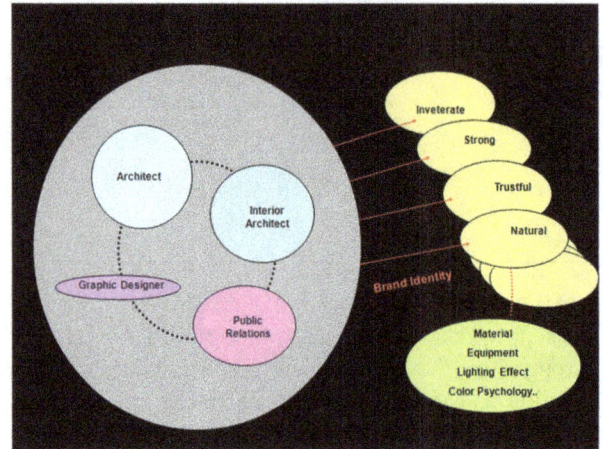

Figure 1. Interdisciplinary relation in transferring identity related meaning in design.

Humans receive sensory data that form perception through stimuli in outer environment. Visual sense is faster than auditory, smell and tactile senses in terms of perceiving natural and artificial environment. Transfer of the identity-related message through visual communication established between the space and the user and solution options in line with spatial functions can be achieved easily thanks to graphic design. This way, the completed design does not only have functionality but also differentiation through artistic identity given to the space and visual richness.

Visual perception varies person to person. Psychosocial and cultural characteristics and experience play an active role in perception of the message, which is planned to be transferred by the design. Surfaces defining limits of the volume play an important role in graphic design of the space, forming a visual identity, creating a difference and achieving message transfer. These surfaces composing the shell of the space are walls, horizontal planes and floors. Besides, vertical circulation instruments connecting different planes and reinforcement elements of the space are other architectural elements that play a role in visual identity.

According to Faulkner, walls are surfaces which create volumes in volume through their division function as per the relation between spatial and graphic design and it is also where spatial graphic is mostly used. Walls can be opaque, transparent, semi-transparent or variable transparent depending on materials used. They play an active role in describing the space with expressions like privacy-openness, transparency-closeness, dividedness or unity (Figure 2).

Figure 2. *Graphic Design on a Wall Plane (URL-1<http://www.interiordesign.net. Date accessed: 15.06.2015).*

Figure 3. *Graphic Design on Covering - Floor Plane (URL-2<http://librarytestkitchen.org, Date accessed: 15.06.2015).*

For integrity in design vertical place surface characteristics have to be taken into consideration during the graphic design process. Doors and windows are vertical surfaces pieces that open into the space, thus these are the first building components, which transfer messages via graphic design, stylistic design and construction technique based on the atmosphere of the inner space. Whether these components are opaque, transparent, semi-transparent, wooden, metal or plastic determines graphic design of how message is to be transferred. Functions which are to be added to vertical planes like exhibitions, information boards/ panels or planes to be used for storing purposes have to be in line with the architectural design integrity not only for graphic design but also as for the technology and materials to be used.

Covering and floors are surfaces used intensively in effective spatial graphic for continuity, flow and guidance in differentiating between different materials or elevation, which are in close contact with users like walls. Graphic prints on the floor can be partially extended to the walls or

even to the ceiling having a restrictive or guiding impact as can be seen in Figures 3 and 4.The ceiling, upper horizontal plane bordering the volume, is another important architectural element which completes inner space, closes on the third dimension, doesn't directly contact the user in contrast to walls and floors, is located further away. Humans' field of vision covers an angle of 40° on horizontal and vertical axis and thus architectural elements like floor, wall and ceiling are surfaces, which have an impact on design and can be used for guidance; and where graphic design can be used effectively to create the defined identity-related image (Figure 4).

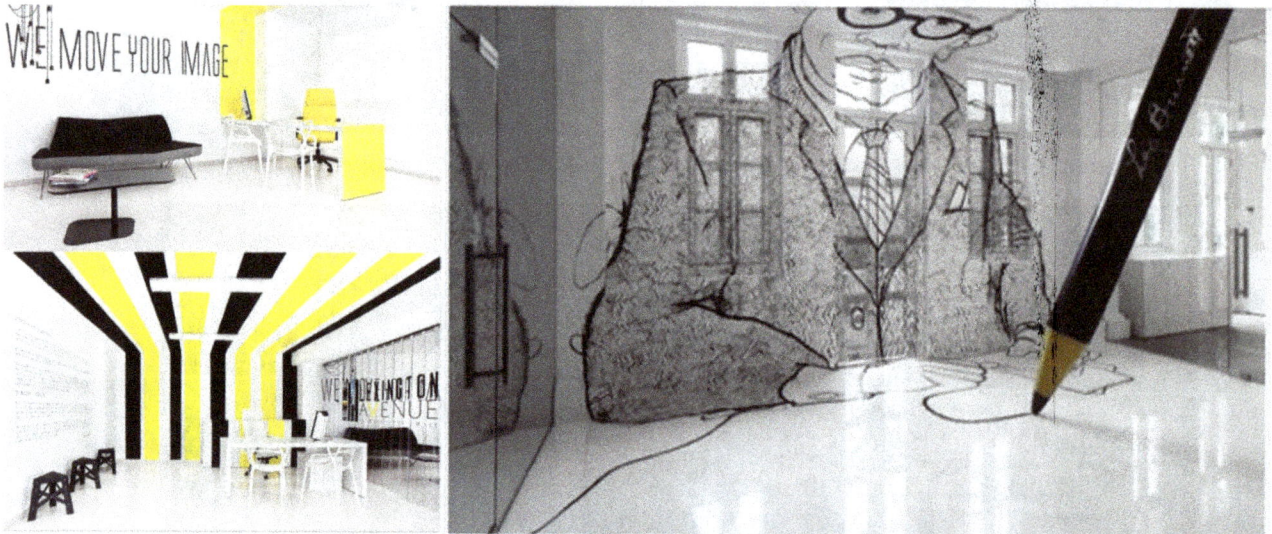

Figure 4. Graphic Design in Floor-Ceiling Plane (URL-3<http:// asphaltartusa.com, Date accessed: 15.06.2015).

Vertical circulation instruments like stair, ramp and elevator first of all serve for continuity and guidance between elevation and floors and can also be used as surface for graphic design the emphasize the abovementioned purposes.

Vertical circulation instuments are not only functional but

they can also be art objects adding esthetic features to the building thanks to their forms, colors and material compositions. Message/information transfer on different vertical circulation elements like stairs, elevators and escalators can be seen in Figure 5.

Figure 5. *Vertical Circulation – Graphic Design (URL-4<http://www.flicker.com, Date accessed: 15.06.2015).*

Figure 6. *Reinforcement Elements - Graphic Design (URL-5<http://www.officefurniturescene.co.uk, Date accessed: 15.06.2015).*

Reinforcement elements can be fixed or moving, horizontal/vertical/inclined in the field of view, at low or high levels. Reinforcements elements, which are designed, and colored, given texture and equipped with materials in line with the spatial design and identity, play an active role in forming a message which matches the identity in the memory of the user in terms of graphic design in addition to fulfilling main design principles like background-object contrast or harmony, rhythm, continuity, consistency, diversity in unity, balance and emphasis (Figure 6).

Visual images like photos, illustration, typography and pictogram, which would emphasize the desired message and space identity defined by the designer on all planes and reinforcement elements composing the space and spatial graphic design compositions which are formed with visual images shall be meticulously discussed in terms of achieving both functional and successful visual communication.

Following adaptation of rapidly developing technology into architectural elements as meaning transfer tools, the relation between architecture and graphic design has gained a new dimension and adapted itself to the new order. Variations of the relation between architectural elements, which will transfer the identity related message through graphic design tools can be seen Figure 7.

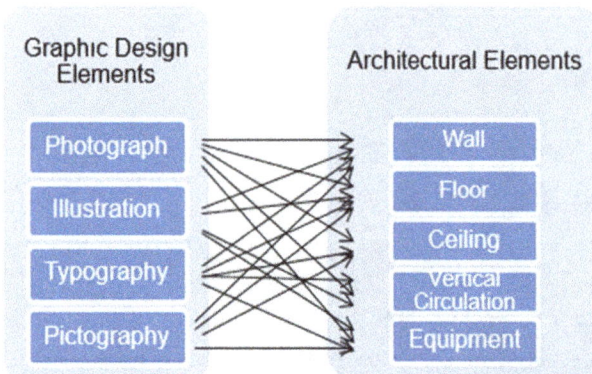

Figure 7. The Relation Between Spatial Meaning Transfer Tools and Graphic Design.

Fixed print or mobile digital arrangements can be done on two dimensional planes using a photograph with high resolution, enough light and a strong graphic language which is used in spatial design to influence visual perception.

Another example for graphic works that is used effectively to give a space an identity is illustrations. Illustration is a common artistic way of expression, which can be understood by everyone independently of language and culture. It is also one of the strong esthetic links strengthening the relation between the space and users.

Typographies, which shapes volume with letters and lines can be either two or three dimensional. They occupy an important position among other graphic design elements in transfer of identity of the space.

Pictogram is defined as international symbol representing a concept, object or a function. It is composed of graphic

drawings and is a strong tool for message transfer. It has esthetic and functional aspects within the space gaining three dimensions in the volume or on two dimensional planes as a strong graphic design tool in identity based message transfer in creating the planned image in the memory of the user. Various graphic design works on architectural elements such as photo, illustration, typography and pictogram can be seen in Figure 8.

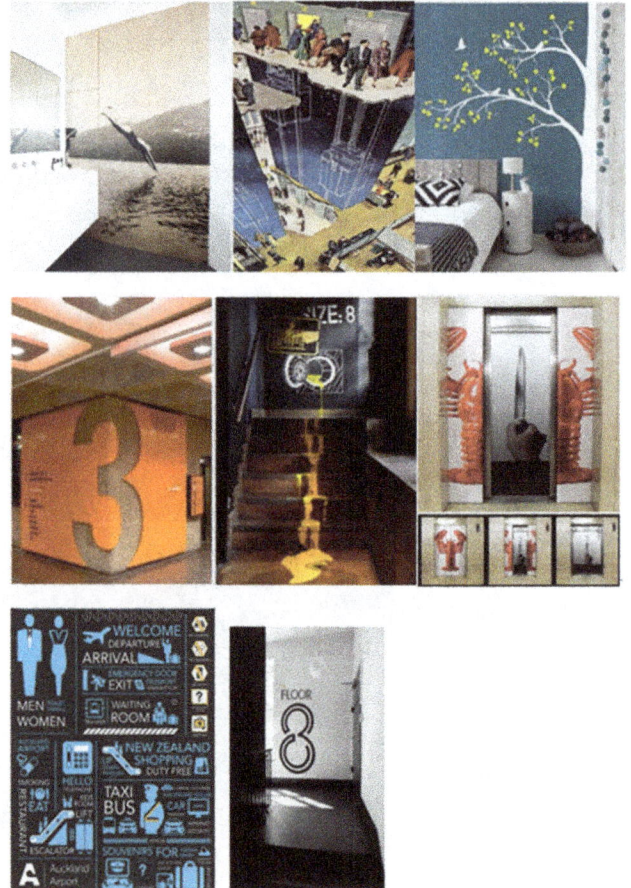

Figure 8. Graphic Design Tools- Photo-Illustration-Typography-Pictogram (URL-6<http://www.design-milk.com, Date accessed: 15.06.2015).

4. Conclusion

Nowadays we see that successful designs are open to multidisciplinary approaches and works. Architecture, which is a place for interaction and experience, is not only a useful artistic production with its functional, esthetic and design features but also a physical environment and a living space where corporate identity works are transferred to the user through this space.

Life of space lasts longer than lives of actions and users involved. Although physical existence of spaces, which have had different functions throughout time and served different groups of users remain the same, the messages it carries or the communication it establishes change in time. Since experience offered by the space directly affects the image, it is one of the most important meaning transfer tools. It covers various design decisions and a wide range of issues like the relation between location of the space and outer environment, details regarding reinforcement elements of inner space, color and illumination designs and graphic expressions. The concept, on which all of these design decisions depend, expresses the overall identity.

The relation between architecture and graphic design should be discussed with a holistic approach in reflecting the corporate identity on the space forming a visual perception. Supporting this interdisciplinary cooperation will have a positive impact on giving the space an identity. Interdisciplinary design approaches have generally proven to bear successful results throughout the history.

Interdisciplinary interaction will make it easier for a designer to reflect his/her talents and authenticity on the space with a broader point of view; nourish the design process and offer chances of new experience through spatial communication.

References

[1]   Anderson, J., (2011), Basics Architecture, Architectural Design 03 AVA Publishing SA, Switzerland.

[2]   Baudrillard, J., (2004), The Consumer Society, Myths and Structures, Sage Publications, London.

[3]   Baudrillard, J., (2012), Tüketimin Tanımına Doğru, Mimarlık ve Tüketim, Boyut Yayınları, İstanbul, s: 67-72.

[4]   Bielefeld, B. & Khouli, S. El, (2010), Basics Entwerfen Entwurfsidee, Birkäuser.

[5]   Brooker, G., Stone, S., (2010), What Is Interior Design? Essential Design Hanbooks, Roto Vision SA, Switzerland.

[6]   Dewey, J., (1997), How We Think. USA: Dover Publication.

[7]   Fielden, Geoffrey BR. et. al., (1963), Engineering Design. London: HMSO.

[8]   Faulkner, R. (1979). Planning a Home. Holt, Rinehartand Winston, New York.

[9]   Jormakka, K., (2008), Design Methods, Birkäuser.

[10]   Pallasmaa, J., (2005), The Eyes of the Skin: Architecture and Senses, Wiley, Chichester.

[11]   Pallasmaa, J. (2009). The Thinking Hand Existential and Embodied Wisdom in Architecture. West Sussex, UK: John Wiley & Sons Ltd.

[12]   URL-1<http://www.interiordesign.net.   Page   accessed: 15.06.2015.

[13]   URL-2<http://librarytestkitchen.org,   Page   accessed: 15.06.2015.

[14]   URL-3<http:// asphaltartusa.com, Page accessed: 15.06.2015.

[15]   URL-4<http://www.flicker.com, Page accessed: 15.06.2015.

[16]   URL-5<http://www.officefurniturescene.co.uk, Page accessed: 15.06.2015.

[17]   URL-6<http://www.design-milk.com,   Page   accessed: 15.06.2015.

Best Bet Technology Package Development to Improve Sorghum Yields in Ethiopia Using the Decision Support System for Agro-Technology Transfer (DSSAT) Model

Fikadu Getachew, Gizachew Legesse, Girma Mamo

Ethiopian Institute of Agricultural Research (EIAR), Climate and Geospatial Research Directorate (CGRD), Addis Ababa, Ethiopia

Email address:

fikeget@gmail.com (F. Getachew)

Abstract: Sorghum is grown mainly in the semi-arid areas. In spite of the fact that there was observed high climate variability in the last few decades, rain fed sorghum [Sorghum bicolor (L.) Moench] production is still an important source of food and feed in the semiarid regions of Ethiopia. Although sorghum is realized as crop tolerant to water deficit, compared with other semiarid crops in Ethiopia, climate variability and change has been challenging its production and no intensive crop simulation modeling was done as it was desired. In this study the CERES-Sorghum Model of Decision Support System for Agro-Technology Transfer (DSSAT) has been tested over the north Rift Valley of Ethiopia. We have checked what would be the best combination of management options under research and farmers' practice conditions for each sites for the historical climatological periods (1980-2010) in which we have found that the model performs well in assimilating the real situation in our sentinel sites in both research and farmers' management practices. The potential grain yield from the DSSAT model would go up to 2.5T/ha under best scenario rainfall seasons without applying the developed technology package application (which we call it farmer's condition). The same sorghum variety has a potential yield of 6.2 T/ha if one can apply the recommended best bet technology packages (planting date, planting population, sowing data, fertilizer application rate and time) within the same season. Hereby we can assert that the application of the developed technology packages would make a difference of up to 3.7 T/ha of grain sorghum yield under the same season. Even though applying the technology packages according to the prevailing seasons would significantly matter the expected grain yield, the worst possible grain yield lose would be minimized by applying the best bet technology packages that fits the specific season. Moreover, the selected sentinel sites were few, the result can be extrapolated using the calibrated crop simulation modeling to larger areas to develop strategic plans to improve grain yield of sorghum in Ethiopia.

Keywords: Crop Simulation, DSSAT, Sorghum, Technology Packages

1. Introduction

Sorghum is the fifth largest cereal crop in the world, after wheat, maize, rice and barley. It is cultivated in wide geographic areas in the Americas, Africa, Asia and the Pacific. It is the second major crop (after maize) across all agro-ecologies in Africa (Taylor, 2003). It is universally considered to have first been domesticated in North Africa, possibly in the Nile or Ethiopian regions around 1000 BC (Kimber, 2000).

Sorghum is a singularly viable food grain crop for many food insecure people in sub-Saharan Africa (ICRISAT, 1994) because it is rather drought resistant among cereals and can withstand heat stress. Those parts of Africa, where sorghum is a significant arable crop are semi-arid and include the highlands of east Africa where bi-modal rainfall is intermittent. Sorghum is not only drought resistant but can also withstand periods of water-logging. The precise reasons for sorghum's environmental tolerance are not fully understood, and are undoubtedly multi-factorial (Doggett, 1988).

Over the past 25 years sorghum production has increased steadily in Africa, from 11.6 M T in 1976 to 20.9 M T in 2001, with most of this due to increased crop area not to improved rate of production. Average yields remain below 1 T/ha due to the applied subsistence farming practices; with low inputs (no inorganic fertilizer or pesticides) and traditional cultivar varieties (ICRISAT, 1996). Without mechanization and large-scale operations the consequent low yields leave no surplus sorghum, without which farmer's food security and processing industries cannot be created. However, where intensive agriculture is practiced with improved technology on varieties or hybrids, yields are much higher and comparable with other major cereals (Belton, 2004).

Integration of optimal technology packages and marketing could lift the livelihood of subsistence small-scale sorghum farmers and the adaptive capacity of the encompassing community.

The spread of cultivation areas into environmentally sensitive areas with great bio-diversity is highly damaging and unsustainable, and efforts must be aimed at intensifying sorghum agricultural practice in Ethiopia. Higher yields are essential, not only for rural food security but also for increasing population density and market commercialization.

To determine the best land-use practices for higher crop production, one must take into consideration the economic sustainability of the farmer as well as the ecological conditions of the environment. Among the tools that can be used to help solve some of these issues are computer-based biophysical simulation models. These crop simulation models have become more widely used in the past few decades by scientists to hypothesize ways to improve agricultural production under seasonal and daily weather variability. The models capture much of what we know about crop growth response to factors of temperature, solar radiation, rainfall, soil traits and crop management (Boote et al., 1998). Crop models have been used to evaluate management practices to improve yield for a given climatic region (Boote et al., 1996; Singh et al., 1994a, 1994b), to plan irrigation (Hook, 1994), and to evaluate climatic yield potential for different regions (Aggarwal and Kalra, 1994) or different costs (Alagarswamy et al., 2000).

Management practices such as sowing date, row spacing, sowing density, cultivar choice (both seasonal length and genetic traits), soil water availability, and fertilizer application are factors to enhance productivity. Variability of rainfall (onset, intensity, and cessation) as well as temperature, day length, and solar irradiance are important climatic factors that also impact management practices in a given region.

The models have been evaluated extensively and applied in agriculture to problems such as estimating the sensitivity of crop production to climate change (Williams et al., 1988; Alexandrov and Hoogenboom, 2000; Mall et al., 2004).

In this regard, the Ethiopian Institute of Agriculture Research (EIAR) has conducted research in sorghum technology development to improve productivity for small-holder farmers. Various technology applications have been documented, from place to place and time to time. From this we have distilled a combination of technologies into a 'package' via agricultural simulation models such as DSSAT v4.5, so that sorghum farmers can obtain useful advice to enhance their potential yield, regardless of the prevailing weather conditions.

Here we report on such a study to critically assess an optimal technology package; including stakeholder feedback on its practical application in the field.

2. Study Area

The dry land sorghum growing areas of Ethiopia can be characterized as areas receiving 350-800 mm of rainfall with a broad unimodal distribution. The rainfall has a coefficient of variability > 30% and displays multiple onsets and cessations (Hailu and Kidane, 1988). Sorghum in Ethiopia performs best with average temperatures 24 to 26°C.

Our simulation modelling was conducted at three specific sites in the northern Rift Valley of Ethiopia (Figure 1) i.e. Fedis, Miesso and Kobo districts located at 42.03E,9.08N; 39.38E,12.08N and 40.46E,9.14N respectively. Their corresponding elevations are 1700, 1470 and 1400 m.

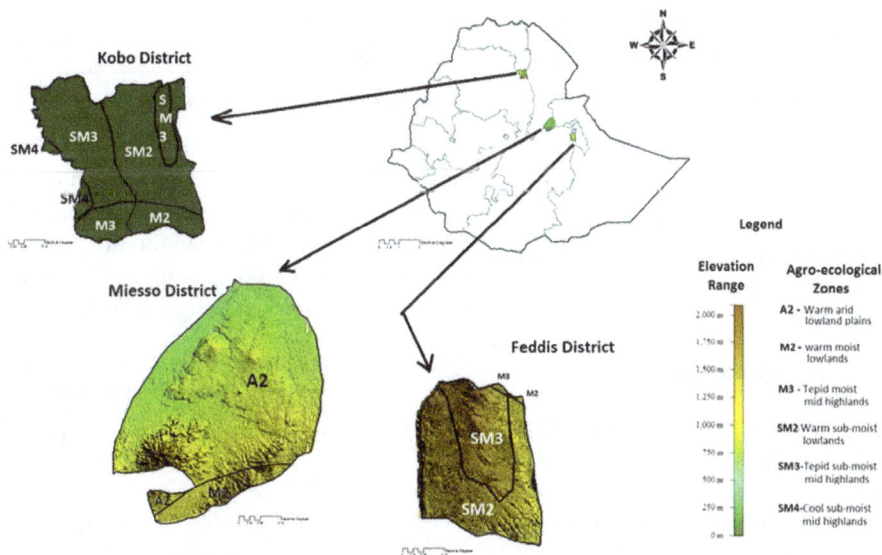

Figure 1. Selected sites for Sorghum modelling in northern Rift Valley of Ethiopia.

The agro-ecological zones of the study sites fall under A2, M2, M3, SM2, SM3 and SM4[i] with mean annual rainfall for Fedis, Meisso and Kobo of 614 mm, 765 mm and 691 mm respectively. The average monthly rainfall during the Sorghum growing period is shown in Table 1.

Table 1. Average monthly rainfall (mm) for growing period at the three sites.

Experiment Sites	June	July	August	September
Fedis	51.4	65.6	77.1	96
Kobo	36	137	200	65
Miesso	40.9	135.7	145.6	95

3. Data and Methodology

The DSSAT v4.5 Crop growth model to hypothesize improvement in production of Sorghum in Ethiopia used 32 years of observed weather data. The following biophysical and management options were employed to arrive at the optimal technology package.

3.1. Plot Design

Three field experiments with the sorghum cultivar MEKO were conducted at Fedis, Kobo and Miesso sites under rain fed condition in 2010 and 2011. In all experiments, both research and farmer conditions were separately observed. Each plot had 50 m length x 50 m width, 44 rows and 0.75 m row spacing. Recommended crop management such as fertilizer

rate application, sowing date, sowing depth, farm implement technologies and row spacing information was derived from earlier research works that has been conducted in the selected experiment sites (give peer-reviewed published reference).

3.2. Soil and Weather Data

As stated by Hailu and Kidane, 1988, sorghum can grow in different soil types from light sands to heavy clays if they are well drained. The pH should be above 5 but it performs best on deep, fertile sandy loams. Good yields are also possible on heavy but well drained soils. In fact, good fertility, drainage and optimum temperature are most important considerations in the successful culture of sorghum. This crop can tolerate considerable quantities of alkali or salts. Under rain fed conditions it performs well in the soils of high water retention capacity.

Major soil characteristics for the three sites namely; Cation exchange capacity, pH, percent of total N, percent of clay and silt, percent of organic carbon, Bulk density, hydraulic conductivity, root growth factors, percent of saturation, drained upper limit and lower limit, root growth factors and percent saturation were collected from each experiment site (Table 2). Monthly (from daily) weather data included solar radiation, maximum and minimum temperature and rainfall over the period 1973-2011 were obtained from meteorological stations at each site from the National Meteorology Agency of Ethiopia.

Table 2. Soil data used for this simulation just for the Miesso site which is one of the three study sites.

Profile depth cm	Lower limit	Drained upper limit	Saturation	Root growth factor	Hydraulic conductivity	Bulk density (g/cm³)	Organic carbon %	Clay %	Silt %	Total N %	pH	CEC cmol/kg
10	0.341	0.483	0.683	1	0.06	0.74	1.5	58	28	0.06	7.8	48.9
30	0.352	0.488	0.69	1	0.06	0.72	1.04	60	24	0.08	8	41.8
60	0.394	0.528	0.643	0.407	0.06	0.85	1.04	66	24	0.03	7.9	41.8
90	0.368	0.506	0.651	0.223	0.06	0.83	1.04	62	26	0.01	7.9	41.8
120	0.38	0.515	0.68	0.122	0.06	0.75	1.04	64	24	0.04	7.8	45.2
150	0.419	0.548	0.625	0.067	0.06	0.9	1.04	70	22	0.03	7.8	40.5
180	0.404	0.533	0.618	0.037	0.06	0.92	1.04	68	20	0.04	7.8	39.2

Table 3. The genetic coefficient data used for the simulation.

Genetic coefficient	P1	P2O	P2R	P5	G1	G2	PHINT	P3	P4	P2	PANTH
Value	294	12.5	30	399	2.9	6	49	152.5	81.5	102	617.5

3.3. Cultivar Selection

The Sorghum cultivar used in this study is called MEKO. It is an early maturing variety which fits well to the dry semi-arid areas. The data required for genetic coefficient code definition are depicted in the Table 3.

3.4. Planting Date and Fertilizer Application

The planting time and fertilizer application (rate and timing) were analyzed under two scenario groups. The first group is under farmer condition and the second is under research condition. The planting time for the first group is under normal farmer condition and for the second was based on the analysis

of mean rainfall data. The time of fertilizer application depended on the respective planting time and the application rate taken from previous research recommendations for the sites. The type of fertilizer used in this study is Diammonium Phosphate (DAP) (100 kg at the time of sowing) and Urea (50 kg when plants are knee height) (Table 4).

Table 4. Fertilizer application time and rate in all sites.

Scenarios	Fertilizer application time and rate	
	DAP (100 Kg/Ha)	Urea at knee height (50 kg/Ha)
Farmer condition	None	None
Research condition	20-Jul	20-Aug

3.5. Tillage and Management Practices

The farm implement technologies used for this simulation was classified as farmer and research based conditions. It is known that the level of depth that farm implements can be till have an influence on the productivity of crops. Accordingly the following type of farm implements and associated depth were taken in the simulation activity (Table 5 and Table 6).

Table 5. *Tillage practice used for all implementation sites.*

scenario	Tillage Practice	Depth (cm)
Farmer condition	Cultivator field	13
Research condition	Cultivator, ridge till	20

Table 6. *Management options used for the simulation of sorghum at Meisso.*

Scenarios		Planting date	Emergence date	Planting	Planting distribution	Plant population at seedling (plant/m2)	Plant population at emergence (plant/m2)	Row spacing (cm)	Planting depth (cm)
Farmer condition	Meisso	25-Jun	02-Jul						
	Kobo	16-Jun	24-Jun	Dry	Broadcast	9	9	50	3
	Feddis	26-Jun	03-Jul						
Research condition	Meisso	20-Jul	27-Jul						
	Kobo	12-Jun	19-Jun	Dry	Row	9	9	75	5
	Feddis	23-Jun	30-Jun						

3.6. Planting Date

Experimental results under dry land farming conditions have clearly revealed that dry or early sowing gives substantially higher yield compared to the traditional late planting after two or three effective rainfall (Hailu and Kidane, 1988). At Kobo dry sown sorghum (1-15 June) produced 2.3 T/ha, whereas, sowing after one, two and three effective rains gave 1.8, 1.4, 1.1 T/ha (Table 7). In the same way, in our simulation study we found that the date that gave the highest yield was June 15 when compared with late planting (July 20). However, an earlier planting date i.e. June 5 gave smaller yields than June 15. It would of course depend also on the fluctuations of rainfall during the growth period.

Table 7. *Effect of planting date on grain yield of sorghum at Nazareth (Hailu and Kidane, 1988).*

Planting Dates	Yield (T/ha)			
	1983	1984	1985	Mean
Dry planting	3.455	0.611	2.746	2.268
After one Effective rain	2.466	0.618	2.506	1.771
After two Effective rain	1.53	0.463	2.383	1.407
After three Effective rains*	1.041	0.319	1.897	1.086

3.7. Agronomic Practices Adopted in This Study (Tied Ridge)

Soil water stress is the 'bottleneck' of sorghum production in dry land areas. Several research activities were carried out to develop soil management practices, which store and conserve as much rainwater as possible by reducing runoff and improving infiltration and water storage in the soil profile. To this effect, tiled ridges have been found to be very efficient in storing the rain water and lead to substantial grain yield improvement for sorghum. According to Kidane and Rezene (1989), a 45% increase of yield can be obtained for sorghum when compared to the traditional practice; depending also on soil type, slope, rainfall and cultivar (Table 8).

Table 8. *Effect of soil conservation methods (tiled ridges) on grain yield of sorghum in the semi-arid areas of Ethiopia (Kobo and Melkassa) (Ridge height = 35 cm, Ridge spacing 75 cm, Ridges tied at 5 m interval, Numbers in Parenthesis are percentage of grain yield increase over control. (Source Kidane and Rezene 1989).*

Soil conservation method	Average grain yield T/ha		
	Kobo	Melkassa	Mean
Flat planting (control)	1.6	0.80	1.20
Tiled Ridges planting in furrow	2.9 (81%)	3.0 (150%)	2.95 (145%)

3.8. Crop Simulation Model

CERES (Crop-Environment-Resource-Synthesis)-sorghum module in DSSAT v4.5 model is PC based crop simulation model which integrates all factors into a Cropping System Model (CSM) in a modular approach. The CSM uses one module for simulating soil water, nitrogen and carbon dynamics, while crop growth and development are simulated with the CERES, CROPGRO, CROPSIM and SUBSTOR modules. These components simulate the changes over time in the soil and plants that occur on a single land unit in response to weather and management practices.

4. Results and Discussion

4.1. Validation of Model Performance

The CERES-Sorghum Model of DSSAT has been tested over the selected sites in the northern Rift Valley of Ethiopia. The result shows that this model performs well under different 'packages' over two seasons. The statistical correlation is 97%. (see figure 2 below)

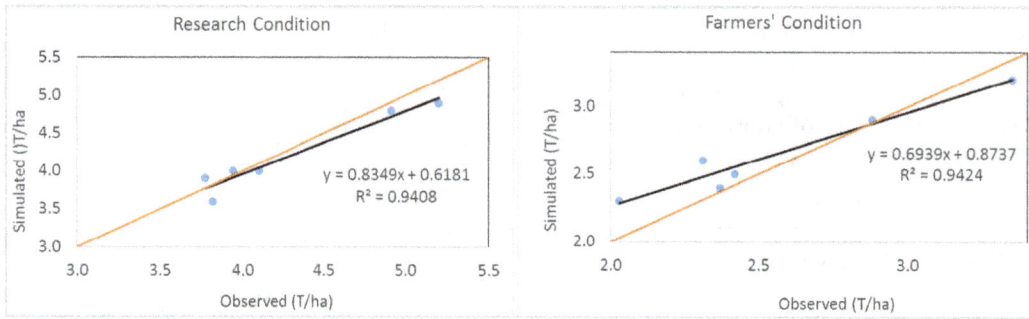

Figure 2. *Validation of the CERES-Sorghum model under farmers and research condition in the study areas during the experiment years.*

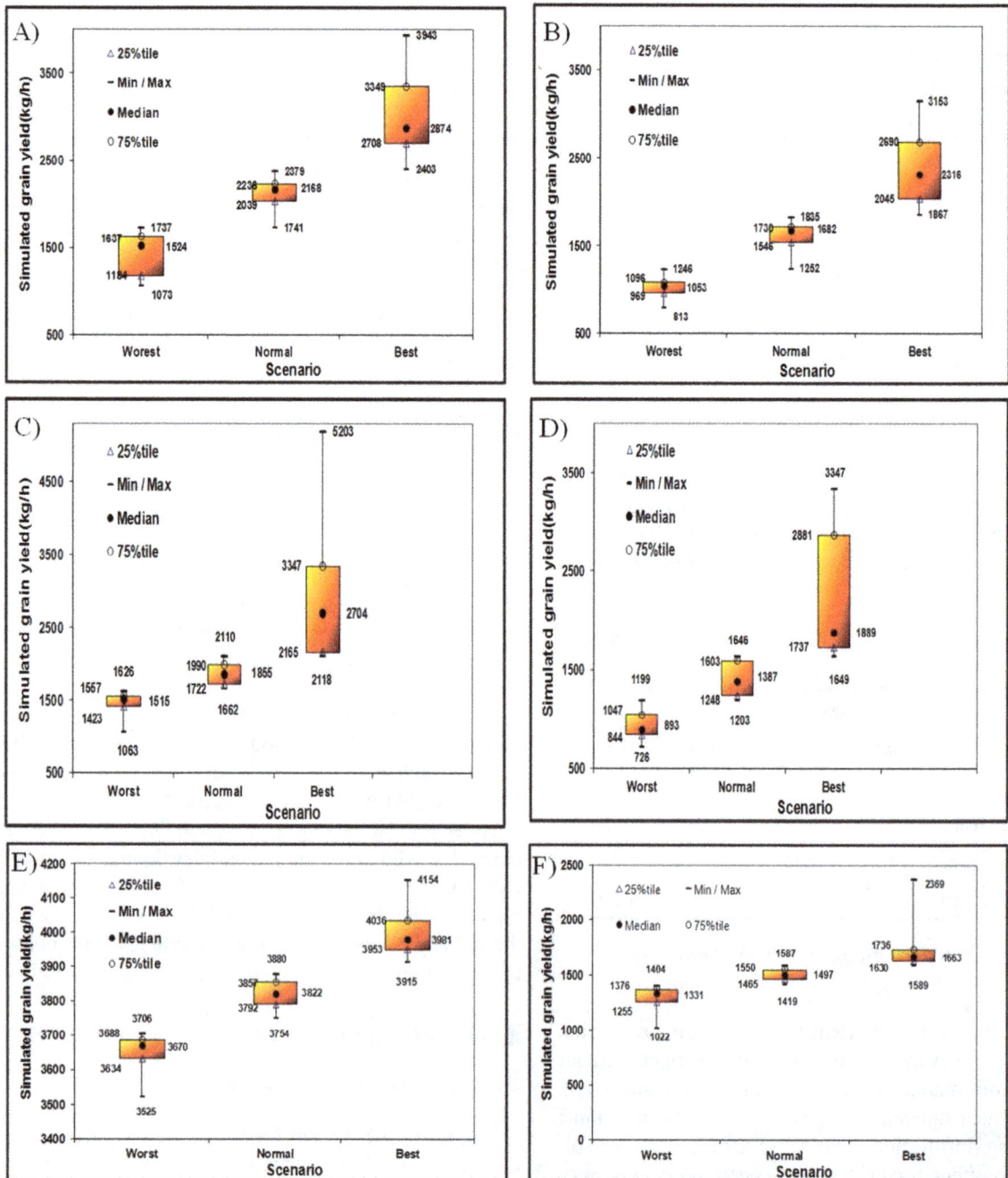

Figure 3. *Box plot of potential grain yield of sorghum under research (Right) and farmers' (Left) management practice condition at Meisso (A and B), Kono (C and D), Feddis (E and F).*

4.2. Simulated Sorghum Grain Yield

In this study sorghum grain yield has been simulated depending on soil type and weather data. Mean simulated grain yield for these three sites was found to be comparable to the national grain yield of sorghum. The models realistically simulated the potential grain yield data. Values for plant parameters and soil parameters described in this paper offer user's reasonable inputs for simulating sorghum grain yields semi-arid areas of Ethiopia.

As it can be seen in figure 3 the mean sorghum grain yield under best scenarios would be as high as 2.3 T/h at Meisso in which the same sorghum variety would be yield as high as 3.9 T/h at Feddis by applying best bet technology packages. In other hands even if the worst scenario would happen the grain yield sorghum would be 1.06 T/h at Kobo under best bet management options were applied. Meanwhile the situation would be worsen when we didn't apply those technologies and the yield would be get down to 0.72 T/h at Kobo.

In our annual analysis, a management decision like crop and cultivar selection, planting density and spacing, planting date, timing, amount and types of fertilizer application and other options were evaluated to compare model output under expert research and subsistence farmer conditions. This allows the evaluation of management options (Tsuji et al., 1998). In view of this, the result depicted in box plot (Figure 3) revealed that the average yield of sorghum is 2.3 and 1.0 T/ha at Kobo under research and farmer condition, respectively. In the 75 percentile > 3.0 T/ha is obtained under research condition while in farmer condition 1.2 T/ha. The figure also gives similar results for other percentiles, but clearly shows the research condition outperforms the farmer condition by a factor of about 3.

5. Conclusion and Recommendations

The result of the validation suggests that the CERES-Sorghum model, as applied to the Meko cultivar was good at all experiment sites (i.e. Meisso, Kobo and Feddis). Moreover this is a suitable tool for optimizing management decisions to improve the potential grain yield of sorghum in Ethiopia.

The use of a crop simulation model incorporating biophysical factors can be used to explore possibility of options that help to efficiently utilize the existing resource of the area, while reducing the risk associated with climate. At the same time, it offers the possibility of saving time and resources required for the development of crop technologies. In this regard, DSSAT can provide possibility of different technological package for any combination of sowing date, varietal choice, soil type and crop management. Here we have given evidence that model is validated for local growing condition. Even though the simulation output is promising, it should be realized that the availability and quality of existing soil and weather data is a key element for DSSAT. Therefore, in order to verify crop outputs for different technological packages, it is recommended that data on genetic coefficient should be collected.

The experiment sites in this study, both for calibrating and validating the CERES-Sorghum model, were located in the low latitudes, which limits the use of the model for sites in other latitudes. Future research should therefore include studies to calibrate the model in sites other than in the low latitudes.

Acknowledgements

The authors would like to thank the Association for Strengthening Agricultural Research in East and Central Africa (ASARECA) for providing support to this study, and the Ethiopian Institute of Agricultural Research for providing full material and technical support. The National Metrological Service Agency is thanked for provision of weather data at the experiment sites. We thank the staff at Meisso, Feddis and Kobo Municipal Agricultural Department for their expertise for field implementation; and also we thank the Farmers for their part in the study.

References

[1]   Aggaewal PK, and Kalra. N. (1994). Analyzing the limitations set by climatic factors, genotype, and water and nitrogen availability on productivity of wheat: II. Climatic potential yield and management strategies. *Field Crops Res.* 38: 93-103.

[2]   Alagarswamy, G and Singh, P and Hoogenboom, G and Wani, S P and Pathak, P and Virmani, S M (2000) Evaluation and application of the CROPGRO-Soybean simulation model in a Vertic Inceptisol. Agricultural Systems, 63 (1). pp. 19-32. ISSN 0308-521X.

[3]   Alexandrov, V. A., Hoogenboom, G., 2000. The impact of climate variability and change on crop yield in Bulgaria. Agricultural and Forest Meteorology, Volume 104, Pages 315-327. Pages 315-327.

[4]   Belton, P. S., & Taylor, J. R. (2004). Sorghum and millets: protein sources for Africa. Trends in Food Science & Technology, 15(2), 94-98.

[5]   Boote, K. J., Jones, J. W., & Pickering, N. B. (1996). Potential uses and limitations of crop models. Agronomy Journal, 88(5), 704-716.

[6]   Boote, K. J., Jones, J. W., Hoogenboom, G., & Pickering, N. B. (1998). The CROPGRO model for grain legumes. In Understanding options for agricultural production (pp. 99-128). Springer Netherlands.

[7]   Doggett, H. (1988). Sorghum. Harlow, Essex, England: Longman Scientific & Technical.

[8]   Gebre, H., & Georgis, K. (1988). Sustaining crop Production in Ehe Semi-Arid areas of Ethiopia. Ethiopian Journal of Agricultural Sciences.

[9]   Hook, J. E. (1994). Using crop models to plan water withdrawals for irrigation in drought years. Agricultural Systems, 45(3), 271-289.

[10] International Crops Research Institute for the Semi-arid Tropics, Agriculture Organization of the United Nations. Commodities, & Trade Division. (1996). The world sorghum and millet economies: facts, trends and outlook. Food & Agriculture Org.

[11] International Crops Research Institute for the Semi-arid Tropics. (1994). ICRISAT now: sowing for the future. Patancheru, Andhra Pradesh, India: ICRISAT.

[12] Kidane, G. and R. Fesahays. 1989. Dry land research priorities to increase crop Productivity, pp.57-64. In Proceedings of 21st NCIC. Addis Ababa, Ethiopia.

[13] Kimber, C. T. (2000). Origins of domesticated sorghum and its early diffusion to India and China. *Sorghum: Origin, history, technology, and production*, 3-98.

[14] Mall, R. K., Lal, M., Bhatia, V. S., Rathore, L. S., & Singh, R. (2004). Mitigating climate change impact on soybean productivity in India: a simulation study. *Agricultural and forest meteorology*, *121*(1), 113-125.

[15] Parry, M. L., Carter, T. R., & Konijn, N. T. (Eds.). (2013). *The Impact of Climatic Variations on Agriculture: Volume 1: Assessment in Cool Temperate and Cold Regions*. Springer Science & Business Media.

[16] Singh, P., Boote, K. J., & Virmani, S. M. (1994). Evaluation of the groundnut model PNUTGRO for crop response to plant population and row spacing. *Field Crops Research*, *39*(2), 163-170.

[17] Singh, P., Boote, K. J., Rao, A. Y., Iruthayaraj, M. R., Sheikh, A. M., Hundal, S. S., ... & Singh, P. (1994). Evaluation of the groundnut model PNUTGRO for crop response to water availability, sowing dates, and seasons. *Field Crops Research*, *39*(2), 147-162.

[18] Taylor, J. R. N. (2003, April). Overview: Importance of sorghum in Africa. In Proceedings of AFRIPRO Workshop on the Proteins of Sorghum and Millets: Enhancing Nutritional and Functional Properties for Africa, Pretoria, South Africa (Vol. 9).

[19] Tsuji, G. Y., Hoogenboom, G., & Thornton, P. K. (1998). Understanding options for agricultural production (Vol. 7). Springer Science & Business Media.

[20] Williams, G. D. V., Fautley, R. A., Jones, K. H., Stewart, R. B., & Wheaton, E. E. (1988). *Estimating effects of climatic change on agriculture in Saskatchewan, Canada* (pp. 219-379). Kluwer Academic Publishers.

---

i * A2=Warm arid lowland plains, M2=warm moist lowlands, M3=Tepid moist mid highlands, SM2=Warm sub-moist lowlands, SM3=Tepid sub-moist mid highlands and SM4=cool submost high land

Isolation of *Pseudomonas fluorescens* Species from Rhizospheric Soil of Healthy Faba Bean and Assessed Their Antagonistic Activity Against *Botrytis fabae* (Chocolate Spot Diseases)

Fekadu Alemu

Department of Biology, College of Natural and Computational Sciences, Dilla University, Dilla, Ethiopia

Email address:
fekealex@gmail.com

Abstract: Crop protection is an important area of agriculture which needs attention because of most of the hazardous inputs added into the agricultural system are in the form of chemicals. Production of the crop is, however, constrained by several disease infections including fungal diseases. The present study, was isolate twelve *Pseudomonas fluorescens* isolates from rhizospheric soil of faba bean and were tested for their antagonistic activity against *Botrytis fabae* that is known to attack faba bean crops. All *Pseudomonas fluorescens* isolates are shown successfully employed in controlling chocolate spot diseases of plant. *Pseudomonas fluorescens* isolates 10 (88.1%) showed highest antagonistic activity against *Botrytis fabae*. All isolate of *Pseudomonas fluorescens* are indicated successfully employed in controlling chocolate spot diseases of plant due to their antifungal metabolites. Therefore, these isolates can be used as potential of biocontrol agents.

Keywords: Biocontrol, *Botrytis fabae*, Chocolate Spot Diseases, Faba Bean, *Pseudomonas fluorescens*

1. Introduction

Faba bean (*Vicia fabae* L.) is one of the most important food legumes due to its high nutritive value both in terms of energy and protein contents (24-30%) and is an excellent nitrogen fixer (Sahile *et al.*, 2008). Ethiopia is the third largest producers of faba bean in the world, next to china and Egypt (Torres *et al.*, 2006) and its share is only 6.96% of world production and 40.5% of Africa (Asfaw *et al.*, 1994). Faba bean is grown on 370,000 hectares in Ethiopia with an annual production of about 450,000 tonnes (ICARDA, 2006). Despite its wide cultivation, the average yield of faba bean is quite low in Ethiopia, because of many biotic and abiotic constraints (Sahile *et al.*, 2008). Common names of *Vicia fabae* L is Ackerbohne in (Germany, Austria), Bob obecny (Czech Republic), Broad bean, Faba bean, Field bean, Horse bean, (United Kingdom), Favetta (Italy), Féverole (France), Haba (Spain), Hestebønne (Denmark), Põlduba (Estonia).

In many parts of Ethiopia broad beans are a daily part of the diet of the population. They are an important source of dietary protein, especially valuable during the numerous days of fasting that are observed (Asfaw, 1979) and also providing minerals (iron, zinc, calcium) and vitamins (B1, B2, C) in human diet and livestock rations and a source of biological nitrogen fixation in cereal rotation systems (Khalil and Erskine, 2001). Broad beans may be consumed green, either raw, roasted, or boiled; as dry seed, having been soaked, roasted, or boiled; in a preparation with a hot sauce called "wot"; ground and mixed with barley, wheat, or teff flour to form "injera" (a kind of pancake-type bread); or used in the preparation of various sauces (in a mixture with mustard and spices that is fermented for 4-5 days) (Asfaw, 1979). Therefore, increasing the crop production is one of the most important targets of agricultural policy in several countries.

The first description of chocolate spot disease came from Berkley during 1849 to 1875 in a series of articles in Britain. However, he was unable to associate a pathogen to the disease (Gaunt 1983). Sardina (1930, 1932) described the pathogen *Botrytis fabae* during 1928-1929 when the faba bean crops were attacked by the disease in different areas of Spain. The disease has been reported from Tunisia, Algeria, Morocco, Libya, Ethiopia, England, Spain, Norway, Germany, Scotland, Russia, Japan, China, Canada, North and South America, and Australia (Abdelmonem, 1981; Conner, 1967; Hebblethwaite, 1983; Ikata, 1933; Mengistu, 1979; Sardiña, 1929; Tupenevich and Kotova, 1970; United States Dept. of Agri., 1960).

Losses caused by chocolate spot are due mainly to a decreased number of pods per plant (Williams, 1978), damage the foliage, limit photosynthesis activity, and reduce faba bean production globally (Torres *et al.*, 2004). Other workers showed that faba bean leaves approaching maturity are more susceptible than the younger ones (Deverall and

Wood, 1961 and Mansfield and Deverall, 1974).

The chocolate spot disease (*B. fabae*) occurs mainly on leaves, but stems and flowers may also be infected under sever conditions as indicated fig. 1. Disease severity is favoured between 92-100% relative humidity and 15-20°C (Harrison, 1980 and 1984), other factors including inoculum density, waterlogging, high plant density and host physiology (Griffiths and Amin, 1977; Ingram and Hebblethwaite, 1976; Moore and Leach, 1968) have been shown to be closely related with disease development. Also, under prolonged wet conditions, the disease may be epidemic with heavy crop losses (Bernier *et al.*, 1993). The spots on leaves and stems enlarge and develop a grey, dead centre with a red-brown rim or margin. Chocolate spot can kill flowers and stems. Spores will form on this dead tissue ((Noorka, and El-Bramawy, 2011).

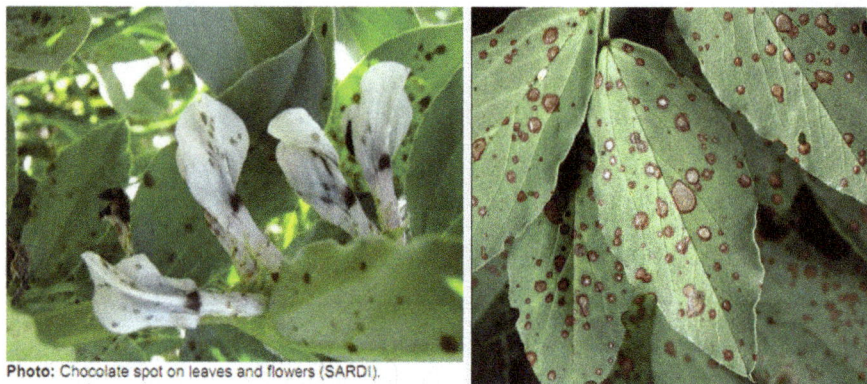

Photo: Chocolate spot on leaves and flowers (SARDI).

Figure 1. Chocolate spot disease on the leaves and flowers of faba bean.

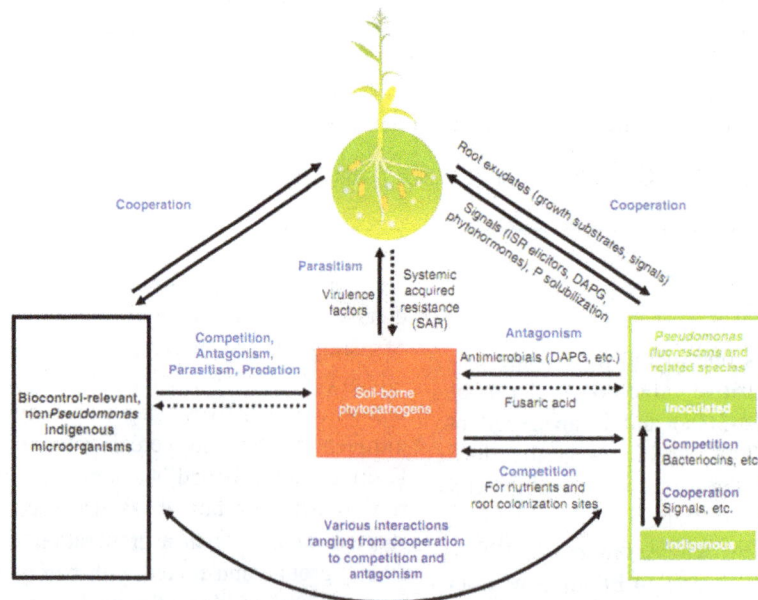

Figure 2. *Overview of plant-protection mechanisms in biocontrol agents from Pseudomonas fluorescens and closely-related species of fluorescent Pseudomonas. These pseudomonads may act directly on the plant, noticeably via production of various signals (2, 4-diacetylphloroglucinol (DAPG), phytohormones, etc.) and / or by triggering induced systemic resistance (ISR) pathways, and the plant provides them with organic exudates and molecular signals. They may also inhibit the phytopathogens by competition and / or antagonism mediated by secondary metabolites such as DAPG. In addition, these effects are modulated by the action of certain non Pseudomonas members of the microbial community, which may also have direct or indirect (i.e. via the plant) biocontrol effects and / or interfere with the functioning of biocontrol agents from Pseudomonas fluorescens and related species. As for Pseudomonas inoculants, their ecology and plant-beneficial properties can be influenced positively (via signalling and cooperation) or negatively (via competition) by indigenous root-colonizing pseudomonads. Dashed lines are used to indicate possible feedback responses of partners subjected to negative interactions, noticeably inhibition of DAPG production in Pseudomonas by fusaric acid from Fusarium oxysporum phytopathogens, and systemic acquired resistance in plant in response to infection (Couillerot et al., 2009).*

A strain of *Pseudomonas fluorescens* showed antagonistic property against *Rhizoctonia solani* (Howell and Stipanovic, 1979). A number of *Pseudomonas fluorescens* strains isolated from the rhizosphere of potato plants were reported to be antagonistic to *Rhizoctonia solani* in vitro and effectively reduced stem canker under laboratory conditions (Chand and Logan, 1984). Several strains of siderophore producing *P. fluorescens* have been shown to inhibit *Fusarium oxysporum f.sp.cubense, Fusarium oxysporum f.sp. vasinfectum, Rhizoctonia solani, Acrocylindrium oryzae, Xanthomonas campestris pv oryzae* and *P. syringiae pv phaseolicola* (Sakthivel *et al.*, 1986) as shown in fig. 2.

P. fluorescens isolated from rhizosphere of rape seedlings has been reported to inhibit *Fusarium roseum* and *Pythium ultimum* (Dahiya *et al.*, 1988). Bare root dip treatment or soil drench with *Pseudomonas fluorescens* CHA0 significantly suppressed *Rhizoctonia solani* of tomato (Siddiqui and Shaukat, 2002). *Pseudomonas fluorescens* strain A 506 is a commercially available biological control agent (Available in the name of blight Ban 506; N. farm Americas Inco., Sugar Land, TX) used for the suppression of fire blight on pear and apple trees (Temple *et al.*, 2004).

Biological control is the strategy for reducing disease incidence or severity by direct or indirect manipulation of microorganisms (Papavizas, 1985; Paullitz and Bekanger, 2000; Tesfaye Alemu and Kapoor, 2004; Zhang *et al.*, 1994). Biological control has attracted great interest because of increasing regulation and restriction of fungicides or unnecessary control attempts by other means. It is especially attractive for soil borne diseases because it needs critical evaluation of economics of the country and the pathogens are difficult to reach with specific fungicides (Montealegre *et al.*, 2003). The application of biological controls using antagonistic microorganisms has proved to be successful for controlling various plant diseases in many countries (Sivan, 1987). Outside the host, the biocontrol agent may be antagonistic and thereby reduce the activity, efficiency, and inoculum density of the pathogen through antibiosis, competition and predation/hyper parasitism. This leads to a reduction in inoculum potential of the pathogens (Baker, 1977).

There is an urgent need to improve *Vicia faba* yield, since this plant remains an important part of the diet of both humans and domestic animals in many parts of the world, because of its high nutritive value in both energy and protein contents. Furthermore, faba bean supplies an important benefit to the crop by fixing atmospheric nitrogen in symbiosis with *Rhizobium leguminosarum* thus, reducing costs and minimizing impact on the environmental, that is why increasing the plant production is one of the major targets of the agricultural policy in several countries (El-Fallal and Migahad *et al.*, 2003; Mahmoud *et al.*, 2004).

However, this crop is subjected to many abiotic and biotic stresses that seriously compromise the final yields. Among the menacing biotic stresses, chocolate spot, caused by *botrytis fabae*, is a worldwide disease capable of devastating the unprotected faba bean, result in harmful effects on growth, physiological activities, and yield. Chocolate spot disease of faba bean is the most widespread and destructive disease in Ethiopia with yield reductions of up to 61% on susceptible cultivars (Dereje and Beniwal, 1987). The problem of adequately protecting plants against the fungus by using fungicides has been complicated by development of fungicidal resistance and many chemicals traditionally used to control chocolate spot disease is less effective (Harrison, 1988), giving only partial disease control, high cost of their use and /or adverse effects on growth and productivity of faba bean as well as on the accompanying microflora (Khaled *et al.*, 1995). Therefore, controlling *B. fabae* by biocontrol agents seemed to better and preferred than the chemical control (El-Fallal and Migahad *et al.*, 2003; Mahmoud *et al.*, 2004). The present study is an initiative and also helps in better understanding and utilization of P. fluorescens isolates as biocontrol agent. It can be also assist further exploited for the commercial production of an inoculum to use a bio-primed seed as biocontrol and biofertilizers are an efficient approach to replace chemical fertilizers, fungicide, and its incorporation in the production system of *Vicia faba*. This is also the first study to control the *Botrytis fabae* plant pathogen through biocontrol. Therefore, the present study was designed to isolate *Pseudomonas fluorescens* species from rhizospheric soil of healthy faba bean will be evaluated for their antagonistic activity against *Botrytis fabae*.

2. Materials and Methods

2.1. Soil Sample Collection

The rhizospheric soil samples were collected in an envelope from fields growing faba bean (*Vicia faba* L.) from five Kebales of Salele zone: Mechale wartsu at altitude of 2560 meters above sea level, Wachale at altitude of 2540 meters above sea level, Gore kateme at altitude of 2590 meters above sea level, Eveno at altitude of 2510 meters above sea level and Gago at altitude of 2520 meters above sea level of North Showa of Oromiya Region of Salele zone, Ethiopia as shown in fig. 3. The soils were brought to Mycology Laboratory, Department of Microbial, Cellular and Molecular Biology, College Natural Sciences, Addis Ababa University.

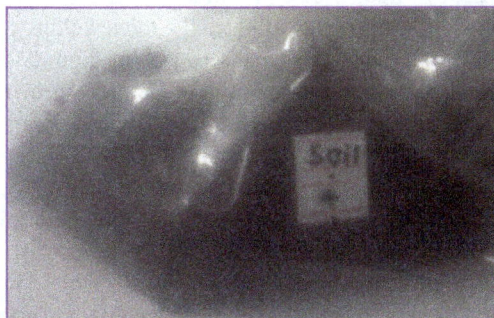

Figure 3. Soil samples.

2.2. Isolation of Pseudomonas fluorescens

Isolation of *Pseudomonas fluorescens* isolates studies were carried out on King's B medium (KBM) (King *et al.*, 1954). 1g of rhizosphere soil sample was suspended in 99 ml of sterile distilled water. Samples were serially diluted and 0.1 ml of sample was spreaded on King's B medium plates. After incubation at 28°C for 48 h the plates were exposed to UV light at 365 nm for few seconds and the colonies exhibiting the fluorescence were picked up and streaked on to the slants for maintenance, purified on King's B medium plates and also desigented as P f1-12 which stands for *Pseudomonas fluorescens* isolates used for further studies as indicated in fig. 4.

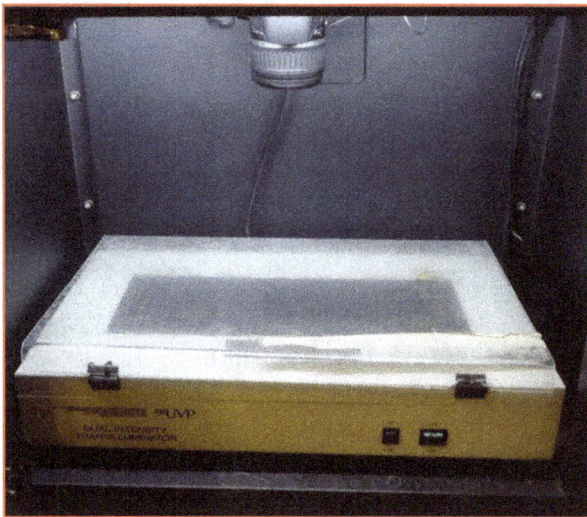

Figure 4. UV lamp apparatus set up.

2.3. Collection of Pathogen

One isolate of *Botrytis fabae* was obtained from Holeta Agricultural Research Centre, Ethiopia. This pathogenic fungus was isolated from the Leaf of infected faba bean leguminous crops.

2.4. In Vitro Tests of Fungal Antagonism

All *Pseudomonas fluorescens* isolates were assessed for potential antagonistic activity against *Botrytis fabae* on King's B agar using dual culture technique (Rangeshwaran and Prasad, 2000). An agar disc (4 mm dia.) was cut from an actively growing (96 hr) phytopathogen, B. *fabae* culture and placed on the surface of fresh King's B agar medium at the center of the Petri plates. A loopful of actively growing *Pseudomonas fluorescens* isolates (each) was placed opposite to the fungal disc and streaking the *Pseudomonas fluorescens* isolates on the plate at four locations, approximately 3 cm from the center. Plates inoculated with phytopathogen and without bacteria were used as control. All in vitro tests of antagonism were performed triplicates, with new coinoculations used each time. Plates were incubated at room temperature for 7 days. Degree of antagonism was determined by measuring the radial growth of pathogen with

bacterial culture and control and percentage inhibition calculated by the following equation (Riungu *et al.*, 2008) as follow as below.

$$\text{Percent inhibition } = \frac{100(C - T)}{C}$$

Where,
C=Radial growth of fungus in control plates (mm)
T=Radial growth of fungus on the plate inoculated with antagonist (mm)

2.5. Data Analysis

All the measurements were replicated three times for each assay and the results are presented as mean. IBM SPSS 20 Version statistical software package was used for statistical analysis of mean in each case.

3. Results and Discussion

3.1. Isolation of Pseudomonas fluorescens

During this research study, 12 *Pseudomonas fluorescens* were isolated from rhizospheric soil of healthy faba bean from five Kebales of Salale zone of Oromiya Region on King's B medium and observed under UV light at 365 nm for few seconds and pigment producer was screened as indicated fig. 5. Then it was purified again on same medium and observed under UV light as shown in fig. 6. All the rhizospheric isolates were named as *Pseudomonas fluorescens* isolate 1=P f1, *Pseudomonas fluorescens* isolate 2=P f2, *Pseudomonas fluorescens* isolate 3=P f3, *Pseudomonas fluorescens* isolate 4=P f4, *Pseudomonas fluorescens* isolate 5=P f5, *Pseudomonas fluorescens* isolate 6=P f6, *Pseudomonas fluorescens* isolate 7=P f7, *Pseudomonas fluorescens* isolate 8=P f8, *Pseudomonas fluorescens* isolate 9=P f9, *Pseudomonas fluorescens* isolate 10=P f10, *Pseudomonas fluorescens* isolate 11=P f11, *Pseudomonas fluorescens* isolate 12=P f12, and maintained on Nutrient Agar slants for further testing.

Figure 5. Pseudomonas fluorescens was isolated based on their pigment production under UV light at 365 nm.

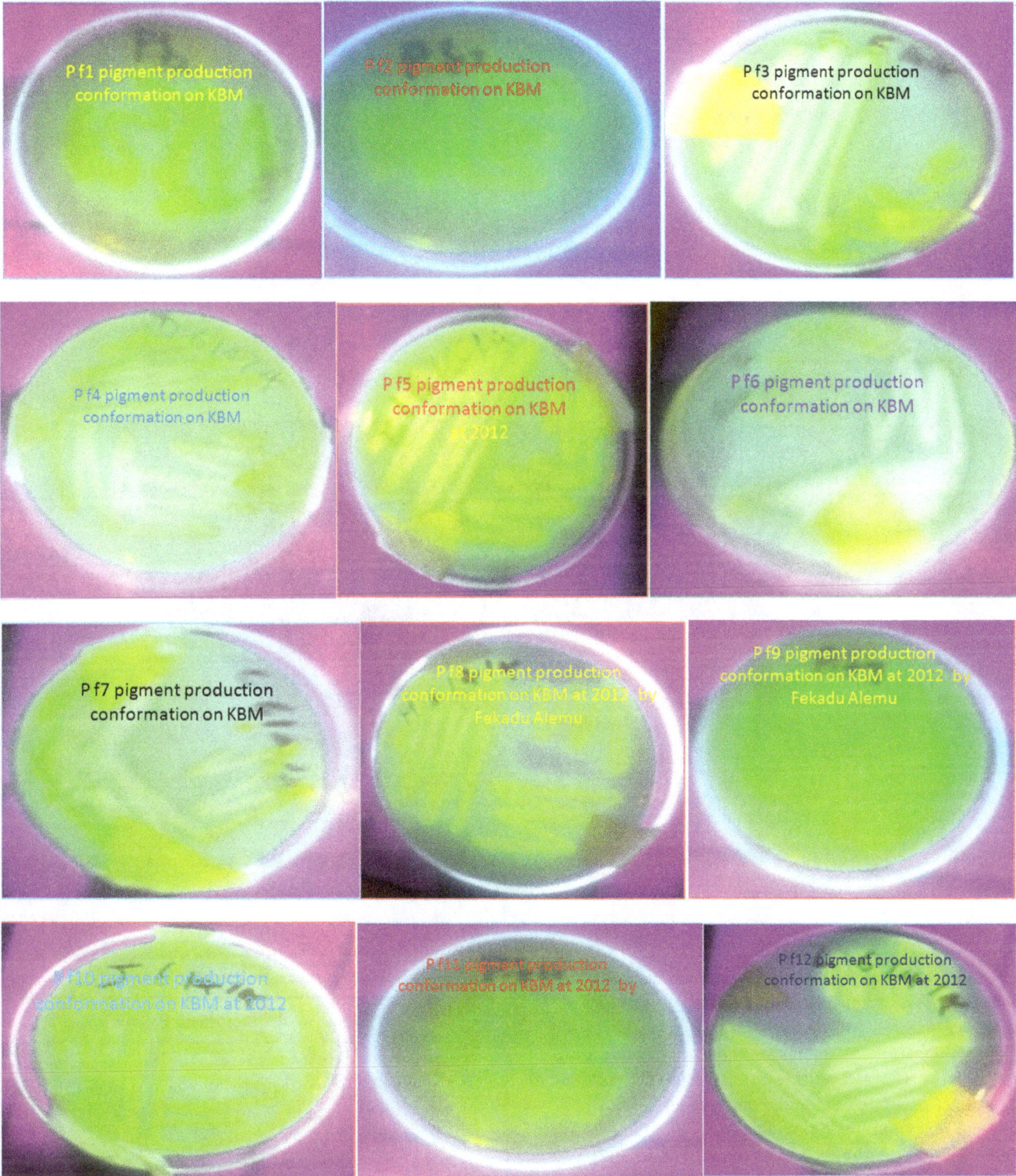

Figure 6. *Pseudomonas fluorescens isolates was confirmed again under UV light at 365 nm.*

3.2. Pathogens

Spore morphology of *Botrytis fabae* attachment to mycelia when slide culture was observed under microscope (Fig. 7 and Fig. 8).

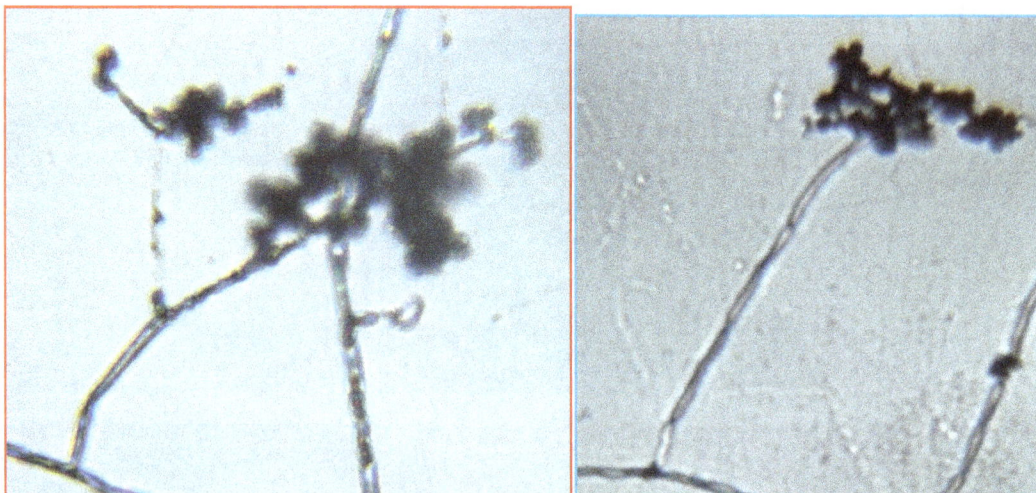

Figure 7. Conidiophore of Botrytis fabae (Branched dichotomously).

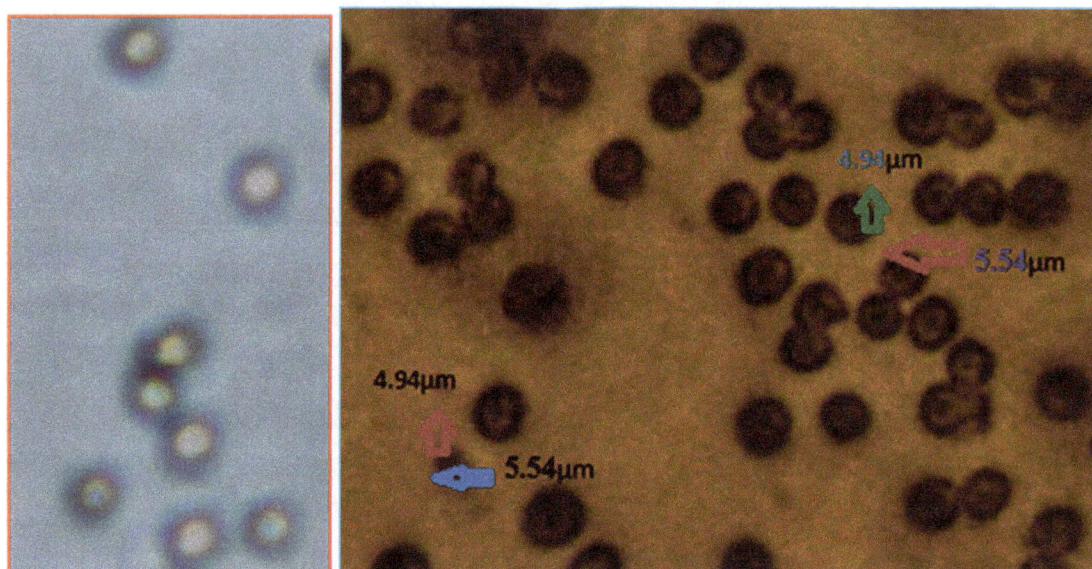

Figure 8. Conidia of Botrytis fabae (Ovoid shape) with measure horizontal and vertical diameter at 400x magnifications.

3.3. In Vitro Antagonism

The results of in vitro tests of antagonism toward the plant pathogen *B. fabae* were shows in Table 1 and fig. 9. Inhibition was clearly discerned by very limited growth of fungal mycelium in the inhibition zone surrounding a bacterial colony. The antagonistic effects of *P. fluorescens* isolates against *B. fabae* were in the range of 84.1- 88.1%. P f10 gave the maximum inhibition about 88.1%, followed by P f 9 (88.0%), P f 7 (87.6), P f 1 (87.0%), P f 8 (86.7%), P f 3 (86.5%), P f11 (86.3%), P f 2 (85.4%) or P f6 (85.4%), P f5 (84.8%), P f 12 (84.8%) and P f 4 (84.1%) respectively. Results as indicated that the mycelial growth diameter of *B. fabae* was with range 2.86-2.13cm as well as the lowest mycelial growth was obtained by (2.13cm) P f10 dual culture. Control plates not treated with *P. fluorescens* isolates were completely covered by the *B. fabae* showing no inhibition.

Table 1. Effect of Psedomonas fluorescens isolates treatments aganist the leaner mycelial growth of Botrytis fabae in vitro tests.

Pseudomonas fluorescens isolates	Antagonistic effect against Botrytis fabae	
	Mycelial diameter (cm) Mean ± SD	Inhibition (%)
P f 1	2.30 ±0.26458	87.0
P f 2	2.63 ±0.32146	85.4
P f 3	2.43 ±0.20817	86.5
P f4	2.86 ±0.11547	84.1
P f5	2.73 ±0.30551	84.8
P f6	2.63 ±0.49329	85.4
P f7	2.23 ±0.25166	87.6
P f8	2.40 ±0.36056	86.7
P f9	2.20 ±0.00000	88.0
P f10	2.13 ±0.15275	88.1
P f11	2.46 ±0.50332	86.3
P f12	2.80 ±0.20000	84.8
Control	9.00 ±0.00000	-

Key: Mean and SD= standard deviation

In the present study, twelve Pseudomonas fluorescens were isolated from the rhizospheric soil of healthy faba bean plants and were tested for their antifungal activity and promising antagonistic against B. fabae. The dual culturing of pathogen with 12 isolates of P. fluorescens showed clearly potential of biocontrol agent as shown in fig. 9.

All P. fluorescens isolates treatments reduced the mycelial growth of B. fabae on King's B medium. These might be due to producing secondary metabolites which inhibited growth of B. fabae. Similarly, the antimicrobial activity of Pseudomonas fluorescens had reported against numerous fungi (Khan and Zaidi, 2002 and Sivamani and Gnanamanickam, 1988). Pseudomonas fluorescens was shown to effectively inhibit R. solani and P. oryzae by agar plate method (Rosales et al., 1995). Vidhyasekaran and Muthamilan (1995) recorded that Pseudomonas fluorescens strains showed inhibitory action against the chickpea (Cicer arietinum) wilt pathogen Fusarium oxysporum f. sp. ciceris under in vitro studies.

As present results showed that all P. fluorescens isolates were success on growth inhibition against B. fabae with highest 88.1%. On other study, according to Sherga (1997), Bacillus isolates can be used as a biocontrol agent against Botrytis fabae and Botrytis cinarea. The highest reduction was caused by Bacillus isolate 115y (64%) against Botrytis fabae (Sahile et al., 2009). Pseudomonas fluorescens isolated from rhizosphere of organic farming area is an effective against Rhizoctonia solani (Anitha and Das, 2011). Pseudomonas fluorescens strain 003 was found to effectively inhibit (85%) the mycelial growth of fungal pathogens tested Rhizoctonia solani (Reddy et al., 2007).

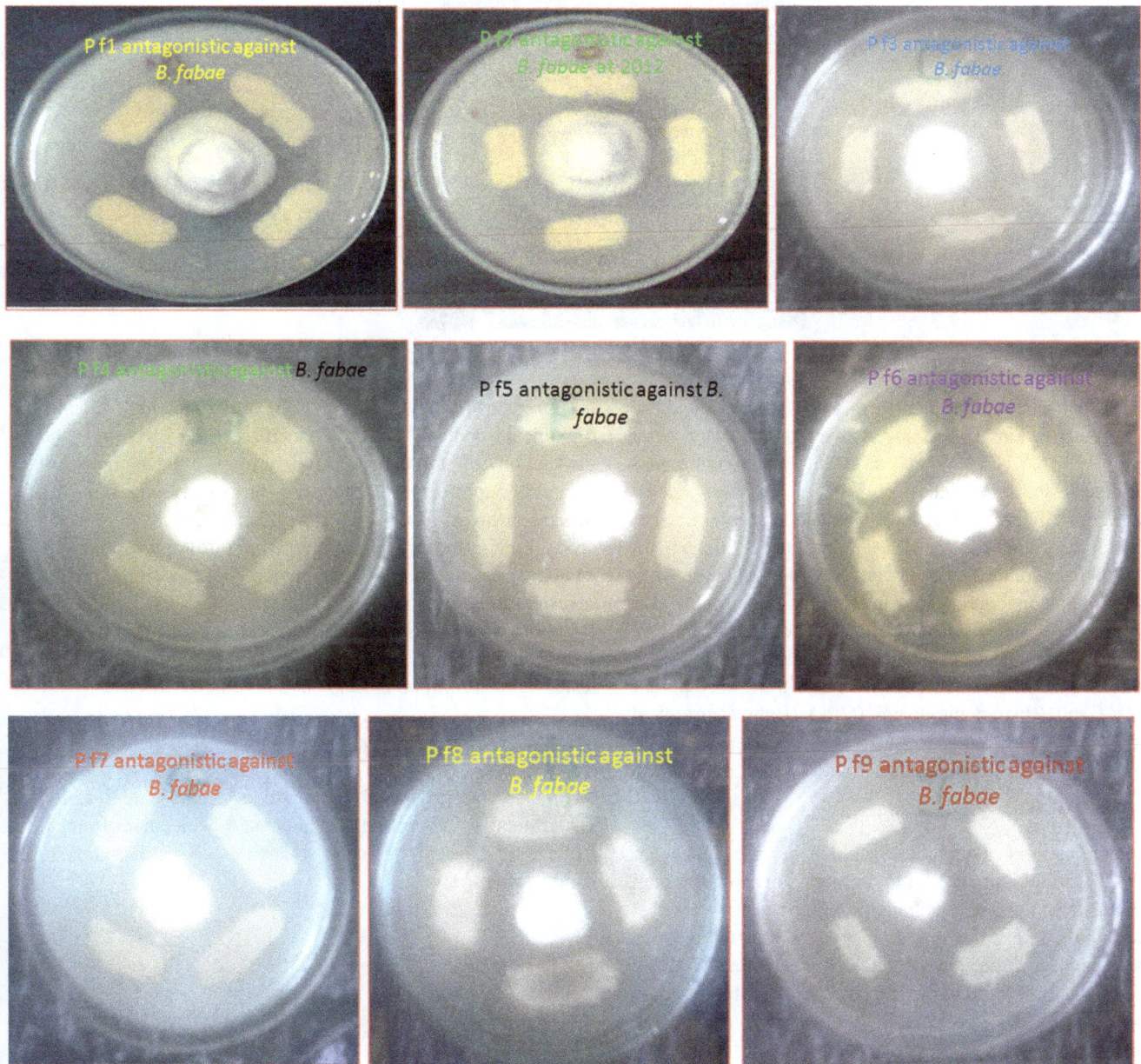

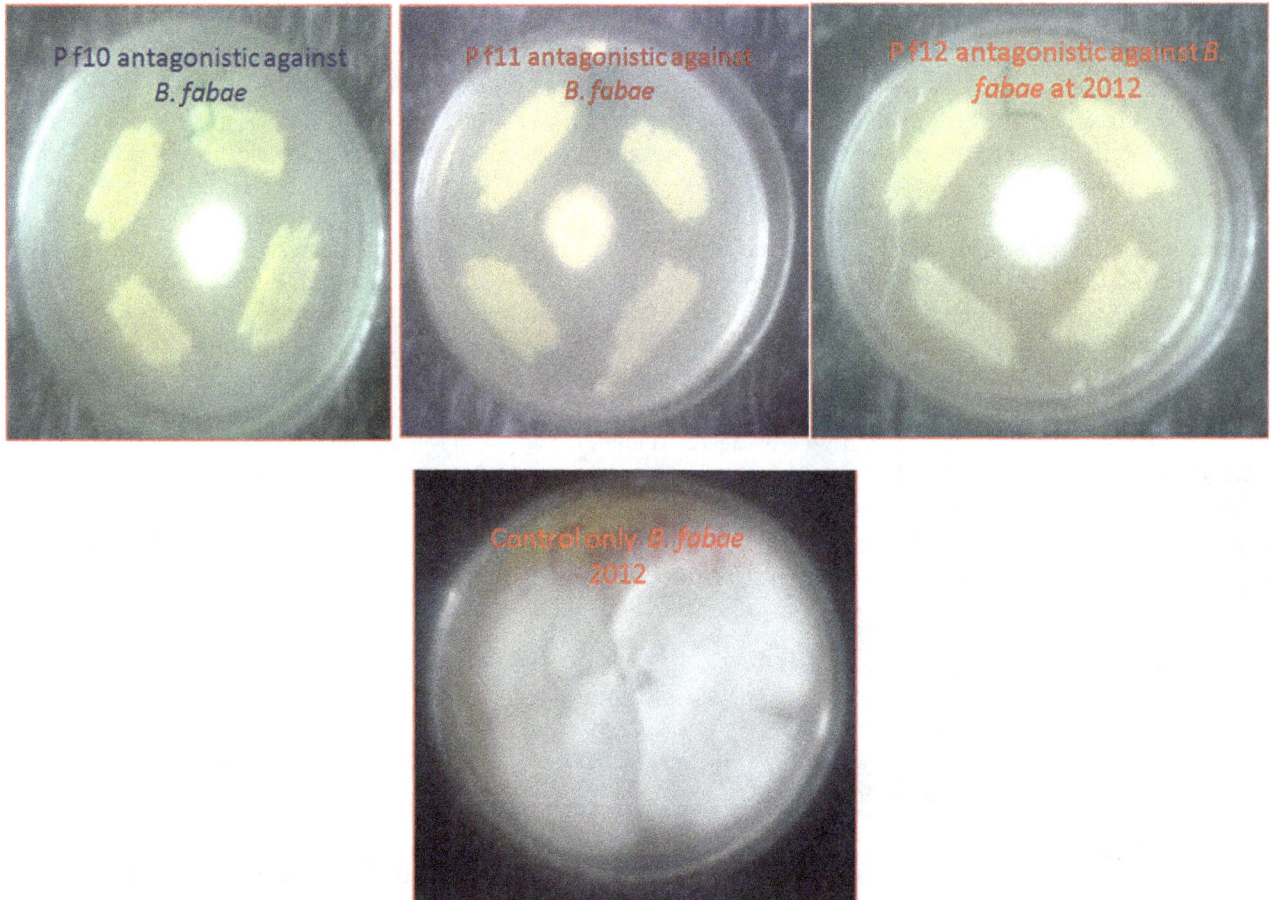

Figure 9. *Dual culture of Pseudomonas fluorescens isolates with B. fabae on King's B medium.*

And also Pseudomonas fluorescens 003 was found to be highly effective in controlling Rhizoctonia solani with inhibition 58% (Reddy et al., 2010). Pseudomonas flourescence showed highest antifungal activity against Penicillium italicum (94%) and was moderately effective against Aspergillus niger (61%) (Mushtaq et al., 2010). Isolate of Pseudomonas fluorescens on co-inoculation with fungal pathogens showed maximum inhibition for phytopathogens of Collectotrichum gleosporioides (58.3%), Alternaria brassicola (50%), Alternaria brassiceae (12.5%), Alternaria alternate (16.66%), Fusarium oxysporum (14.28%) and Rhizoctonia solani (50%) (Ramyasmruthi et al., 2012).

4. Conclusion

The application of fungicides for plants disease control are largely affecting human health, normal flora of soil and environment and also the pathogenic fungi became very fast resistant to them. For this reason, biocontrol inoculation such as *Pseudomonas fluorescens* isolates are needed antagonistic activities against *B. fabae* is an acceptable alternative to chemical fungicides application. *Pseudomonas fluorescens* isolates under investigation possess a variety of promising properties which make them better biocontrol agents that are capable of producing, antifungal substances, and subsequent enhancement of yield of faba bean crop. Such type of study is

necessary as it advocates that the environmental friendly biological control use for increase soil fertility and production of faba bean crop through management of chocolates spot disease. Therefore, further investigations are needed to investigate which type of biochemical production are making P.fluorescens isolates as one of the most suitable candidate in suppressing the phytopathogenic fungi. More investigations are needed to investigate this regard for isolation and characterization of antifungal compounds. *Pseudomonas fluorescens* isolates must be eveluted and tested against *B.fabae* in field condition in near future.

Acknowledgments

I am grateful thanks to Addis Ababa University, Department of Microbial, Cellular, and Molecular Biology, College of Natural Sciences which are giving facilities to conduct this study.

References

[1] Abdelmonem, A. M., 1981. Studies on chocolate spot of broad bean in Libya. Annu. Agric. Sci., 16: 119-131.

[2] Anitha, A., Das, M. A., 2011. Activation of rice plant growth against *Rhizoctonia solani* using *Pseudomonas fluorescens*, *Trichoderma* and salicylic acid. Res. Biotechnol., 2: 07-12.

[3] Asfaw T., Beyene, D., Tesfaye, G., 1994. Genetic and Breeding of Field Pea. First Natinal Cool season Food Legumes Conference. ICARDA, Aleppo.

[4] Asfaw, T., 1979. Broad beans (*Vicia faba*) and dry peas (*Pisum sativum*) in Ethiopia. In: Food Legume Improvement and Development, pp. 80-82, (Hawtin, G. C. and Chancellor, G. J. eds). International Development Research Centre, Aleppo.

[5] Baker, K. F., 1977. Evolving concepts of biological control of plant pathogens. Annu. Rev. Phytopathol., 125: 67-85.

[6] Bernier, C. C., Hanounik, S. B., Hussein, M. M., Mohamed, H. A., 1993. Field manual of common faba bean diseases in the Nile Valley. International Center for Agricultural Research in the Dry Areas (ICARDA), Aleppo.

[7] Chand, T., Logan, C., 1984. Antagonists and parasites of *Rhizoctonia solani* and their efficacy in reducing stem canker of potato under controlled conditions. Transanctions British Mycol. Soc., 38: 107-112.

[8] Conner, L. L., 1967. An annotated index of plant diseases in Canada. Can. Dept. of Agric. Publication 1251.

[9] Couillerot, O., Prigent-Combaret, C., Caballero-Mellado, J., Moë¨nne-Loccoz, Y., 2009. *Pseudomonas fluorescens* and closely-related fluorescent pseudomonads as biocontrol agents of soil-borne phytopathogens. Lett. Appl. Microbiol., 48: 505–512.

[10] Dahiya, J. S., Woods, D. L., Tewari, J. P., 1988. Control of *Rhizoctonia solani*, causal agent of brown girdling root rot of rapeseed by *Pseudomonas fluorescens*. Botanical Bulletin of Academic Sinica, 29: 135-141.

[11] Deverall, B. L., Wood, R. K. S., 1961. Infection of bean plants (*Vicia faba* L.) with *Botrytis cinerea* and *Botrytis fabae*. Annu. Appl. Biol., 49: 461-472.

[12] El-Fallal, A. A., Migahed, F. F., 2003. Metabolic change in broad bean infected by *Botrytis fabae* in response to mushroom spent straw. Asian J. Plant Sci., 2: 1059-1068.

[13] Gaunt, R. E. 1983. Shoot diseases caused by fungal pathogens. In: The Faba Bean (*Vicia faba* L.), pp. 463-492, (Hebblethwaite, P. D., ed). Butterworths, London.

[14] Griffiths, E., Amin, S. M., 1977. Effects of *Botrytis fabae* infection and mechanical defoliation on seed yield of field beans (*Vicia faba*). Annu. Appl. Biol., 86: 359-367.

[15] Harrison, J. G., 1984. Effect of humidity on infection of field bean leaves by *Botrytis fabae* and germination of conidia. Trans. Br. Mycol. Soc., 82: 245-248.

[16] Harrison, J. G., 1988. The biology of *Botrytis* spp. on *Vicia beans* and chocolate spot disease a review. Plant Pathol., 37: 168–201.

[17] Harrison, J. G., 1980. Effects of environmental factors on growth of lesions on field beans by *Botrytis fabae*. Annu. Appl. Biol., 95: 53-61.

[18] Hebblethwaite, P. D., 1983. The Faba Bean. Butter worths, London.

[19] Howell, C. R., Stipanovic, R. D., 1979. Control of *Rhizoctonia solani* on cotton seedlings with *Pseudomonas fluorescens* and with an antibiotic produced by the bacterium. Phytopathol., 69: 480-482.

[20] ICARDA, 2006. Technology Generations and Dissemination for Sustainable Production of Cereals and Cool Season Legumes. International Center for Agricultural Research in the Dry Areas, Aleppo.

[21] Ikata, S., 1933. Studies on red spot disease of *Vicia faba*. Rep. Agric. Expt. Sta. Okayamaken.

[22] Ingram, J., Hebblethwaite, P. D., 1976. Optimum economic seed rates in spring and autumn sown field beans. Agric. Prog., 51: 1-32.

[23] Khaled, A. A., Abd El-Moity, S. M. H., Omar, S. A. M., 1995. Chemical control of some faba bean diseases with fungicides. Egypt J. Agric. Res., 73: 45-56.

[24] Khalil, S. A., Erskine, W., 2001. Combating disease problems of grain legumes in Egypt. Rain legumes no. 32 – 2nd quarter.

[25] Khan, M. S., Zaidi, A., 2002. Plant growth promoting rhizobacteria from rhizosphere of wheat and chickpea. Annu. Plant. Protec. Sci., 10: 265-271.

[26] King, E. O., Ward, M. K., Raney, D. E., 1954. Two simple media for the demonstration of pyocyanine and fluorescein. J. Lab. Clin. Med., 44: 301-307.

[27] Mahmoud, Y. A. G., Ebrahim, M. K. H., Aly, M. M., 2004. Influence of plant extracts and microbioagants on physiological traits of faba bean infected with *Botrytis fabae*. J. Plant Biol., 47: 194-202.

[28] Mansfield, J. W., Deverall, B. J., 1974. Changes in wyerone acid concentrations in leaves of *Vicia faba* after infection by *Botrytis cinerea* or *B. fabae*. Annals Appl. Biol., 77: 227-235.

[29] Mengistu, A., 1979. Food legume diseases in Ethiopia. In: Food Legume Improvement and Development, pp.106-108, (Hawtin, G. C. and Chancellor, G. J., eds). ICARDA/IDRC, Ottawa.

[30] Montealegre, J. R., Perez, L. M., Herrera, R., Silva, P., Besoain, X., 2003. Selection of bioantagonisic bacteria to be used in biological control of *Rhizoctonia solani* in tomato. Environ Biotechnol., 6: 1-9.

[31] Moore, K. G., Leach, R., 1968. The effect of 6-benzyl- aminopurine (benzyladenine) on senescence and chocolate spot (*Botrytis fabae*) of winter beans (*Vicia faba*). Annu. Appl. Biol., 61: 55-76.

[32] Mushtaq, S., Ali, A., Khokhar, I., Mukhtar, I., 2010. Antagonisitic potential of soil bacteria against food borne fungi. World Appl. Sci. J., 11: 966-969.

[33] Noorka, I. R. El-Bramawy, M. A. S., 2011. Inheritance assessment of chocolate spot and rust disease tolerance in mature faba bean (*Vicia faba* L.) plants. Pak. J. Bot., 43: 1389-1402.

[34] Papavizas, G. C., 1985. Trichoderma and Gliocladium: biology, ecology and potential biocontrol. Annu. Rev. Phytopathol., 23: 23-54.

[35] Paullitz, T. C., Bekanger, R. R., 2000. Biological control in greenhouse systems. Annu. Rev. Phytopathol., 39: 103-133.

[36] Ramyasmruthi, S., Pallavi, O., Pallavi, S., Tilak, K., Srividya, S., 2012. Chitinolytic and secondary metabolite producing *Pseudomonas fluorescens* isolated from Solanaceae rhizosphere effective against broad spectrum fungal phytopathogens. Asian J. Plant Sci. Res., 2: 16-24.

[37] Rangeshwaran, R., Prasad, R. D., 2000. Biological controls of *Sclerotium* rot of sunflower. Indian Phytopathol., 53: 444-449.

[38] Reddy, B. P., Rani, J., Reddy, M. S., Kumar, K. V. K., 2010. Isolation of siderophore- producing strains of rhizobacterial fluorescent *Pseudomonads* and their biocontrol against rice fungal pathogens. Int. J. Appl. Biol. Pharm. Technol. 1: 133-137.

[39] Reddy, K. R. N., Choudary, K. A., Reddy, M. S., 2007. Antifungal metabolites of *Pseudomonas fluorescens* isolated from rhizosphere of rice crop. J. Mycol. Plant Pathol. 37(2).

[40] Riungu, G. M., Muthorni, J. W., Narla, R. D., Wagacha, J. M., Gathumbi, J. K., 2008. Management of Fusarium head blight of wheat and deoxynivalenol accumulation using antagonistic microorganisms. Plant Pathol. J., 7: 13-19.

[41] Rosales, A. M., Thomashow, L., Cook, R. J., Mew, T.W., 1995. Isolation and identification of antifungal metabolites produced by rice associated antagonistic *Pseudomonas* spp. Phytopathol., 85: 1029-1032.

[42] Sahile, S., Fininsa, C., Sakhuja, P. K., Ahmed, S. 2009. Evaluation of pathogenic isolates in Ethiopia for the control of chocolate spot in faba bean. Afri. Crop Sci. J., 17: 187-197.

[43] Sahile, S., Fininsa, C., Sakhuja, P. K., Seid, A., 2008. Effect of mixed cropping and fungicides on chocolate spot (*Botrytis fabae*) of faba bean (*Vicia faba*) in Ethiopia. Crop Protection, 27:275-282.

[44] Sakthivel, N., Sivamani, E., Unnamalai, N., Gnanamanickam, S.S., 1986. Plant growth promoting rhizobacteria in enhancing plant growth and suppressing plant pathogens. Curr. Sci., 55: 22-25.

[45] Sardina, J. R., 1930. A new species of Botrytis attacking broad beans. Rev. Appl. Mycol., 9: 424.

[46] Sardina, J. R., 1932. Two new diseases of broad beans. Rev. Appl. Mycol., 11: 346-347.

[47] Sardiña, J. R., 1929. Una nueva especie de *Botrytis* que ataca a las habas. Real Sociedad Española de Historia Natural Memorias, 15: 291-295.

[48] Sherga, B. M., 1997. *Bacilus* isolates as potential biocontrol agents against chocolate spot on Faba beans. Can. J. Microbio. 43: 915-924.

[49] Siddiqui, I. A., Shaukat, S. S., 2002. Resistance against the damping-off fungus *Rhizoctonia solani* systemically induced by the plant growth promoting rhizobacteria *Pseudomonas aeruginosa* (IE-6S+) and *Pseudomonas fluorescens* (CHA0). J. Phytopathol., 150: 500-506.

[50] Sivamani, E., Gnanamanickam, S. S., 1988. Biological control of *Fusarium. oxysporum* in banana by inoculation with *Pseudomonas fluorescens*. Plant Soil, 107: 3-9.

[51] Sivan, A., 1987. Biological control of *Fusarium* crown rot of tomato by *Trichoderma harzianum* under field conditions. Plant Dis.71: 587-592.

[52] Temple, T. N., Stockwell, V. O., Zoper, J. E., Johnson, K. B., 2004. Bioavailability of iron to *Pseudomonas fluorescens* strain *A506* on flowers of Pear and apple. Phytopathol., 94: 1286-1294.

[53] Tesfaye Alemu and Kapoor, I. J., 2004. In Vitro evaluation of *Trichoderma* and *Gliocladium* spp against Botrytis corm rot (*Botrytis gladiolorum*) of Gladiolus. Pest Mgt. J. Ethiopia, 8: 97-103.

[54] Torres, A. M., Roman, B., Avila, C. M., Satovic, Z., Rubiales, D., Sillero, J. C., Cubero, J.I., Moreno, M. T., 2006. Faba bean breeding for resistance against biotic stresses: towards application of marker technology. Euphytica, 147: 67-80.

[55] Tupenevich, S. M., Kotova, V. V., 1970. Biological features of the causal agent of bean brown patch. Trudy veses. Inst. Zashch. Rast., 29: 143-150.

[56] United States Department of Agriculture, 1960. Index of plant diseases in the United States. Agric. Handbook, 165: 275-276.

[57] Vidhyasekaran, P., Muthamilan, M., 1995. Development of formulations of *Pseudomonas fluorescens* for control of chickpea wilt. Plant Dis., 79: 782-786.

[58] Williams, P. F., 1978. Growth of broad beans infected by *Botytis fabae*. J. Hort. Sci., 50: 415-424.

[59] Zhang, B., Ramonell, K., Somerville, S., Stacey, G., 1994. Characterization of early, chitin induced gene expression in Arabidopsis. M. P. M. I., 15: 963–970.

16

The Impact of Gas Flaring in Nigeria

Omoniyi Omotayo Adewale, Ubale Mustapha

Department of Petroleum Engineering, Abubakar Tafawa Balewa University, Bauchi State, Nigeria

Email address:

omotosimple4u@gmail.com (O. O. Adewale)

Abstract: Nowadays, petroleum hydrocarbons are widely used as fuels for energy and power generation. However, the production of such fuels complements the flaring of gas in Nigeria. Therefore, this research investigates the various ways in which gas flaring affects the populace of the Niger Delta region of Nigeria, which is the bedrock of Nigeria's current oil and gas resources. Also, the research work attempts to offer solutions on how a feasible and reliable gas industry and market can be developed in Nigeria. The research makes use of literatures relevant to gas flaring, and a questionnaire was distributed to three states namely; Akwal bom, Rivers and Bayelsa. To figure out the environmental, social and economic impacts of gas flaring in those areas, a thorough research led to findings that gas flaring causes health problems for the people, a damaged and unsustainable environment, as well as socio-economic problems. Furthermore, the research shows recommendations for solutions on the development of a viable gas industry, and also economic prospects in relation to the development of a proper gas plan in Nigeria.

Keywords: Petroleum Hydrocarbons, Oil and Gas, Gas Flaring, Environment, Income, Health

1. Introduction

Nigeria is a nation highly endowed with natural resources such as mineral deposits, natural gases and petroleum. These have been a blessing in some instances, and in others it is nothing but a continuous source of pain and sorrow to the countries that are located where these natural resources may be found. Some of these natural sources are a windfall for these nations, while others struggle to deal with social afflictions such as pollution, disease and environmental consequences that come with the resources (Human Development Report 2011)[13].

The extensive inventory of natural resources, consisting of natural gas and petroleum, in the nation of Nigeria is situated in a region referred to as the Niger Delta. With access to the Atlantic Ocean and Lake Chad by means of the many rivers, the Niger Delta is a diverse ecosystem of marine organisms recognized as a prime source of aquatic food production (Awosika et al 2001)[5].

As a result of decades of overextended petroleum and natural gas exploration and drilling, the lack of foresight to strategically extract these natural resources with minimal impact on the environment, and the primary and collateral pollution associated with these activities has had an adverse impact on coastal and oceanic inhabitants in jeopardy of

being eradicated (Argo 2001)[3]. Occurrences of the flaring of natural gas and sanitary sewage overflow/ run-off are two of the many factors that contribute to the degradation of the surrounding environment, and increase pollutants contaminating the air and water for residents of the Niger Delta region.

The Niger Delta is located in the southern area of the southern region of Nigeria (Aghalino 2000a, b)[1,2] and is surrounded by towns and villages of Escravos, Ekpan, and Batan with fishing and farming as the prime industries that support the regional economy (Kamalu and Wokocha 2010)[15].

The magnitude of the oil exploration industry in the Niger Delta, and the subsequent environmental damage and pollution that is associated with it have impacted the way of life and economies of the local inhabitants who earn an income by way of fishing or agricultural farming. But the effects of natural gas flaring and petroleum spillages have tainted the fish supply with toxins as well as the fruits and vegetables that are harvested as well as taking a toll on the health and well-being of the residents too (Ashton et al. 1999)[4]. As the production of petroleum takes precedence over the interests of the local residents, their rights towards the alternatives of being gainfully employed in other sectors of the economy have been largely ignored. The exploration of petroleum comes with a price to be paid, and that price is the severe and detrimental effects on the environment, which is viewed as a source towards fostering climate change which within the past decade garnered a lot of attention (Vatn 2005)[23], though the failure to respect property ownership rights can amount to problematic conflicts and subsequent losses (Awosika et al. 2001)[5].

The petroleum industry has actually come under attack in a physical sense from local residents who claim they are yet to realize any compensation for seizure of their land and loss of income and livelihood, with damage to petroleum industry infrastructure. In some instances, the petroleum industry has engaged in isolated combat warfare against local residents with the taking of hostages to further emphasize their claims for compensation (Dibia 2011)[8]. A systematic plan has to be devised with a framework that ensures that the residents of the Niger Delta will be compensated for their loss of income and livelihood and not be subjected to unlawful violent tactics by the petroleum corporations. A possible solution towards sufficient compensation would be employment within the petroleum industry, unfortunately though, the education and skills of the local resident is insufficient as well as instituting standard, regulated policies and procedures to minimize impact on the communities (Kamalu and Wokocha 2010)[15].

1.1. Understanding Gas Flaring

Onshore and offshore wells both conduct the procedure of flaring of natural gas (Farina 2010[11], GGFR 2002[12]); with the intent of facilitating the reduction of pressure in the well and viewed by some as a safety precaution, but in most instances is conducted as a means of disposing excess natural gas, as illustrated in Figure 1 below. Flaring of natural gas in Nigeria adds approximately 1% to the worldwide CO_2 emissions which poses extensive issues for the environment (World bank, 2010)[24], and is viewed as a pathway to issues associated with absence of consumer goods, cultural and ethnic perceptions towards flaring as waste and insufficient opportunity for economic benefit (World bank, 2010)[24].

Over the last two decades, the quantity of natural gas that has been flared has remained at a consistent rate of 100bcm/year with less than 15 countries accounting for nearly 80% of the volume discharged into the atmosphere, contaminating the environment and a needless exhaustion of valued natural resources (Svalheim 2005)[22]. Collaborative efforts amongst governments and the petroleum industry to curb or eliminate natural gas flaring include the Kyoto Protocol, The Vietnam Rang Dong Project and efforts of the World Bank in conjunction with the Norwegian government (World Bank 2010)[24].

Figure 1. Sample Picture of a Gas Flaring Activity (The World Bank 2011)[25].

1.2. Statement of the Problem

Many problems have become prevalent over the last several decades within the Niger Delta region as a result of excessive of flaring Natural Gas, such as:
- Environmental Damage
- Economic Strife and
- Socio-Ecological Issues.

1.3. Objectives

The objectives of this research are as follows:
1. Examine contemporary information concerning gas flaring in Nigeria.
2. Assess the economic and socio-political factors behind gas flaring in Nigeria.
3. Investigate probable economic prospects from the

elimination of gas flaring in Nigeria.

4. Propose resolution towards developing a sustainable gas industry in Nigeria.

1.4. Significance of the Project

With petroleum's inability to be considered a clean burning fuel as a result of the amount of toxins and hazardous compounds that it releases into the atmosphere, it is no longer a viable source of energy, and instills economic distress on individuals and businesses (Davies 2001)[7]. Hence, the quest for other sources of energy which is cost affordable, clean and economic friendly makes natural gas seem as a natural alternative to fuel our residences, businesses, and transportation needs over the long term (Natural Gas 2011)[18].

The use of natural gas as an alternative to petroleum or fossil fuels offers an array of economic benefits. In comparison to petroleum, natural gas is lower in terms of cost to produce; and burns with negligible amounts of gas emissions (Madueme 2010)[16, 17] and despite these facts, Nigeria continues to flare approximately 80% of its annually produced natural gas inflicting health issues upon individuals

and damaging the environment. Therefore, this study aims to concentrate on the economic impacts of gas flaring in Nigeria, and also the abilities of the Nigerian gas industry in bringing about viable development in the country.

2. Methodology

Research Questions

The questions posed for this research were developed from the framework of the research with the aim of determining the socio-economic and environmental impacts resulting from gas flaring on the populace of the Niger Delta region. The research questions are presented as follows:

1. What is the availability of literature relevant to natural gas flaring in Nigeria?
2. What are the environmental, economic and political impacts of gas flaring in Nigeria?
3. What are the economic prospects in relation to the development of a proper gas plan in Nigeria?
4. What are the opportunities for developing extensive infrastructure for natural gas in Nigeria?

Table 1. Research Methods, Analysis and Selection.

Technique	Advantages	Disadvantages
Interview	• Issues associated with the subject are addressed. • No future relationship between the interviewer and the interviewee • It gives the interviewer the opportunity to improve the research by asking other questions that are relevant to the research	• There may be bias due to interviewer not understanding the interviewee • Conducting interviews is costly as it requires relocation • Difficult in recalling interviewee statements
Questionnaires	• Questionnaires are inexpensive method • It can be completed confidentially • Many responses can be made before analysis • Easy to compare and assess • High rate of dependability	• High probability of responder bias • Low response rate • Data received not relevant to questions • Feedback is useless • Problems assessing collected data
Focus groups	• Preferred means of data generation • Means of gathering major and detailed information without bias • Spans broad range of interest	• Mandates presence of trained facilitator prior to data being generated • Scheduling difficulties to accommodate everyone
Observations	• Monitoring events as they occur • Create accurate data and information	• Data collected may not match the research topic
Case studies	• Offers adequate data methodology and results for project • High rate of dependability for data collection	• Difficulty in acquiring data • Analysis of collected data is time consuming

3. The Process of Data Collection

To conduct this research in an unbiased, independent manner, a research statement was drafted in detail explaining the scope of the research, its purpose and the objectives that were to be met. A consent form was sent to all relevant government agencies of Nigeria and petroleum organizations that either is active or having pending operations with regard to petroleum production. Questionnaires (see Appendix) were distributed to verified residents of pre-selected towns and villages that are in near proximity to petroleum production wells. A large sampling of residents was necessary to obtain a true representative view.

100 individuals were selected to receive a questionnaire with their responses recorded electronically or in the presence of a research poll representative. The questionnaires were issued to individuals explaining the purpose of the research along with the assurance that any personal

information recorded would remain confidential.

The critical need for the research of this project necessitated a personal trip to the Niger Delta region to see firsthand the existing conditions that may be attributed to natural gas flaring. The visit insured verified distribution of the questionnaires and allowed for the potential respondent to be personally qualified to participate in the research. With a personal visit in lieu of having others administer the questionnaire, it insured a timely response by all 100 individuals, secure collection of all recorded data and completed within the parameters of the original research statement.

A percentage of questionnaires were distributed to the petroleum organizations by means of email. When potential respondents did not submit a response as it approached the deadline, this researcher dispatched made an unannounced visit to inquire as to the status of their completion. The researcher took the opportunity to speak to willing officials and staff members as to gas flaring activities and the eventual phasing out of such practices in accordance with government mandates. Critical information provided by those interviewed served as the foundation for the analyzing of attained data.

To support the assessment and analysis of data received; this researcher reviewed electronically retrievable literature and databases relevant to natural gas flaring. Examination was made regarding the social, economic and environmental impacts annually on the inhabitants of the Niger Delta region and the Nigerian government as a result of natural gas flaring. Documents such as conference papers, position/opinion papers, government findings, academic journals as well as internal communications of various agencies (like CBN, NNPC and OPEC) and commercially available publications such as newspapers and magazines were reviewed. Information and data accumulated from research was substantial and supported the intent and mission of this study, and states that type of method utilized for the collection of data clarifies between effects arising from independent variables and those effects that are induced by reactive measurements.

The populace of the Niger Delta region of Nigeria was the focus or 'target' of the research conducted due to its association with natural gas flaring from petroleum exploration and production. Of the 36 states that make up the nation of Nigeria, 9 states are situated in the region referred to as the Niger Delta region. Geodynamics of the Niger Delta region are consistently similar with extensive depositories of petroleum and natural gas throughout Nigeria, though the quality and amounts verified in the Niger Delta region may differ as a result of formation processes. A sample from two or more locations in the Niger Delta will yield the same result as any samples selected from the remaining states. This is due to the fact that the majority of samples is exposed to and suffers from similar social, economic and environment threats because of the uncontested, unregulated flaring, petroleum exploration and production.

3.1. Research Limitation

- The accessibility to gas flaring information relevant to petroleum organizations operating in the Niger Delta region is non-existent.

- Illiteracy amongst the populaces of the Niger Delta. This proved the completion of the questionnaires difficult, necessitating personal assistance to assist in their completion.
- The response rate of questionnaires distributed via email was very low. These may be due to lack of access to the Internet or an association with the unwillingness of the petroleum organizations to return their responses.
- The importance of the research had to be emphasized to the staff of the petroleum organizations and the populaces of the Niger Delta who expressed concerns about maintaining confidentiality
- The research was time and labor intensive.

3.2. Results and Discussion

Chapter 4 presents the analysis and results stemming from the questionnaire survey collected from three states; Eket of Akwa Ibom, Ogba Land of Rivers state and Imiringi of Bayelsa so as to determine the socio-economic and environmental impact of natural gas flaring on the people that reside in the Niger Delta region. A pie chart is also used to demonstrate the responses received from those returning completed questionnaires.

100 questionnaires were distributed to individuals, with 90 returned fully completed and 22.2% of the individual respondents were female and 77.8% were male. Furthermore, the result shows that the questionnaire grouped respondents' classes as:

Age
- 19-29 (47.8%)
- 30-39 (26.7%)
- 40-49 (21.1%)
- 50 + (4.4%)

Education
- WASSCE/GCE (36.7 %)
- OND (31.1%)
- BSc/HND (27.8%)
- PhD &Msc (4.4%)

Residency
- Niger Delta region (93.3%)
- Outside of Niger Delta (6.7%)

Responses

Table 2. States where the interview was conducted

Names Of States	Names Of villages/Local govt	Number of questionnaires distributed	Number of questionnaires collected
1.Akwa Ibom	Eket	30%	29%
2.Rivers	Ogba land	25%	25%
3.Bayelsa	Imiringi	25%	24%
4.Federal Ministry Of Environment		10%	7%
5.Department Of Petroleum Resource		5%	2%
6.NNPC		5%	3%

The above table presents the number of questionnaires distributed and those collected in percentage, 29% out of 30% of the respondent are people of AkwaI bom, 25% out of 25% are people of rivers, 24% out of 25 are people of

Bayelsa, 7% out of 10% are staff of the Federal ministry of environment, 2% out of 5% are staff of the Department of Petroleum Resource and 3% out of 5% are staff of the NNPC.

Does gas flaring affect the social life, economy and evnironment of the Niger Delta?

■ Yes ■ No

20%

80%

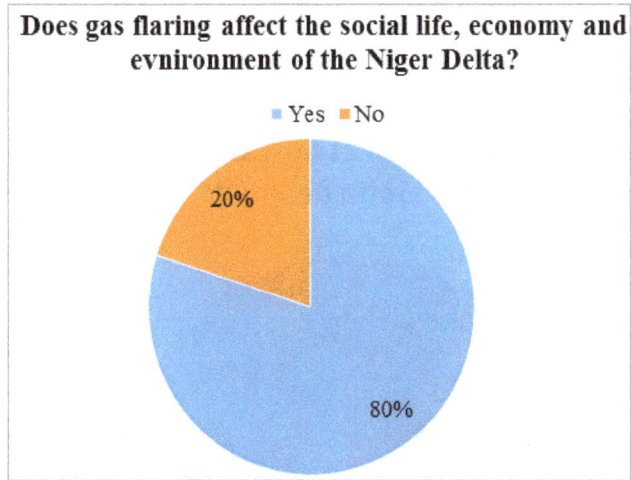

Figure 2. The socio-economic and environmental impacts of gas flaring in the Niger Delta.

80.0% of the questionnaire respondents believe that natural gas flaring did contribute to social and economic threats, while other 20.0% said it was not an issue. The Niger Delta region is agrarian in nature that sustains the local populace by means of fishing and agricultural farming. The harvests from bother industries are sold in local marketplaces. It is widely believed and documented that flaring of natural gas releases hazardous chemical and toxic compounds into the atmosphere, which by means of acid rains percolate into the soils and eventually is absorbed to the rooted fruits and vegetables. These toxic materials also poison waterways and estuaries killing untold amounts of marine life, which serves as a means of support for many residents as fishermen. Looking at the results of the respondents, it can be deduced that 80.0% agree wholeheartedly that under production of crops which denies them the opportunity to sell their goods, which stymied local economy's growth is directly correlated to natural gas flaring in the local area.

A review of literature during investigation revealed that discharges of nitrous and sulfur oxides along with atmospheric conditions are likely to induce acid rain. It causes agricultural solids to become acidic; this destroys seeds and rooted vegetables and fruits, It is just as it is devastating to marine life. It is believed that climate change can be attributed to exorbitant quantities of methane and carbon dioxide in the atmosphere. As natural gas flaring continues, the production of methane gas along with carbon dioxide accumulates in such massive amounts that it has a devastating impact and in some instances initiates changes in weather patterns that causes drought, coastal erosion, desertification and flooding. Flooding is just as powerful a force as drought is in the eradication of crops, which impacts the Niger Delta's economy. This confirms the findings of Ishione (2004) [14], which concluded that acid rains, climatic change and temperature elevations have impacts on the economy of the Niger Delta region.

As the natural gas flaring continues to disrupt the economy of the Niger Delta, social disequilibrium has had an impact on social matters of concern, such as; poverty, hunger and more seriously, violence. Community leaders have reached out for assistance from petroleum organizations for assistance to offset the hardships endured as a result of the natural gas flaring. The petroleum organizations are, in some instances slow to respond in a timely manner with relief. Temperaments in the community run rampant when delays are perceived as stalling and avoidance tactics by the petroleum organizations. Abductions, social unrest, rioting and criminal activity occur and more often than not petroleum companies collaborate with the government to forcefully eliminate the demands of the communities that possess the oil.

Seeking infrastructure improvements from the Nigerian government and petroleum organizations, the populace of the Niger Delta has come to the realization that constitutes a political bloc, with some clout in government regulation and legislation. Specific individuals are chosen to act as liaisons between the Nigerian government, the petroleum organizations and for community affairs. These individuals (now in a different social capacity) have increased earnings and in a different social standing that the community represents. This again presents issues of class differentiation and issues pitting the petroleum organizations against local communities. Sonibare(2006) [21] suggests that community interaction with political forces resulting from petroleum issues are corrupt and does not serve the community in a beneficial way.

Approximately 45,800 gigawatts of heat is released into the atmosphere on a daily basis as a result of gas flaring in Nigeria. According to the World Bank, it would satisfy, approximately 25.0% of the energy demands of the United Kingdom. If the Nigerian government would enforce strict regulations on gas flaring, the government will generate about $2.4 billion in revenue every year, and also be able to provide a cleaner and sustainable environment. Moreover, employment opportunities will be provided to the populaces of the region

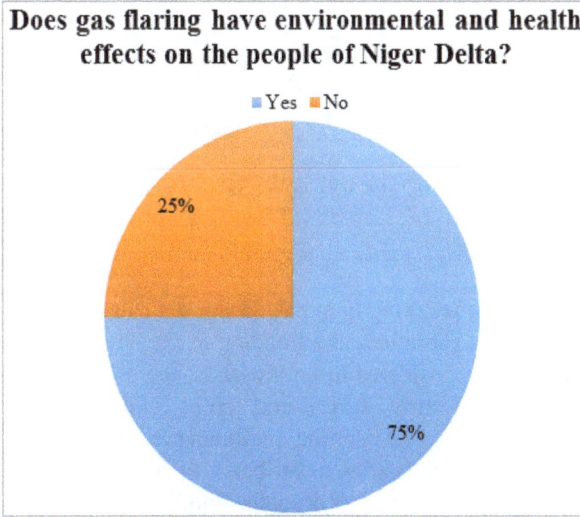

Does gas flaring have environmental and health effects on the people of Niger Delta?

■ Yes ■ No

25%

75%

Figure 3. Environmental and Health Effects of Gas Flaring.

75.0% of the respondents agree that consequence to an

individual's health and surrounding environment can occur as a result of Niger Delta natural gas flaring with the remaining 25.0% seeing no consequence at all. The toxic compounds released into the atmosphere have been shown to cause devastating damage to skin cells, onset of different types of cancers, blood disorders, bronchitis, types of anemia in the communities in proximity to the flaring discharges into the atmosphere. These discharges are comprised of toxic compounds and elements, such as Benzene and enter the food chain. Adverse health conditions substantially reduce one's life expectancy. This makes the life expectancy of the Niger Delta people to be no more than 40 years while in the rest of the country life expectancy is approximately 45 years. Hence a decrease in greenhouse gas emission benefits all human health (Ishione 2004)[14]. In comparison to costs associated with flaring of natural gas, the cost to capture associate gas is 4 times greater (ESMAP 2001)[9]. As a result, petroleum organizations flare the gas. The deposits discharged from the gas have drastic impacts on the environment (e.g. climate change). As noted in fig.3, 75% of the sampling respondents agree that gas flaring impacts the environment since it releases hazardous and toxic substances like methane, carbon monoxide and sulfur dioxide (Ezzati and Karmmen 2002)[10]. These substances hinder environmental conservation by causing change in climate, flooding and erosion of the shoreline (Ishione 2004)[14].

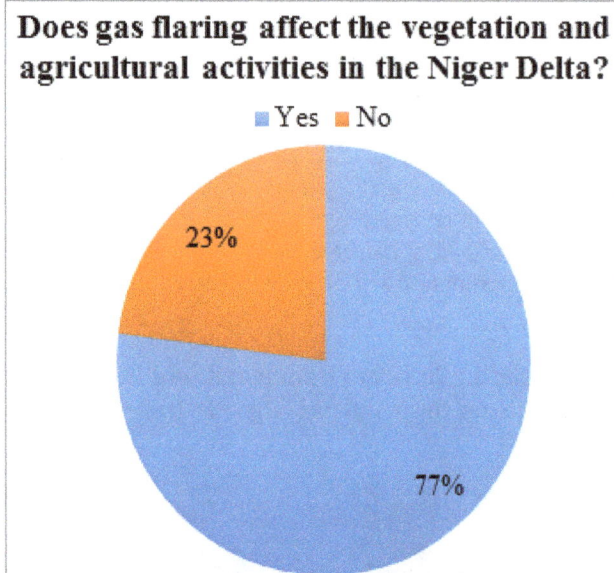

Figure 4. *The effect of gas flaring on agriculture and vegetation.*

Climate change resulting from gas flaring can cause severe erosion of coastal shorelines and flooding in the Niger Delta region which are considered to be lowlands; the devastating climatic events destroy agricultural crops and lead to outbreak of diseases. 77.0% of the questionnaire respondents agree that natural gas flaring contributes to destruction of agricultural crops by means of drought, flooding and drastic variations in temperatures.

Acidic soil as a result from acid rain serves to only inhibit the development and growth of agricultural crops and the

industry itself. Acidity of the soil depletes essential nutrients need, and can be attributed to the flaring of natural gas, as confirmed by investigative research conducted by Ishione (2004)[14].

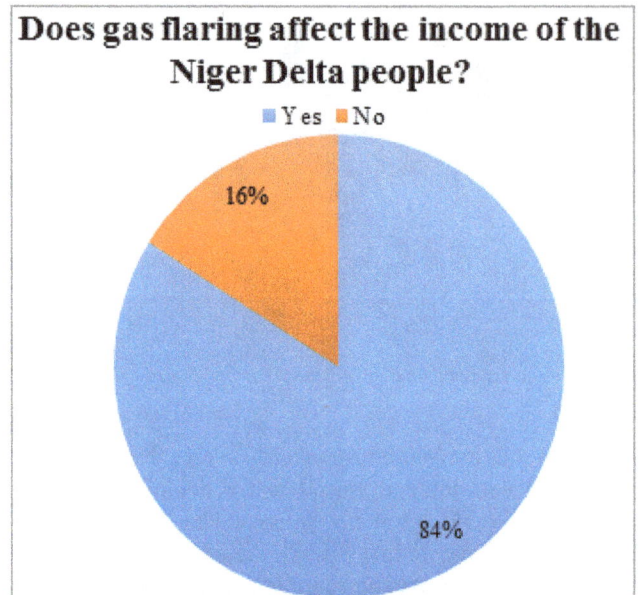

Figure 5. *The effect of gas flaring on income generation.*

Results from the questionnaire indicate that 84.0% of the respondents believe that flaring of natural gas has a direct impact on the income of the population of the Niger Delta region, while the remaining 16.0% indicate that gas flaring felt no impact at all. This amounts to approximately 1 out of 5 respondents stating that natural gas flaring has an impact on the industries they are employed in and has a direct correlation to their annual incomes. A majority of the population of the Niger Delta region are farmers or workers in the agricultural sector in addition to fisherman. Many of the respondents do not understand the scientific or technical reason why their crops and fishing hauls are being destroyed or depleted but they do know that prior to the installation of petroleum rigs and extensive petroleum exploration, that they harvested bumper agricultural crops and netted tons of fish that were sold in local and exported market places for substantial monetary gain.

Of the income generated by the local population, much of it is spent on medical treatments to treat medical conditions, ailments and chronic diseases resulting from the flaring of natural gas. In some instances, farmers pool some of their incomes to initiate erosion control, vegetation diseases, and flaring eradication to protect their agricultural harvests and incomes. These 'investments' by farmers and fishermen to protect the incomes severely decrease the available income to feed their families, which supports the conclusions of the World Bank (2011)[25] that the series of events that arise from gas flaring purposely reduce the income of the Niger Delta inhabitants.

Are the government and oil companies making efforts to stop gas flaring in the Niger Delta region?

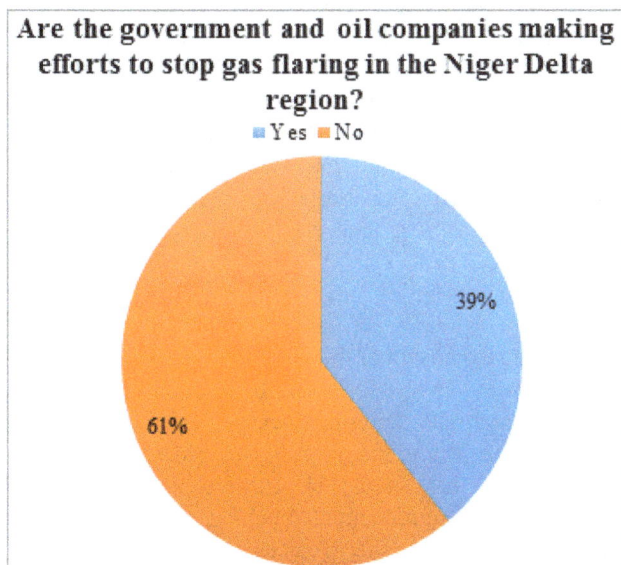

Figure 6. *Government and Oil company efforts to eradicate gas flaring.*

Should the rules guiding oil production in Nigeria be reviewed?

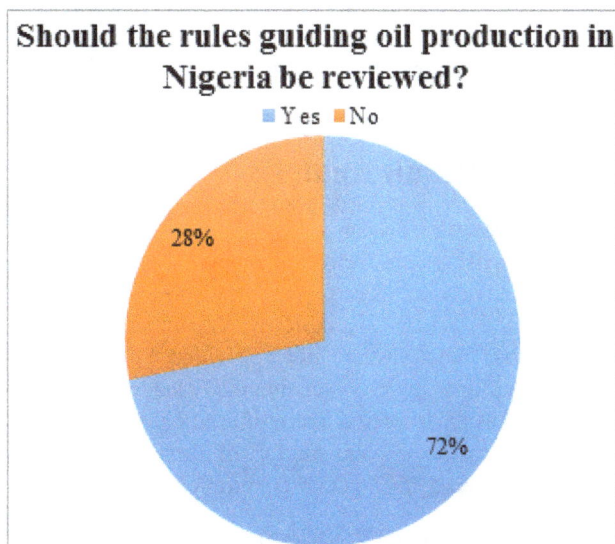

Figure 7. *The need for review of laws guiding oil production in Nigeria.*

Is corruption and other unethical practices encouraging continuous gas flaring in Nigeria?

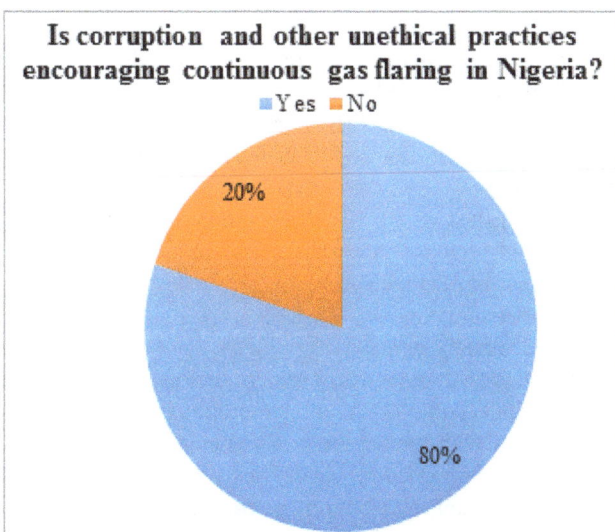

Figure 8. *The effect of corruption and unethical practices on gas flaring.*

39.0% of the respondents to the questionnaire believe that the Nigerian government along with the petroleum organizations have a credible plan established to eliminate natural gas flaring, while the remaining 61.0% believe that no such plan exists. Further investigation and analysis of the respondents' answers indicate that the majority of the 39.0% that believe that petroleum organization and the government have plans to eliminate natural gas flaring are in fact workers for the petroleum organizations and the government. This leads to the belief that the responses are biases, with the fact that the other 61.0% of respondents are residents of the Niger Delta with no affiliation or allegiance with the government or petroleum organizations. The establishment of the national Gas Master plan in 2008 was to focus on maximizing the conversion of waste gas into a beneficial product that can be monetarily capitalized on. Creation of measures, standards, regulations and legislation are fantastic, but enforcement and monitoring by the general public is often difficult if not nonexistent, due to the fact that the residents lack discretionary income to purchase radio, televisions or computers to stay in touch with activities that are of importance to them, such as natural gas flaring. Chevron has initiated the capture of associated gas while other petroleum organizations have started designing development of new platforms that will help them comply with the government's regulations to eliminate gas flaring in the country by the designated time.

A part of the questionnaire ascertains the opinion of respondents toward eliminating the flaring of natural gas in Nigeria. 72.0% of the respondents indicated that established measures, standards, regulations and legislation should be examined and consideration given to updating them and eliminating unnecessary one. 28.0% of the respondents believe that the already established measures, standards, regulations and legislation are sufficient. These 28.0% of respondents, it should be noted are employed by the government and petroleum organization.

80.0% of the respondents agree that corruption plays a vital role in the everlasting practice of gas flaring in Nigeria. The remaining 20.0% disagree with that claim. There are many hidden agendas behind the continuous gas flaring in the Niger delta, and this keeps the Niger Delta people in suspense.

The majority of the people that agree that corruption and ethical issues are some of the major driving forces of gas flaring in Nigeria are the civil servants and the populaces of the Niger Delta. But the remaining 20.0% that disagree with the argument are mostly the workers of the multinational oil companies. The survey was crafted to uncover the prime reason why flaring of natural gas in Nigeria continues in direct violation of measures, standards, regulations and legislation. Meanwhile the 20.0% of the respondents believe that the measures, standards, regulations and legislation are being enforced in an honest and legal manner. There is much political influence in the natural gas and petroleum industries in Nigeria and have the ability to influence decision makers in government to achieve whatever is necessary, such as continuously extending the government imposed deadline for the cessation and elimination of natural gas flaring in Nigeria (Ishione 2004)[14].

4. Interview Result Analysis

Members of the staff of senior management at The Federal Ministries of Environment, Petroleum and Department for Petroleum Resource were questioned in a follow up to the questionnaires they return uncompleted. The intent was to uncover answers that were not provided. 7 senior management staffers and 2 community leaders were interviewed and as an incentive, were offered paid positions in association with the implementation of the Gas Master plan that was adopted by the Federal government in 2008.

4.1. What are the Economic Prospects in Relation to the Development of a Proper Gas Plan in Nigeria

The analysis of the interviewees responses indicate that the government has proposed a Gas master plan, which is a policy used to improve the local demand of natural gas, and also develop better facilities for gas production. This policy (if implemented) will develop the weak sectors of the Nigerian economy.

The primary thrust of the Gas Master plan is to increase and enhance the demand for natural gas in Nigeria. Key points of the master plan include;

- Export of Natural Gas
- Increase Domestic demand
- Economic growth other than business sectors

The promising future of the elimination of natural gas flaring is viewed with the prospect of financial gain, which will be utilized in a myriad of sectors and purposes, such as power generation and distribution, petrochemical industry, fertilizer production throughout the country. Economic prosperity is anticipated with employment opportunities multiplying throughout the country.

With the prospering economy, Nigeria will benefit from an increase in the foreign earning exchange, which will enable industrial, and agricultural diversification to support the country instead of being continuously dependent on the sale of petroleum exclusively. Attempts will be made to legislate government transparency regarding the Gas master plan. Other critical information disclosed by interviewees include

- Increase in revenue allocated for treating diseases arising from gas flaring
- Cleaner and sustainable environment
- Improved agricultural sector

4.2. What are the Prospects for Developing Extensive Infrastructure for Natural Gas in Nigeria

Opinions of the interviewees reflect the prospects and hopes for the development of thriving natural gas industry in Nigeria as a result of the stringent implementation and enforcement of the National Gas Master plan. As result of the

implementation and enforcement, the interviewees foresee promising economic development, growth of the country and enrichment or betterment of the lives of the Nigerian citizens. Due to continuous exploration and production of petroleum, fields are becoming increasingly depleted, natural gas is an inexpensive energy source alternative. This allows for the development of new and advanced technologies and contemporary engineering concepts for this burgeoning industry in Nigeria.

Regulatory complexities and commercial legislative polices of the government are instrumental in the development and expansion of the natural gas industry and is in accordance with observation made by Aghalino (2009)[2], which is concerned with the critical analysis of the viability of projects, the integrity of engineering designs and concepts and other factors that should be considered before the commencement of the production of natural gas in Nigeria. Odumugbo(2011)[19] asserts that if adequate attention is paid to those factors, the Nigerian economy shall experience extensive economic opportunities.

- Without an established and self-sustaining power grid in Nigeria, many entrepreneurs are relegated to providing necessary power to operate their business. The associate cost of energy production is inherently passed on to the consumer. With a revamped petroleum process, natural gas is used for power generation instead of being flared into the atmosphere. This is in concert with positions taken by Aghalino (2009)[2], and Odumugbo (2011)[19] that nation can reach a level of consistent power for domestic and industrial uses via the conversion of flared gas. According to Ishione (2004)[14], this can only be attained by aggressive construction of power transmission and distribution and reception facilities.
- Large-scale production of natural gas must be preceded with an established infrastructure for the containment and storage of natural gas, countrywide. Demand for gas will increase due to its inexpensiveness and will be utilized as a new source of energy in regions that were strictly dependent on petroleum. New modes of delivery will facilitate the expansion of the use of natural gas and create new marketplaces for its use. Foreign and domestic investment opportunities will become available to maintain existing infrastructure and encourage the conceptual design for the eventual replacement of the current infrastructure with cost effective, technologically advanced infrastructure in projects such as the Nigerian Liquefied Natural Gas Plant in Bonny. These enhancements elevate Nigeria's standing as a contemporary in terms of energy production and allow for fiscal gain through elevated foreign exchange.
- The opinions of the interviewees are in agreement with the inferences of Odumugbo (2010)[19], which claim that natural gas re-injection restores the pressure of the reservoir and prevents the emission of hazardous and toxic materials into the environment. This provides a cleaner and more sustainable environment (Ayodele

1998)[6]. Introducing the re-injection process of natural gas to stimulate petroleum will create opportunities of increasing the wells economic life cycle. The opinions of the interviewees is in compliance with the inferences of Odumugbo (2010)[19], which claims that natural gas re-injection stimulates petroleum reservoirs and spars the environment to be susceptible to toxic compounds, a by-product gas flaring (Ayodele 1998)[6].

- Human health and wellbeing along with the environment are the innocent victims of natural gas flaring. Greenhouse gases being emitted into the atmosphere contribute to a continual deterioration and decline of the quality of air, soil and water resources while inflicting horrific diseases and ailments on individuals.

- The populations of the Niger Delta region will be afforded the opportunity to breathe clean air, drink fresh water and be able to harvest edible fruits, vegetables and fish abundantly with concern for marine life being tainted with toxic compound and chemicals, with the development of the gas industry in Nigeria. The quality of health will increase in generations to come and rejuvenate surrounding forests devastated by soil toxicity and acid rains (Opukri and Ibaba 2008)[20].

- The process of exporting the Natural Gas requires skilled technicians. As the development of the natural gas industries takes to a larger scale, a substantial amount of employment opportunities will be created in the petroleum and natural gas sectors. This employment boom will boost the economy, creating job in other sectors, which is in accordance with the investigation and research of Ayodele (1998)[6] that the proliferation of the production of natural gas in Nigeria will stimulate employment opportunities for Nigerian citizens.

5. Conclusions

With worldwide awareness of the ramifications of climate change and its annihilating wrath of devastation it can wreak upon the environment, it is enough for many to realize that preventive measures are in order. One of those preventive measures is the use of a clean natural resource, such as natural gas.

For nearly 55 years, the petroleum industry in Nigeria wields with as iron first, so to say how procedures with regard to petroleum exploration and production are handled. They believe that flaring of natural gas is an effective means of waste disposal with little regard for the impact it is having on the economic conditions and the personal health of individuals in the Niger Delta. Even though continuous attempts at eradicating natural gas flaring have failed in the past, they did not have political support and message conveyed in manner of reaping the benefits economically and health wise for individuals.

The conclusion of this project can unequivocally state that the economy, environment and populace of the Niger Delta have endured hardship resulting from the unabated flaring of natural gas. The waste by products of gas flaring is a contributing factor towards initiating climate change, which imperils the air, water and food supplies. Air becomes un-breathable leading to illness as well as waterways becoming toxic, thus impacting fisheries, agricultural farmlands and the economy.

This research also ascertains that the Nigerian government and the Niger-Delta populace can experience several environmental, economic and health benefits if gas flaring were to cease, and points out several measures that can be adopted in order to develop a feasible natural gas business in Nigeria.

Recommendation

1. Strict government legislations on how the gas production projects should be conducted.
2. Regulatory agencies must fulfill their responsibilities and duties of enforcing laws and regulations to cease gas flaring.
3. The government must provide incentives and fiscal policies conducive to large-scale production and demand for gas.
4. Establish research and development of natural gas, initiate workforce-training programs.
5. The escaping natural gas should be captured and re-injected in order to increase the pressure of the reservoir.
6. Good transportation network should be provided
7. Government should impose penalties against firms continuing to engage in gas flaring.
8. The government should allow the gas sectors to undergo deregulation to improve its productivity and technologies.

Appendix

Questionnaire Sample

The Impact of Gas Flaring in Nigeria

The purpose of this study is to investigate the effects of gas flaring in Nigeria. Participation is voluntary, and all the data/information obtained from this study will be anonymous and confidential. Thank you for your time.

Email: mustinmama@yahoo.com

Section 1: (Bio Data)
1. Gender
Male [ ] Female [ ]
2. Age Category
(a) 19-29 [ ] (b) 30-39 [ ] (c) 40-49 [ ] (d) 50 and Above [ ]
3. Residency
(a) Niger Delta [ ] (b) Other [ ]
(c) State..........., (d) Local Govt ....
4. Qualification
(a) PhD & MSc [ ] (b) BSc/HND [ ] (c) OND [ ] (d) WASSCE/GCE [ ]
5. Employer Name ....

Specify if: (a) Private [ ] (b) Governmental [ ]
6. Position ....

Section 2: Assessment

1. Does gas flaring affect the social life, economy and environment of the Niger Delta?
(a) Yes [ ] (b) No [ ]

2. Does gas flaring have environmental and health effects on the people of Niger Delta?
(a) Yes [ ] (b) No [ ]

3. Does gas flaring affect the vegetation and agricultural activities in the Niger Delta?
(a) Yes [ ] (b) No [ ]

4. Does gas flaring affect the income of the Niger Delta people?
(a) Yes [ ] (b) No [ ]

5. Are the government and oil companies making efforts to stop gas flaring in the Niger Delta region?
(a) Yes [ ] (b) No [ ]

6. Should the rules guiding oil production in Nigeria be reviewed?
(a) Yes [ ] (b) No [ ]

7. Are corruption and other unethical practices encouraging continuous gas flaring in Nigeria?
(a) Yes [ ] (b) No [ ]

References

[1] Aghalino S. (2000) Petroleum Exploitation and Environmental Degradation in Nigeria. In: HI Jimoh, IP Ifabiyi (Eds.): *Contemporary Issues in Environmental Studies.* Ilorin: Haytee Press

[2] Aghalino S (2009) Petroleum Exploitation and the Agitation for Compensation by Oil Mineral Producing Communities in Nigeria. *Geo-Studies Forum*, 1(1, 2): 11-20.

[3] Argo, J (2001) Unhealthy effects of upstream oil and gas flaring [Online], Available from: http://www.sierraclub.ca/national/oil-and-gas-exploration/soss-oil-and-gas-flaring.pdf[Accessed: 28.7.2012]

[4] Ashton, N., Arnott, S., and Douglas, O. (1999) The Human Ecosystems of the Niger Delta – An ERA handbook. Lagos: Environmental Rights Action.

[5] Awosika, F., Osuntogun, C., Oyewo, O. and Awobamise, A, (2001) Development and Protection of the Coastal and Marine Environment in Sub sahara Africa: Report of the Nigeria Integrated Problem Analysis

[6] Ayodele, A. (1998) Improving and Sustaining Power (electricity) 'Supply for Socio-economic Development in Nigeria'. *Bulhon Publication of CBN* 36, 38–58.

[7] Davies, P. (2001) *The New Challenge of Natural Gas.* Paper presented at "OPEC and the Global Energy Balance: Towards a Sustainable Future", Vienna, September 28, 10-11

[8] Dibia, B. (2011) *The Impact of Gas Flaring in Nigeria.* Unpublised MSc thesis. Coventry: Coventry University.

[9] ESMAP (2001) African gas Initiative: main report. [Online], Available from: http://www.worldbank.org/html/fpd/energy/AGI/240-01 %20Africa%20Gas%20Initiative%20Main%20Report.pdf[Accessed: 26.7.2012]

[10] Ezzati, M. and Kammen, D (2002) 'Household energy, indoor air pollution, and health developing countries: knowledge base for effective interventions'. *Annual Reviews Energy and Environment* 27: 233-270.

[11] Farina, M. F. (2010) Flare Gas Reduction: Recent Global Trends and Policy Considerations. Oak Park: General Energy Company.

[12] GGFR (2002) Report on Consultations with Stakeholders. World Bank Group in collaboration with the Government of Norway [online] Available from: http://www.worldbank.org/ogmc/files/global_gas_flaring_initiative.pdf[Accessed: 01.08.2012]

[13] Human Development Report (2011) [online] available from: http://hdr.undp.org/en/ [Accessed: 29.7.2012]

[14] Ishione, M (2004) GasFlaring in the Niger Delta: the Potential Benefits of its Reduction on the Local Economy and Environment [Online], Available from: http://nature.berkeley.edu/classes/es196/projects/2004final/Ishone.pdf[Accessed: 24.7.2012]

[15] Kamalu, O.J., and Wokocha, C.C (2010) 'Lan resource Inventory and Ecological Vulnerability; Assessment of Onne Area in Rivers State, Nigeria'. *Research Journal of Environmental and Earth Science* 3(5) 438-447

[16] Madueme, S (2010) "Economic Analysis of Wastages in the Nigerian Gas Industry". *International Journal of Engineering Science and Technology* 2(4), 618-624

[17] Madueme S (2010) Gas flaring activities of major oil companies in Nigeria: An economic investigation. *InternationalJournal of Engineering and Technology*, 2(4): 610-617.

[18] Natural gas (2011) History [online], Available from: http://www.naturalgas.org/overview/history.asp[Accessed: 25.7.2012]

[19] Odumugbo, C (2010) Natural Gas Utilisation in Nigeria: Challenges and Opportunities. *Journal of Natural Gas Science and Engineering* 2(6) 310-316.

[20] Opukri, O., and Ibaba, S. (2008) "Oil Induced Environmental Degradation and Internal Population Displacement in the Nigeria's Niger Delta".*Journal of Sustainable Development in Africa* 10 (1), 173-193

[21] Sonibare, A., and Akeredolu, A (2006) "Natural Gas Domestic Market Development for Total Elimination of Routine Flares in Nigeria's Upstream Petroleum Operations". *Energy Policy*34 (6), 743-753.

[22] Svalheim, S. (2005) Norwegian initiative for responsible, environmentally-friendly gas management, Published by the Norwegian Petroleum Directorate [online] Available from: http://www.npd.no/cgibin/MsmGo.exe?grab_id=24&EXTRA_ARG=&CFGNAME=MssFindEN%2Ecfg&host_id=42&page_id=4285184&query=gas+flaring&hiword=gas+flaring+[Accessed: 25.7.2012]

[23] Vatn, A. (2005) *Institutions and the Environment*. Cheltenham: Edward Elgar Publishing Limited

[24] World Bank (2010) [online], Available from: http://web.worldbank.org/WBSITE/EXTERNAL/TOPICS/EX TSDNET/0,,contentMDK:22679372~menuPK:64885113~pag ePK:7278667~piPK:64911824~theSitePK:5929282,00.html[A ccessed: 25.7.2012]

[25] World Bank 2011, October 2003 [online], Available from: http://go.worldbank.org/ARNU3Z3BR0 [Accessed: 25.7.2012]

17

Effect of Diatomaceous Earths on Mortality, Progeny and Weight Loss Caused by Three Primary Pests of Maize and Wheat in Kenya

Ngatia Christopher Mugo*, **Mbugua John Nderi**, **Mutambuki Kimondo**

Kenya Agricultural Research Institute (KARI), Nairobi, Kenya

Email address:

chrisngatia@gmail.com (Ngatia C. M.)

Abstract: The Kensil fine (KF) dust was evaluated under laboratory conditions for the control of three important storage insect pests of maize and wheat. Serial concentrations of KF, Dryacide (DA) and Wood ash (Ash) were admixed with 100g of maize or wheat in ventilated glass jars. Mortality of *S. zeamais*, *P. truncatus* and *R. dominica* was assessed at 7, 14, 28, 56 and 84 days interval after grain treatment. At 28 days, all the three dusts effectively controlled *S. zeamais* with 95% - 100% mortality while only DA was effective against *P. truncatus*. Both KF and Ash, with 84% and 92% mortality, did not reach the threshold required for *P. truncatus*. Mortality in *R. dominica* only peaked after 56 days but again only DA treatment was effective at 84 days. The delayed effect of the Diatomaceous earths (DE) and ash treatments appear to contribute to the higher damage inflicted; hence more weight loss than was expected. At 28 days mean sample weight loss by *S. zeamais* was 4.5% while *P. truncatus* and *R. dominica* caused 4.2% and 3.5% respectively. The emerged progeny after 14 days exposure to the three dusts was different for each pest with DA producing the least and KF the most. These results formed the criteria on which to base future trials under simulated farmer storage practice.

Keywords: Diatomaceous Earths, Delayed Effect, Grain Storage, Farmer Practice, Pest Control

1. Introduction

Storage insect pests have been linked with reduced income and food insecurity at farm household level (Stathers, 2002). Studies show the *Sitophilus zeamais* Motschulsky and the *Prostephanus truncatus* Horn are among the main storage insect pests of maize, (Golob et al., 1996; Brice et al., 1996, Marshland & Golob 1996; Donaldson et al., 1996) while the *Rhizopertha dominica* prefers smaller grains like sorghum and wheat (Navarro and Donahaye, 1976). In their effort to mitigate losses, farmers either sell their harvested grain early or use different means of grain protection including traditional methods (Golob, et al., 1983). The use of chemicals among farmers has been on the rise, though the impact of insect infestation in terms of grain damage and weight loss also appears to display an upward trend (de Lima 1979; Muhihu and Kibata, 1985; Mutambuki and Ngatia, 2006).

Coupled with inherent problems like the development of resistance, risks of exposure to toxic pesticides, environmental contamination and zero tolerance in the grain trade, the urge to replace chemical pesticides with effective, safe and environmentally friendly pest control products seems imminent. Kuronic, (1997) noted that diatomaceous earths (DEs), the fossilized skeletons of diatoms, have the greatest potential to replace pesticides. They have low mammalian toxicity (Golob 1997 and Kuronic 1998) and can control a wide range of stored products insect pests (Barbosa et al., 1994; Subramanyam et al., 1998; Mewis & Ulrichs 2001; Arthur, 2002; Arthur & Throne 2003; Athanassiou et al., 2005; Wakil et al., 2010). Their potency does not expire and can therefore protect grain during a storage season (Stathers, et al., 2004). Ebeling (1971) explains that DEs work by absorbing the epicuticular lipids which leads to excessive water loss and death of insects.

DEs are found in different parts of the world, and based on their physical properties and the diatom species, Kuronic (1997, 1998) found significant differences in their efficacy against insects. Athanassiou, et al., (2003) added the grain type (Fields & Kuronic 2002; Nikpay 2006; Athanassiou et al., 2006) and grain moisture, temperatures and relative humidity also influence their efficacy. The fact that DEs can combine with chemical pesticides, (Stathers, 2002; Ceruti and Lazzari, 2005), or bio-control organisms to enhance potency (Lord, 2001; Akbar et al., 2004) appear to attract local commercial interests aimed at broadening the areas of use for the DEs. The main hindrance is the stringent regulatory requirements enforced by the Pest Control Products Board (PCPB) which recommends local efficacy trials before any pest control product can be registered. Towards this requirement, the African Diatomite Industries Limited (ADIL), requested the Kenya Agricultural Research Institute (KARI) to evaluate Kensil F before seeking its registration for use in the storage sector. This paper describes the laboratory evaluation process for Kensil Fine (KF) dust which was compared with Dryacide (DA) from Australia and Wood ash (Ash) for their control of three important stored products insect pests.

2. Materials and Methods

2.1. Grain Conditioning and Treatment

A 90kg bag of freshly harvested maize and 45 kg of wheat were fumigated in metal drums using phosphine gas generating tablets for 7 days. The grain was then screened over a sack sieve (maize) and 1mm aperture test sieve (wheat) to remove dust and non-grain material. A 100g of maize were put in each of the 720 glass jars of 300cc capacity, half of which were closed with wire gauze and the remaining with watman filter papers. The jars closed with wire gauze were then grouped into 4 lots of 90 and again divided into 5 batches of 18 jars of which a set of 3 replicates was treated with 0, 0.1%, 0.2%, 0.3%, 0.4% and 0.5%w/w serial doses of Kensil F, Dryacide or Wood Ash. The same was repeated for the jars closed with filter paper. A similar quantity (100g) of wheat in another 360 jars closed with filter paper was treated in the same way.

2.2. Introduction of Test Insects

Sitophilus zeamais (Motschulsky) *Prostephanus truncatus* (Horn) and *Rhizopertha dominica* (F.), all ex laboratory cultures maintained on whole maize and wheat grain at 25±5°C and 70%±2% relative humidity (r.h) were used. Twenty unsexed but active adult *S. zeamais* were introduced into each of the 90 jars with maize treated with KF and repeated for jars treated with DA and Ash respectively. A similar number of *P. truncatus* adults were introduced into the maize jars closed with wire gauze (to check escape) and treated with the three dusts. Finally, the process was repeated for *R. dominica* in wheat jars. All the jars were randomly placed in the temperature control room (TCR) set at 25±5°C and 70±2% relative humidity. Mortality of the parent population was assessed after 7 days exposure for each of the post-treatment intervals of 7, 14, 28, 56 and 84 days. Any

increase especially after 28 days was classified as the F_1 progeny.

2.3. Progeny Monitoring

Progeny monitoring was done only in the jars exposed for 14 days. After accounting for the parental population, the jars were incubated in the same CTR for a further 21days before the contents were sieved and the status of the recovered adult insects noted. Sieving to remove emerged insects was repeated at 2-day interval until all the jars failed to produce any adults for three consecutive attempts when the total emergence per jar was noted.

2.4. Estimate of Weight Loss in Samples

Grain weight reduction occasioned by insect feeding was calculated from the differences between the original and the final weight in each jar and the results expressed as percentage using Harris and Lindblad (1978) derivative formula below:

$$\%wt\ loss = \frac{w_1 - w_2 \times 100}{w_1}$$

Where
W_1 = Original weight without inert dust;
w_2 = Final weight without inert dust.

2.5. Data Handling and Analysis

Data on mortality, F_1 emergence and weight loss was managed with the Excel and analyzed using the statgraphic softwares. ANOVA indicated the main factors that influenced insect mortality while Least Significant Difference (LSD) separated treatment means that significantly contributed to the difference at 95% level of confidence.

3. Results and Discussion

3.1. Results

3.1.1. Mortality of Test Insects Exposed to DE on Treated Maize or Wheat

The ANOVA showed that post-treatment period, pest species and applied dust treatments significantly (P=0.0000) influenced mortality of test insects (Table 1). Insect mortality was different from one interval to the next and for *S. zeamais* and *R. dominica* it increased with storage period. Mortality of *P. truncatus* fluctuated between intervals and Dryacide was the most effective among the three dusts.

Table 2 shows the effect of the three dusts on the mortality of the test insects. At 7 days, Kensil F controlled only 37% of the *S. zeamais,* which increased to 97% at 28 days. Dryacide and Ash were significantly (P =0.0003) better at 69%, rising to 100% and 96% respectively for same period. After 28 days, insect mortality dropped but increased again at 84 days.

Mortality in *P. truncatus* was much lower, at between 23% and 44% at 7 days across dust treatments and thereafter fluctuated up and down between intervals. Again, Kensil F performed poorly and was not significantly (P=0.05) different from Ash. Between 7 and 14 days, mortality in Dryacide treatment was significantly different from the control. At 28 days, the three dusts controlled 84%, 96% and 92% respectively showing great improvement. After 28 days, pest mortality declined but Dryacide maintained a clear lead from the other dusts.

When applied on wheat to control *R. dominica,* the three dusts recorded even lower mortality compared with that in *S. zeamais* and *P. truncatus* in maize. At 7 days, Kensil only achieved 11% compared with 17% and 25% for Ash and Dryacide respectively. At 28 days, mortality in Kensil treated maize had risen almost by four-folds to 43% while that in

Dryacide treatment had increased by more than three-folds to 92%. Dryacide then progressed to effective level (>95%) at 84 days while both Kensil F and Ash only attained 68% and 81% respectively at 56 days. The data portrays a close relationship between Ash and Kensil and that the three dusts were a better alternative to no control. Figure 1 shows the performance of individual dusts based on average mortality as compared with the control. Among the dusts the superiority of Dryacide as a grain protectant was confirmed. *S.* zeamais was the most susceptible followed by *R. dominica* and *P. truncatus*. All the dust treatments indicate the benefits to be gained if farmers could use them for protecting stored grain.

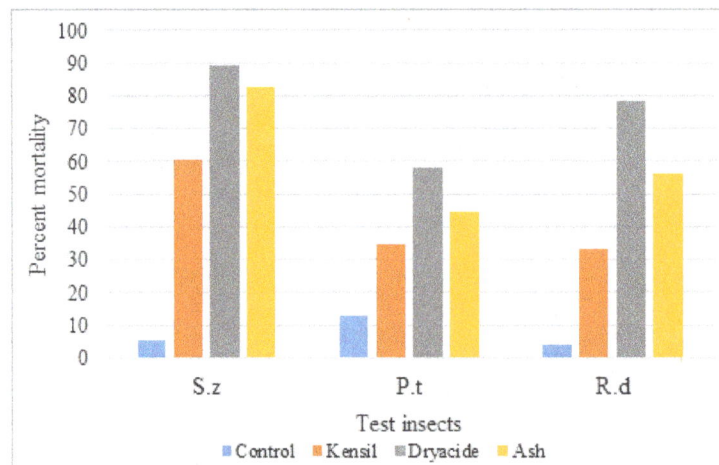

Fig. 1. *The influence of dust treatments on the mortality of three test insects as compared with the control. Key: S.z = Sitophilus zeamais; P.t = Prostephanus truncatus; R.d = Rhizopertha dominica.*

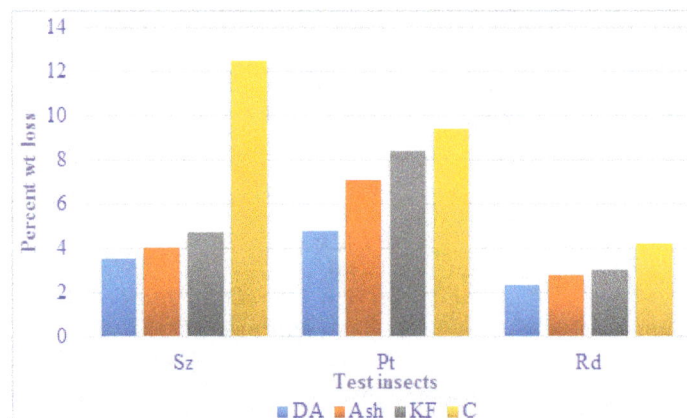

Fig. 2. *The influence of dust treatments on percent weight loss in grain samples exposed to three test insects. Key: S.z = Sitophilus zeamais; P.t = Prostephanus truncatus; R.d = Rhizopertha Dominica; DA = Dryacide, KF = Kensil fine, C = control.*

3.1.2. The Potential of DE to Protect Stored Grain from Damage by Insect Pests

The potential of Kensil Fine to protect stored grain was assessed from the level of grain damage and its subsequent weight loss. Table 3 shows the level of sample weight loss in both maize and wheat when compared with the control. At 14 day interval, dust treatments recorded between 2% and 3% in *S. zeamais* infested maize compared with almost 7% for the control. Dryacide had the lowest while there was no statistical difference between Ash and Kensil. At 28 days,

both Ash and Kensil recorded 4.6% and 5.1% weight loss, an indication that their effectiveness was weak, a situation which persisted to 84 days. With 2.1% - 4.4% weight loss, Dryacide was the only effective protectant against *S. zeamais*. The benefit of applying protection can be worked out from comparing treatment figures against the control at each interval. *P. truncatus* appear to be the most damaging, and despite the dust application, weight loss between 4.6% and 10.3% was recorded at 7 days across the treatments. However, weight loss dropped by a factor of between 0.4 and 0.8 (20%

and 59%) at 28 days, but gradually increased to 6.5% - 13.7% at 84 days. At every interval, Dryacide had the lowest figures and compared with Kensil, the differences were highly significant (P=0.0000), only comparable with the control. Wheat suffered between 2% and 6% weight loss from *R. dominica* infestation in spite of the treatments applied. At 14 days, all treatments recorded between 1.9% and 2.3% weight loss as compared with 3.1% for the control. Highest weight loss (of between 3.1% and 3.8%) was recorded at 28 days with Dryacide having the lowest and Kensil the opposite. From 28 days, sample weight loss progressively dropped and at 84 days only Kensil reached 3.3% level. However, all treatments were significantly (P <0.05) better than the control. Figure 2 illustrates the protective benefit accorded by the applied dusts on stored grain against the damage from test insects. It is clear that more benefit would be realized if the pest was *S. zeamais* and not others.

3.1.3. Emergence of F1 Progeny of Test Insects from DE Treated Grain

Varying numbers of F_1 progeny emerged from the eighteen jars that were used from each treatment (Table 4). More progeny emerged from maize than wheat and *S. zeamais* had consistently lower numbers (between 7 and 30), compared with 19 – 87 for *P. truncatus*. It was not clear why *R. dominica* produced negligible progeny of between 1 and 2 adults across the treatments including the control. Figure 3 shows the influence of the dust treatments on the emerged progeny. Although all treatments significantly (P=0.0000) suppressed progeny emergence, Dryacide was markedly better with below 20 adults. Kensil and Ash could not effectively suppress *P. truncatus,* the most destructive of the three test insects, indicating reduced protection of grain.

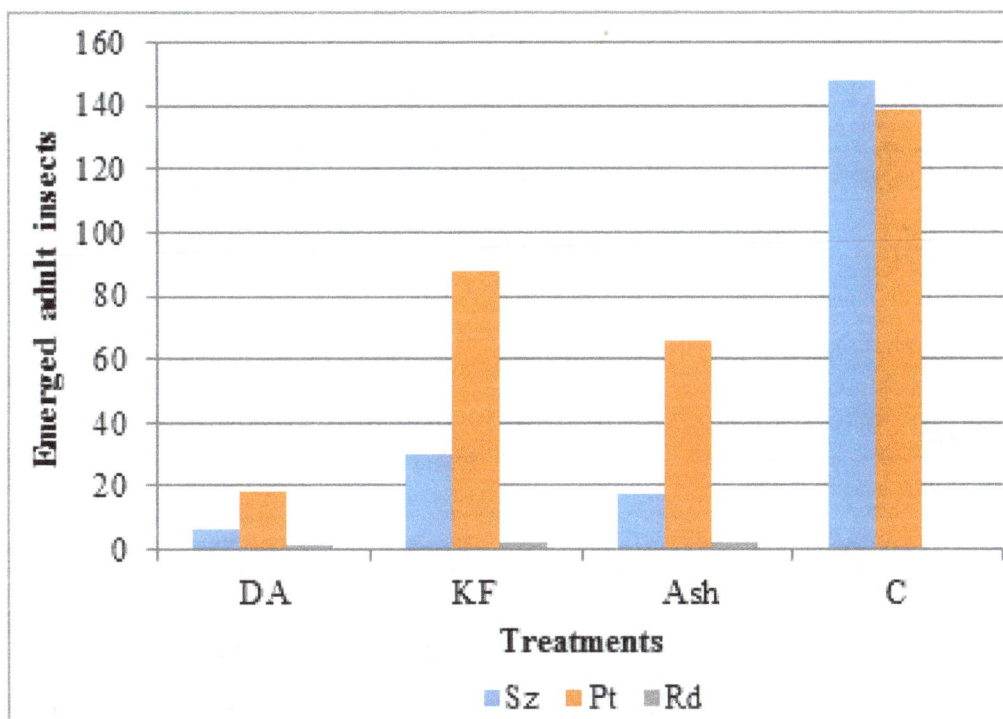

Fig. 3. *Progeny that emerged from Sitophilus zeamais (Sz), Prostephanus truncatus (Pt) and Rhizopertha dominica (Rd) after 14 day exposure on maize/wheat treated with Dryacide (DA), Kensil fine (KF) and Ash compared with untreated control.*

3.2. Discussion

The lab bioassay was set up to generate various data sets that could give a trend in efficacy of Kensil F, a locally mined diatomaceous earth. From the results, the mode of action clearly emerged where mortality was gradual, reaching the peak in 28 days for *S. zeamais* and *P. truncates* respectively. The peak for *R. dominica* was at 56 days. This contrasts with findings in some lab studies where DEs recorded 100% mortality between 7 and 14 days (Collins and Cook, 2006; Athanassiou et al., 2004). In this study the delay was probably due to the absorptive process of the epicuticle described by Ebeling (1971), a clear difference from the almost instant mortality normally observed in chemical trials. Although mortality trend was similar for all test insects, the

level of control showed different pest responses with *Sitophilus zeamais* being more susceptible than the grain borers. Kuronic (1997 and 1998) had stated pest species as a factor in DE efficacy, which could partly be due to specific pest behaviour (borers, in the open tend to be docile and less active while they actively bore into the grain almost immediately after introduction). Such behaviour tend to reduce the chances of abrasion from the dusts which would improve if the insect moved about among DE treated grain. With delayed mortality, test insects had ample time to bore into grain, feed and reproduce, a fact confirmed by a second grain damage peak at 12 weeks. The impact of insect damage directly relates to percent weight loss noted. Grain damage is therefore the main criterion used to judge the protection capacity of a candidate product. The Pest Control Products Board, a Kenya Government regulatory body considers >5% as the ceiling for acceptance to license and register chemical pesticides for use in the storage sector. Higher damage levels also translate to higher weight loss which has a bearing on market value and trade. In this trial, the extensive damage by *P. truncatus* was thus responsible for higher (>5%) weight loss especially in maize treated with either Ash or Kensil F. This again was a reflection of the pest behaviour, and between the two borers, *P. truncatus* was the more voracious feeder and quickly bored into the grain thereby avoiding too much contact with grain protectant. These assumptions will be considered during the simulation trial. The data also allows the calculation of differential loss farmers would suffer by applying a less effective dust. Using Dryacide as the best option during this trial, a farmer would suffer 1.21% for applying Kensil F on maize infested with *S. zeamais*, or 3.6% if *P. truncatus* was the pest. For wheat, the difference was much smaller (0.35%) but the proportional loss could be higher considering the grain size compared to that of maize.

4. Conclusion

The rising interest in DEs has been responsible for many studies some of which have led to commercial exploits. This pilot trial will contribute greatly to the understanding of how the DEs work before formal registration for use in the storage sector in Kenya. It will also help to expand the areas of use for Kensil, while farmers will have a novel alternative to chemical pesticides. Though DEs have a definite role to play in the management of storage insect pests, the study has shown that they are not instant in their efficacy as would be for chemical pesticides. They require actively mobile insects among the treated grain to abrade the waxy layer which prevents excessive water loss from their bodies. These were the behavioral differences between *S. zeamais* and *P. truncatus*. The data generated promising trends and it was left to the proprietor to consider refining the physical properties (size) of Kensil before the research body could plan advanced trials.

Table 1. ANOVA for mortality of test insects exposed to Kensil F, Dryacide and Ash

Source	Sum of Squares	df	Mean Square	F- Ratio	P-Value
MAIN EFFECTS					
A: Weeks	171997	4	42999.2	82.79	0.0000
B: Pests	113318	2	56658.9	109.09	0.0000
C: Treatment	233782	3	77927.0	150.04	0.0000
D: Reps	640.631	2	320.316	0.62	0.5400
RESIDUAL	367721	708	519.38		
TOTAL (CORRECTED)	88745	719			

All F – ratios are based on the residual mean square error.

Table 2. Percent mortality of test insects when exposed to three inert dusts applied on stored maize or wheat

Maize weevil (Sitophilus zeamais)						
Treatments Post treatment period (weeks)						
1	2	4	8	12	Mean	
C	8.3a	11.7a	1.7a	2.3a	3.7a	5.54
KF	37.3a	52.0b	97.0b	38.3b	77.6b	60.44
DA	68.7b	94.7c	100b	83.1d	98.6c	89.02
Ash	68.9b	82.7c	95.7b	72.1c	92.9c	82.46
P value	0.0003	0.0000	0.0000	0.0000 0.0000		
Larger grain borer (Prostephanus truncatus)						
C	3.3a	20.0ab	26.7a	5.8a	9.2a	13.0
KF	22.7ab	17.3a	84.0b	36.8ab	12.7a	34.7
DA	43.7c	35.0b	95.7c	63.1c	53.9b	58.28
Ash	30.7b	21.0a	91.7c	50.7bc	27.6a	44.34
P value	0.0004	0.0174	0.0000	0.0024	0.0000	
Lesser grain borer (Rhizopertha dominica)						
C	0.0a	3.3a	1.7a	12.4a	3.4a	4.16
KF	11.2a	17.7a	42.6a	68.4a	27.0a	33.38
DA	25.4b	82.4c	91.6c	94.1b	97.6c	78.22
Ash	17.1ab	57.2b	64.1b	61.4a	80.7b	56.1
P value	0.0169	0.0000	0.0000	0.0000	0.0000	

NB: Each datum is a mean of 15 dose reps.
Column numbers followed by same letter were not statistically different at 95% level (DMRT).

Table 3. *Percent weight loss caused by three storage insect pests on maize/wheat treated with three inert dusts*

Sitophilus zeamais					
Post treatment storage period (weeks)					
Treatments	2	4	8	12	Mean
DA	2.12a	3.89a	3.66a	4.38a	3.51
Ash	2.65b	4.55b	4.22a	4.75a	4.04
KF	3.07b	5.10c	4.96c	5.76b	4.72
C	6.93c	12.33c	9.20c	21.27c	12.43
P value	0.0000	0.0000	0.0000	0.0000	
Prostephanus truncatus					
DA	4.55a	3.69a	4.35a	6.50a	4.77
Ash	8.64b	4.57b	5.07b	9.94b	7.06
KF	10.31c	4.28ab	5.32b	13.72c	8.41
C	12.23c	7.17c	5.67b	12.40bc 9.37	
P value	0.0000	0.0001	0.0069	0.0000	
Rhizopertha dominica					
DA	1.85a	3.07a	2.47a	2.04a	2.36
Ash	2.23ab	3.53b	2.73bc	2.58ab	2.76
KF	2.30b	3.79b	2.69b	3.27b	3.01
C	3.13b	4.87c	3.00c	5.87c	4.22
P value	0.0222	0.0000	0.0017	0.0000	

Column means followed by same letter were not statistically different at P=0.05
Each datum is a mean of 15 dose levels

Table 4. *Effect of inert dusts applied on maize or wheat on the emergence of F_1 progeny of three storage insect pests after 14 day exposure*

Pest	Applied treatments				
	DA	Ash	KF	Control	P value
S. zeamais	6.87a	17.07b	29.73c	148.0d	0.0000
P. truncatus	18.6a	65.93b	87.27b	138.0c	0.0000
R. dominica	1.13a	2.0a	2.13a	0.7a	ns

Each datum is a mean of 15 observations
Row numbers followed by same letter were not statistically different at P=0.05.

Acknowledgement

This study could not have progressed to fruitful conclusion without the intervention or participation of certain persons and institutions. The authors would like to sincerely acknowledge the drive and support by the African Diatomite Industries Limited (ADIL) who, initiated the Kensil evaluation process to widen its use from beverage and paint industries to include agriculture. Special mention goes to Mr. Philip Chesang for actively monitoring the progress of the work at all project sites and ensuring continuous project funding which is greatly acknowledged. Many thanks go to the KARI Management through our Centre Director for creating conducive environment under which the work was carried out. The authors greatly appreciate those others who cannot be named for playing different roles.

References

[1] Akbar, W., Lord, J. C., Nichols, J. R. and Howard, R. W., 2004: Diatomaceous earth increases the efficacy of *Beauveria bassiana* against *Tribolium castaneum* larvae and increases conidia attachment. Journal of Economic Entomology. 97 (2): 273 – 280 (2004).

[2] Arthur, F., 2002: Survival of *Sitophilus oryzae* (L.) on wheat treated with diatomaceous earth: impact of biological and environmental parameters on product efficacy. Journal of Stored Product Research. 38 (2002) 305 – 313.

[3] Arthur, F. H. and Throne, J. E., 2003: Efficacy of Diatomaceous Earth to control internal infestation of rice and maize weevils (Coleoptera: Curculionidae). Journal Economic Entomology, 96 (2): 510 – 518.

[4] Athanassiou, C. G., Kavallieratos, N. G., Vaiyas, B. J., Dimizas, C. B., Buchelos, C. Th. and Tsaganou, F. C., 2003: Effect of grain type on the insecticidal efficacy of SilicoSec against *Sitophilus oryzae* (L.) (Coleoptera: Curculionidae). Crop Protection. 22: 1141 – 1147.

[5] Athanassiou, C. G., Vaiyas, B. J., Dimizas, C. B., Kavallieratos, N. G., Papagregoriou, A. S. and Buchelos, C. Th., 2005: Insecticidal efficacy of diatomaceous earth against *Sitophilus oryzae* (L.) (Coleoptera: Curculionidae) and *Tribolium confusum* du Val (Coleoptera: Tenebrionidae) on stored wheat: influence of dose rate, temperature and exposure interval. Journal of Stored Products Research 41 (2005) 47 – 55.

[6] Athanassiou, C. G., Vaiyas, B. J., Dimizas, C. B., Kavallieratos, N. G. and Tomanovic, Z., 2006: Factors affecting the insecticidal efficacy of the diatomaceous earth formulation SilocoSec® against adults of the rice weevil, *Sitophilus oryzae* (L.) (Coleoptera: Curculionidae). Applied Entomology. Zool. 41 (2): 201 – 207 (2006).

[7]   Barbosa, A., Golob, P. and Jenkins, N., 1994: Silica aerogels as alternative protectants of maize against *Prostephanus truncatus* (Horn) (Coleoptera: Bostrichidae) infestations. In: Highley, E. J., Banks, H. G., Champ, B. R. (Eds) Stored Product Protection, Vol. 2. CAB International, Wallingford, UK.

[8]   Brice, J., Moss, C., Marshland, N., Stevenson, S., Fuseini, H., Bediako, J., Gbetroe, H., Yeboah, R. and Ayuba, I., 1996: Post-harvest constraints and opportunities in cereal and Legume production systems in northern Ghana. NRI Research Report, 85pp.

[9]   Collins, D. A. and Cook, D. A., 2006: Laboratory studies evaluating the efficacy of diatomaceous earths, on treated surfaces, against stored-product insect and mite pests. Journal of stored Products Research 42 (2006), 51 – 60.

[10] Ceruti, F. C. and Lazzari, S. M. N., 2005: Combination of diatomaceous earth and powder permethrin for insect control in stored corn. Revista Brasileira de Entomologia 49(4): 580 – 583, dezembro 2005.

[11] de Lima, C. P. F., 1979: The assessment of losses due to insects and rodents in maize stored for subsistence in Kenya. Tropical Stored Products Information, 1979. 38, 21 – 26.

[12] Donaldson, T., Marange, T., Mutikani, V., Mvimi, B., Nenguwi, N., Scarbrorough, V. and Turner, A. D., 1996: Household food security study: rapid rural appraisal of villages in three communal lands of Zimbabwe. NRI Research Report R2315(s). iv +50pp +annexes.

[13] Ebeling, W., 1971: Sorptive dusts for pest control. Annual Review of Entomology 16, 123 – 158.

[14] Fields, P. G. and Kuronic, Z. 2002: Post-harvest insect control with inert dusts. In: Pimentel, D. (Ed.), Dekker Encyclopedia of Pest Management. Marcel Dekker, New York, pp 650 – 653.

[15] Golob, P., Dunstan, U. R., Evans, N., Meik, J., Reed, D. and Magazini,I., 1983: Preliminary field trials to control *Prostephanus truncatus* (Horn) in Tanzania. Tropical Stored Products Information. 45, pp 15 – 18.

[16] Golob, P., Stringfellow, R. and Asante, E. O., 1996: A review of the storage and marketing systems of major food grains in northern Ghana. NRI Research Report, 64 pp.

[17] Golob, P., 1997: Current status and future perspectives for inert dusts for the control of stored product insects. Journal Stored Products Research. Vol. 33, No.1, pp 69 -79, 1997.

[18] Kuronic, Z., 1997: Rapid assessment of insecticidal value of diatomaceous earths without conducting bioassays. Journal of Stored Products Research. Vol. 33, No. 3, pp 219 – 229, 1997.

[19] Kuronic, Z., 1998: Review: diatomaceous earths, a group of natural insecticides. Journal of Stored Products Res. Vol. 34, No. 2/3, pp 87 – 97, 1998.

[20] Lord, J. C., 2001: Desiccant dusts synergize the effect of *Beauveria bassiana* (Hyphomycetes: Moniliales) on stored grain beetles. Journal of Economic Entomology. 94 (2): 367 – 372 (2001).

[21] Marshland, N. and Golob, P., 1996: Rapid rural appraisal of post maturity issues in the central region of Malawi. NRI Research Report R 2358(S) 36pp.

[22] Mewis, I. and Ulrichs, Ch., 2001: Action of amorphous diatomaceous earth against different stages of the stored product pests *Tribolium confusum, Tenebrio molitor, Sitophilus granaries* and *Plodia interpunctella.* Journal of Stored Products Research. 37 (2001) 153 – 164.

[23] Muhihu, S. K. and Kibata, G. N., 1985: Developing a control programme to combat an outbreak of *Prostephanus truncatus* (Horn) (Coleoptera: Bostrichidae) in Kenya. Tropical Science, 25: 239 – 248.

[24] Mutambuki, K. and Ngatia, C. M., 2006: Loss assessment of on-farm stored maize in semi-arid areas of Kitui District, Kenya. In: Lorini, I., Bacultchuk, B., Beckel, H. and Deckles, D. Proceedings of the 9th International Working Conference on Stored Product Protection, Campinas, Sao Paulo, Brazil, 15 – 23.

[25] Navarro, S. and Donahaye, E., 1976: Conservation of wheat grain in Butyl Rubber/EPDM containers during three storage seasons. Tropical Stored Products Information. 32, pp 13 – 23.

[26] Nikpay, A., 2006: Diatomaceous earths as alternatives to chemical insecticides in stored grain. Insect Science. (2006) 13, 421 – 429.

[27] Stathers, T. E., 2002: Combinations to enhance the efficacy of diatomaceous earths against the larger grain borer, *Prostephanus truncatus* (Horn). In Credland, P. F., Armitage, D.M., Bell, C. H., Cogan, P. M. and Highley, E. (2003) (Eds). Advances in Stored Products Protection. York. UK.

[28] Stathers, T. E., Denniff, M. and Golob, P, 2004: The efficacy and persistence of diatomaceous earths admixed with commodity against four tropical stored product beetle pests. Journal of Stored Products Research, 40 (2004) 113 – 123.

[29] Subramanyam, Bh. Mudamanchi, N. and Norwood, S., 1998: Effectiveness of Insecto applied to shelled maize against stored product insect larvae. Journal of Economic Entomology. 91, 280 – 286.

[30] Wakil, W., Muhammad, A., Ghazanfar, M. U. and Tahira, R., 2010: Susceptibility of stored product insects to enhanced diatomaceous earth. Journal of Stored Product Research. 46 (2010) 248 – 249.

18

Geophysical Investigation of Loss of Circulation in Borehole Drilling: A Case Study of Auchi Polytechnic, Auchi, Edo State, Nigeria

Yusuf Inusa[1], Ogundele Olusegun John[1], Odejobi Yemi[2], Auwal Ishaq Haruna[3]

[1]Department of Minerals & Petroleum Resources Engineering Technology, Auchi Polytechnic, Auchi, Edo State, Nigeria
[2]Leading Edge Geoservices Ltd, Suite 55, EDPA Shopping Complex, Ugbowo, Benin City, Nigeria
[3]National Engineering and Technical Company Ltd, Corporate H/Q, Plot 1460 Ligali Ayorinde Street, Victoria Island, Lagos

Email address:
macsalem14@yahoo.com (Y. Inusa)

Abstract: Geophysical investigation of loss of circulation in Borehole drilling at the Auchi Polytechnic Auchi, Etsako West L.G.A, Edo State, Nigeria was carried out. Electrical resistivity sounding techniques with Schlumberger array for Vertical Electrical Sounding (VES 1) was used while modified Schlumberger array was used for Vertical Electrical Sounding (VES 2) with Australia Energy Market Commission (AEMC) Soil Resistivity meter. The Digital Ground Resistance and Soil Resistivity Tester model 6470-B was used. The Tester was produced by Australia Energy Market Commission. Geographical coordinate were obtained from a Global Positioning System (GPS) device. The quantitative interpretation of the resistivity sounding curves was done to obtain the layer thicknesses and resistivities with the computer assisted iteration techniques. Five and four geoelectric layers were resolved for Vertical Electrical Sounding (VES 1) and Vertical Electrical Sounding (VES 2) respectively. The loss of circulation within this borehole occurs around the depth of 36m. The loss zones are permeable and aerated with very high resistivity values. The void spaces within the soil layer in the horizon are not conducive. Pre-drilling geophysical investigation should not be optional but a necessity for every borehole drilling project in order to know the subsurface geology.

Keywords: Geophysical, Investigation, Circulation, Borehole, Drilling, Loss

1. Introduction

A borehole drilling project that is expected to meet a desired need, for which it is being embarked upon, should be carefully thought out (planned) in advance. Inadequate or outright negligence of planning has led to the outrageous rates of failure, recorded in the past drilling works. These trends have necessitated the absolute need for pre-drilling investigations (Offodile, 1983). Lack of planning in most borehole drilling project also covers the non – inclusion of experts in water resources management at the level of project conception. Furthermore, most drilling sites are managed by artisans instead of professional hydrogeologist. No wonder quality of groundwater resources development projects in Nigeria remains below average.

Another area where non – adherence to professional ethics is hurting the society is the lack of regard for the environment. We can safely assume that an expert that has gained some level of proficiency will be able to manage a borehole drilling project in such a way that the safety of the environment is not undermined. But in the eyes of an artisan, the most important thing is to get water during drilling regardless of the harm done to the environment in the process. It is therefore needful to conduct pre – drilling geophysical investigation before embarking on borehole drilling, and also the entire drilling process should be supervised and documented by a competent hydrogeologist to attain the much desired success in any borehole drilling project.

Geophysical methods have been very useful in determining the geological sequence and structure of the subsurface rocks by the measurement of their physical properties. Although there are varieties of geophysical techniques, which could be used in groundwater exploration, electrical resistivity method has proved reliable in delineating zones of relatively low resistivity signatory of saturated strata in various geologic terrains. (Odejobi, 1999). Electrical resistivity method has been successfully used across all the geological terrains in Nigeria and has drastically stemmed the tide of failure in borehole construction projects (Ako and Olorunfemi, 1989; Olayinka, 1999; Olayinka and Olorunfemi, 1992).

The advantage that pre – drilling geophysics offer is that one is able to anticipate the subsurface condition and prepare adequately to surmount any challenge that may arise during borehole drilling. This saves time and cost as when compared to drilling blindly when one in note sure of the formation characteristics.

The primary objectives of this investigation include:

1 Conduction of Resistivity sounding at the site and interpretation of the field Vertical Electrical Sounding (VES) data to obtain geoelectric parameters.
2 Determination of the hydrogeologic characteristics of the subsurface at the site based on geoelectric and available geologic information.
3 Determination of the geological configuration and extent of the horizon(s) where loss of circulation took place
4 Make an appropriate recommendations for the planning and execution of subsequent borehole drilling projects at the site

2. Site Location and Description

The geophysical exploration was carried out at the site of two failed borehole drilling around the office block of the Department of Mineral Resources Engineering and Electrical Engineering laboratory of the Federal Polytechnic Auchi. Vertical Electrical Sounding (VES 1) is approximately

defined by the geographical coordinates of Latitude N 7^0 03' 50.8"and Longitude E 6^0 16' 26.5" with an average elevation of about 239m above the mean sea level.

Auchi town is within the Anambra hydrogeological Basin. The basin comprises an, almost, triangular shaped embayment covering an area of about 30,000km^2. It stretches from the area just south of the confluence of the River Niger and Benue across to areas around Auchi, Okene, Agbo and Asaba, west of the river, and Anyangba, Idah, Nsukka, Onitsha and Awka area, east of the river. From an elevation of about 300m, the Udu – Idah escarpment slopes gently towards the southwest into the flood plain of the River Niger and across, to the west. The basin is drained mainly by the Anambra river and its main tributaries, the Mamu and Adada (Offodile, 2002).

Field observations and desktop study show that Auchi is underlain by geologic materials belonging to the Ajali formation. The formation bears false bedded sandstone with associated clay shale intervals in the bottom section (Reyment, 1964).

The Ajali formation is successively underlain by materials belonging to the Nsukka and Mamu formations. The Nsukka formation hitherto called the upper coal measure, bears sandstone, shale and coal while the underlying Mamu formation which has compositional similarity with the Nsukka, offers a higher frequency of coal occurrence.

Hydrogeologically, the Ajali formation constitutes a heterogeneous lithological sequence, with a generally deep water table conditions ranging from 30m to over 170m in places (Offodile, 2002). The sandy aquifers within the formation are friable, poorly sorted and typically whitish at depth. Also, the aquifers are often overlain by highly resistive (dry) sandy/sandstones/ clayey layer and deep down, the groundwater is ferruginised.

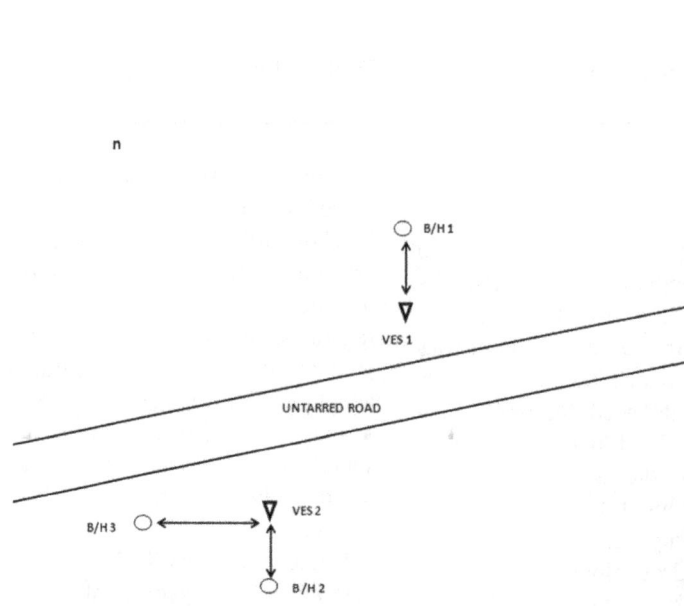

PLS note that the diagramm is not to scale

Figure 1. A schematic diagram showing the site and the ves points

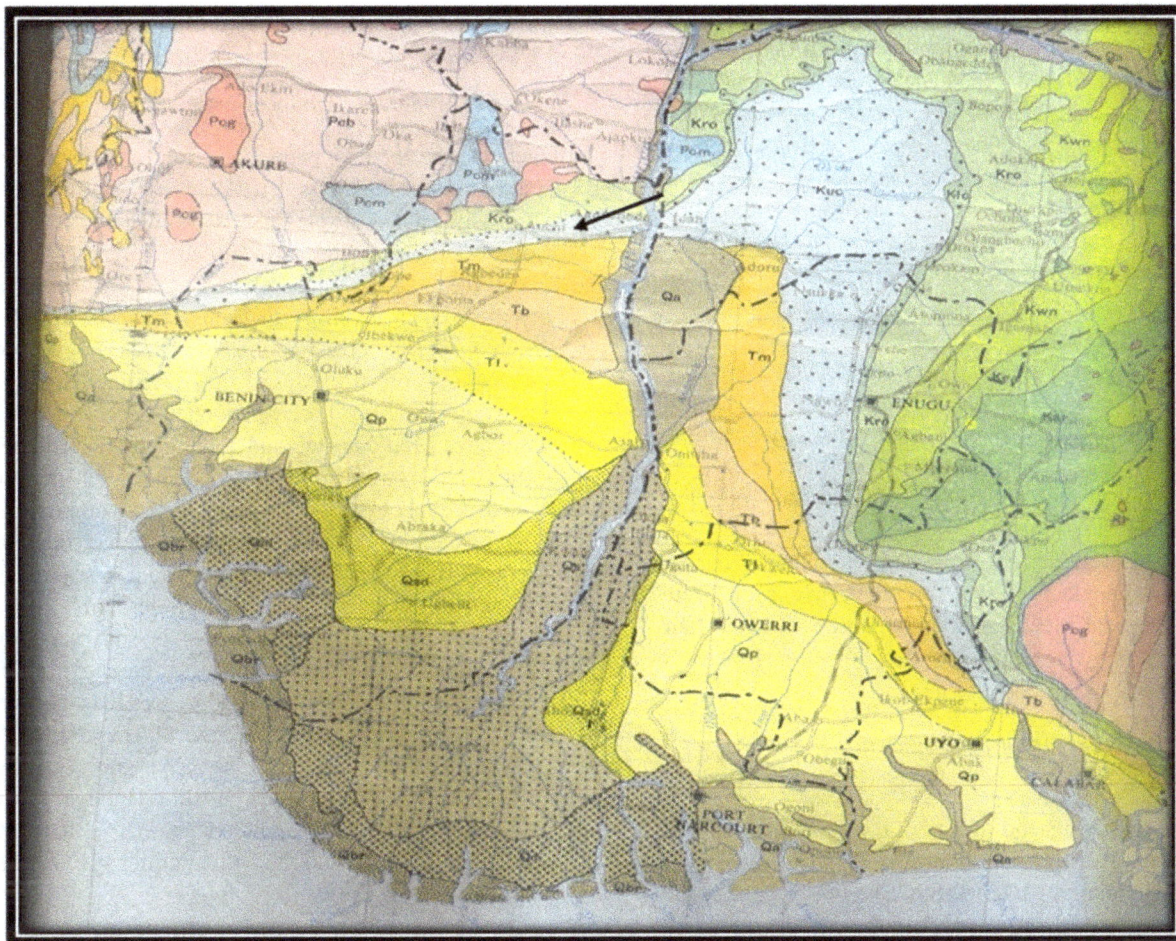

Figure 2. Geological Map of Nigeria Showing The Study Area

Source:NGSA Geological Map of Nigeria 1984 edition Legend

Kuc → Ajali Formation

3. Lost Circulation

For drilling to commence and end seamlessly, the drilling fluid must be in the right proportion and viscous enough to ensure that uninterrupted circulatory process is maintained. Drilling fluid acts as lubricant, cools the bit, removes cutting from the bottom of the hole and takes them to the surface where they can be stored for inspection. The drilling mud is also expected to prevent loss of circulation.

Loss of circulation occurs when the drilling fluid penetrates the permeable and porous formations in the hole. This automatically brings the borehole drilling to a halt. Drillers run into this problem when proper pre – drilling geophysical/geological investigations are not done. This challenge can be anticipated, be fully prepared for and successfully handled with a clear understanding of the specific geology of the proposed drilling site. The loss circulation belt has been found to occur in lignites peaty and carbonaceous shales and behaves sympathetically with it (NFOR, 2006).

In the study area, boreholes 1 and 2 were abandoned due to lost circulation encountered at about 120ft (36m) and 130ft (39m) respectively. However, the third borehole, that is less than 10m away from borehole 2 was successful completed. It should be noted that while permeable and porous formation in the subsurface can cause lost circulation, the content of the permeable rock may not necessarily be liquid. It may be air in which case, the resistivity of that particular substratum becomes very high. This study therefore seeks to unveil the stratigraphical configuration of the loss zones based on geophysical data.

4. Material and Method of Study

Field Procedure

The geophysical exploration carried out at the site was done using electrical resistivity sounding techniques (VES) with Schlumbeger array for Vertical Electrical Sounding (VES 1), while modified Schlumbeger array was used for Vertical Electrical Sounding (VES 2) due to space constraints. This was achieved with the aid of Australia Energy Market Commission (AEMC) soil resistivity meter and other field accessories. Geographical coordinates were obtained from a Global Positioning System (GPS) device.

Two vertical electrical sounding (VES) were done at the site using Schlumberger array. The total spread length (AB/2) attained for Vertical Electrical Sounding (VES 1) and Vertical Electrical Sounding (VES 2) within the limit of the available space and hydrogeological expediency were 200m and 80m respectively. Artificially generated electrical signal is known to penetrate up to AB/2 = 1000m (Kearey and Brooks, 1988). However, the 1000m depth of penetration for borehole drilling represents an outer limit in groundwater exploration and exploitation because it appears to be beyond the boundary of economic exploitation zone in terms of groundwater abstraction.

5. Results and Discussion

The quantitative interpretations of the resistivity sounding curves were done to obtain the geoelectric parameters (i.e. layer thicknesses and resistivities) with the aid computer assisted iteration techniques.

The interpreted results are presented as sounding curves and descriptive geoelectric log. Five geoelectric layers were resolved for Vertical Electrical Sounding (VES 1) while four geoelctric layers were resolved for Vertical Electrical Sounding (VES 2).

5.1. Geoelctric Model for Vertical Electrical Sounding (Ves 1)

Layers one and two stand for lateritic topsoil and sub soil with thicknesses 1.71m and 10.57m respectively. Layer three is the sandy clay/clayey sand horizon. The fourth layer is the dry/resistive sandy horizon. Layer five of unknown thickness is the sandy clay/clayey sand layer. Vertical Electrical Sounding (VES 1) is within the immediate vicinity of the failed borehole 1. The loss of circulation within this borehole occurs around the depth of 120 ft (36m). The geoelectric layer corresponding to the loss zone is layer 4. This layer is designated as a dry/resistive horizon because of its high resistivity value. Ideally, a loss zone is expected to be porous and permeable. The voids within the soil grain are expected to be filled with fluid which could be either air or water. When the void spaces are filled with water which is a conductive fluid, the resistivity of the horizon becomes low. On the other hand, when the void spaces are filled with air which is inert, the resistivity of such horizon goes up.

For Vertical Electrical Sounding (VES 1), we can safely deduce that the loss zone is permeable. But with its high resistivity value, the void spaces within the soil layer in the horizon is not conductive. And considering that the depth to top of this geoelectric layer is about 30m, and its thickness being 19.95m, we can deduce that the horizon is an aerated zone.

5.2. Geoelctric Model for Vertical Electrical Sounding (Ves 2)

The geoelctric model for Vertical Electrical Sounding (VES 2) is of the same trend with Vertical Electrical Sounding (VES 1). The loss zone here is layer 4. The resistivity of this layer too is very high and it is also aerated.

The depth to the top of this layer is 30.55m. The depth to the base of this layer is unknown.

5.3. Geoelctric Model and Loss of Circulation

The geoelectric models generated from the interpreted VES data give a very accurate picture of the stratigraphical composition of the subsurface layers where the lost circulation occurs in the two failed boreholes. This underscores the validity and reliability of the electrical resistivity method in unveiling the subsurface geology. If a thorough geophysical investigation has been carried out before embarking on drilling at the site, the first two boreholes would not have been abandoned because the loss zone would have been detected and if the driller is proficient enough, he would have prepared adequately for the lost circulation and successfully overcome it and complete the drilling. An essential part of the preparation to overcome loss circulation is to use a very viscous drilling mud to seal off the permeable part of the formation within the loss zone. It was also obvious that there was no hydrogeologist supervising the drilling. It is high time the society knew that borehole drilling is not a child's play. It should be taken seriously like any other major construction work because the environmental impact of a failed borehole can be terrifying. A failed borehole that is not properly sealed up can be the receptor of a point source pollutant and contaminant that can subsequently be transmitted to the entire aquifer system of an entire area.

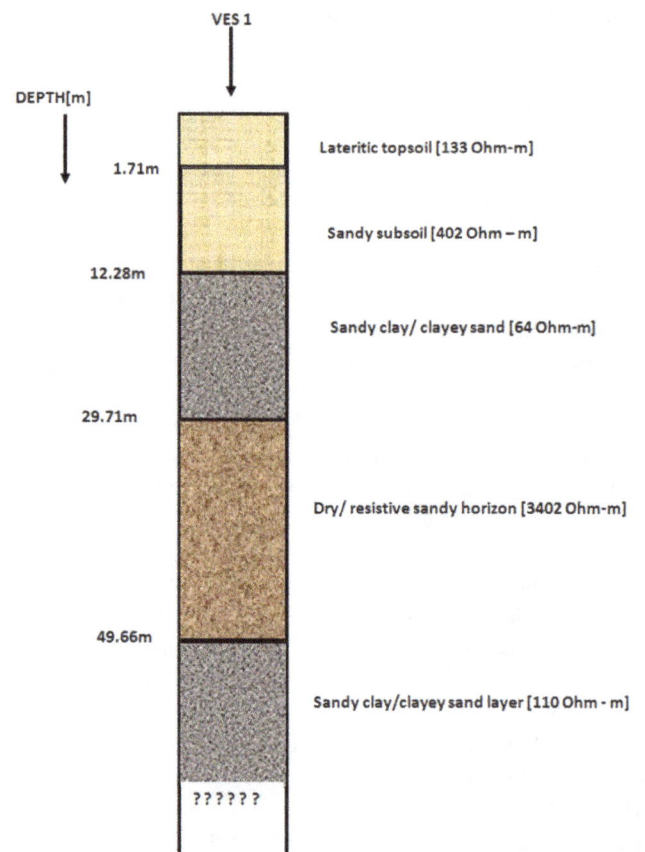

Figure 3. A Descriptive Geoelectric Log For Ves 1 Auchi Poly

Table 1. Geographical Coordinates And Elevations Of Sampled Points

S/N	Description	Position		Elevation (m)
		Latitude [N]	Longitude [E]	
1	VES 1	7⁰ 03' 50.8"	6⁰ 16' 26.5"	239
2	VES 2	7⁰ 03' 52.5"	6⁰ 16' 26.8"	214

Table 2a. Geoelectric Parameters And Inferred Lithology [Ves 1]

Layer NO	APP RES [Ohm-m]	Thickness [m]	Depth [m]	Lithology
1	133	1.71	1.71	Lateritic topsoil
2	402	10.57	12.28	Sandy Subsoil
3	64	17.43	29.71	Sandy clay/clayey sand horizon
4	3402	19.95	49.66	Dry/resistive sandy layer
5	110	-	-	Sandy clay/clayey sand layer

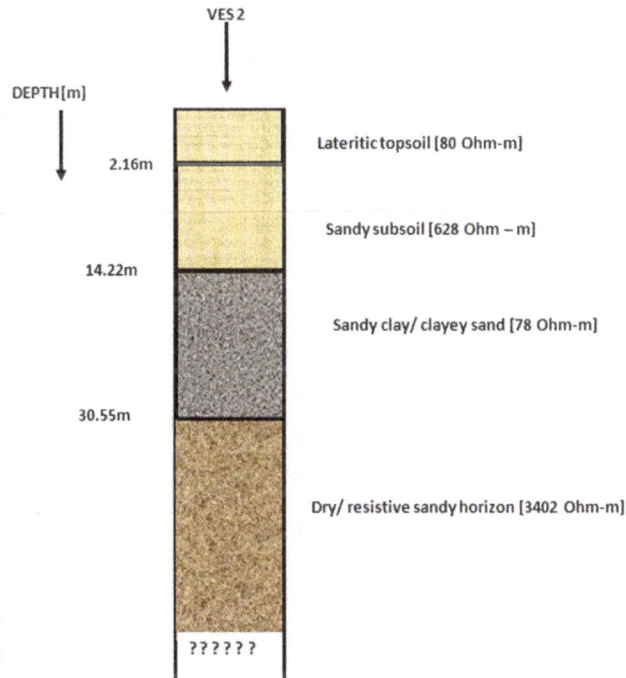

Figure 4. A Descriptive Geoelectric Log For Ves 2 Auchi Poly

Table 2b. Geoelectric Parameters And Inferred Lithology [Ves 2]

Layer NO	APP RES [Ohm-m]	Thickness [m]	Depth [m]	Lithology
1	80	2.16	2.16	Lateritic topsoil
2	628	12.06	14.22	Sandy Subsoil
3	78	16.33	30.55	Sandy clay/clayey sand horizon
4	2919	-	-	Dry/resistive sandy layer

6. Conclusions

The results of the post – drilling geophysical investigations to determine the cause(s) of lost circulation in two abortive boreholes at the Federal Polytechnic Auchi, Etsako West Local Government Area Edo State, is presented in this report.

Hydrogeological/hydrogeophysical deductions made from the interpreted Vertical Electrical Sounding (VES) data shows that the loss zones are permeable and aerated with very high resistivity values. The geoelctric models generated from the interpreted resistivity data agrees with the driller's logs for the abortive boreholes.

The findings of this study serve as the basis for making the following conclusions:

- The formation where the lost circulation occurs is predominantly sandy.
- The formations are permeable and porous thereby allowing the drilling mud to escape inside them
- Although the formations are permeable and porous, the resistivity values are fairly high. This can be attributed to the fluid within the pore spaces of the rock/soil grains which is mainly air.
- Electrical resistivity geophysical method is very useful and reliable in showing the subsurface geology and it also helps in identifying the zone where loss of circulation occurs as well as why it occurs.

Recommendations

The following recommendations have been proffered based on the findings of this study:

- Pre – drilling geophysical investigation should not be optional but a necessity for every borehole drilling project in order to know the subsurface geology.
- It should be made mandatory that a competent hydrogeologist should preside over borehole drilling project and also document the entire process.
- Drilling contractors who did not follow the acceptable best practices in their job execution should be penalized.

References

[1] Ako B.D and Olorunfemi, M.O (1989): Geoelectrical survey for groundwater in the Newer Basals of Vom, Plateau State. Journal of Mining and Geol

[2] Kearey, P. and Brooks, M. (1988): An introduction to Geophysical Exploration. ELBS Edition, pp 199

[3] Odejobi, O.S., (1999): Geoelectric Survey for Groundwater Resources Development in Kudansa, Kaduna State, Nigeria. B.Tech dissertation, Department of Applied Geophysics, Fed. Uni. Of Tech, Akure. Pp 35-36 (unpublished).

[4] Offodile, M.E. (1983): The Occurrence and Exploitation of Groundwater in Nigerian Basement Rocks. Journal of Mining and Geology. Vol. 20, pp 131-145.

[5] Offodile,M.E.(2002):Groundwater study and development in Nigeria. Pp 204-205.

[6] Olayinka, A.I (1990): Case histories of a multielectrodes resistivity profiling array for groundwater in basement complex area of Kwara State, Nigeria. Journal of Mining and Geol, Vol 26 pp 27-38.

[7]   Olayinka A.I and Olorunfemi, M.O (1992): Determination of geoelectrical characteristics in Okene area and implications for borehole siting. Journal of Mining and Geol, Vol 28 pp 403-412.

[8]   Reyment, R. A. (1964): Review of Nigerian Cretaceuos and Cenozoic Stratigraphy. Jour. Min. Geol. Vol 2, no 2, pp 61-80.

[9]   Nfor,B.N.(2006) : Lignite Zone as an Indicator to Loss Circulation Belt-A case study of some locations of Anambra State, Southeastern Nigeria.J.Appl. Sci. Environ.Mgt.Vol10 (3) 31-35

19

Organochloride Pesticides and Polychlorinated Biphenyls in Human Breast Milk: Case Study in the Suburbs of Hue City, Vietnam

Nguyen Van Hop[1, *], Vu Thi Kim Loan[2], Thuy Chau To[3]

[1]Faculty of Chemistry, Hue University of Sciences, Hue city, Vietnam
[2]Department of Chemistry, Hai Phong University of Medicine and Pharmacy, Hai Phong city, Vietnam
[3]Faculty of Resources and Environment, Thu Dau Mot University, Thu Dau Mot city, Vietnam

Email address:
ngvanhopkh@gmail.com (N. V. Hop)

Abstract: Organochlorine pesticides (OCPs, including DDTs and HCHs) and polychlorinated biphenyls (PCBs) were detected in 64 human breast milk samples collected from lactating mothers living in three communes and two wards in the suburbs of Hue city, Central Vietnam in 2010 and in 2011. Assessment of health risk for breast fed babies was conducted, basing on estimation of estimated daily intake (EDI) of the pollutants (OCPs and PCBs) by the babies and then comparing with tolerable daily intake (according to the guideline of Canada Health Agency). The results obtained showed that the EDIs were much lower than the TDIs.

Keywords: OCPs, PCBs, Breast Milk, Hue City

1. Introduction

Diclodiphenyltricloetans (DDTs) and hexacloxyclohexans (HCHs) were two of the widely-used organochlorine pesticides (OCPs) in Vietnam since 1960s. The OPCs and polychlorinated biphenyls (PCBs) (used as transformer oil) belong to 12 persistent organic pollutants (POPs), which are the most dangerous chemicals for environment and human health (Polder A., 2009; Sudaryanto A., 2006, Devanathan G, 2009). For recent years, many places of remains of use-prohibited and/or use-expiry date OCPs were found in many provinces/cities in Vietnam. 28 sites of pesticides remains were found in the rural and urban areas in Hue city of Thua Thien Hue province in 2004. The remains of pesticides estimated was 3,140 kg, of which OCPs, phosphorous pesticides and others occupied 17%, 18% and 65%, respectively (Hung N.V., 2005). OCPs and PCBs collection and treatment after use-expiry date in Vietnam was not tightly regulated and controled. These raised much concern about scatter of the OCPs and PCBs into environment (water, soil, sediment and air) and adverse effect on human health. A study on the levels of DDTs and HCHs in soil samples collected in distance of 50 – 100 m from the OCPs remain sites showed that (Hung N.V., 2005): i) DDTs level in the 14 of 20 soil samples were higher than the requirement of Vietnam regulation TCVN 5941-1995 (100 ppb): in the range of 120 – 470 ppb; ii) HCHs levels in the 12 of 20 soil samples were higher than that of the regulation (100 ppb): in the range of 110 – 500 ppb.

The researches on the OPCs in Tam Giang – Cau Hai lagoon in Thua Thien Hue province, a biggest lagoon in South East Asia, going along the sea with the area of 22,000 ha and receiving river waters from the land and sea water through two inlets (Thuan An and Tu Hien inlets), showed that:

- DDTs content in the lagoon sediment (based on dry weight) was in the range of 9.8 - 33.4 ppb (n = 27 in year 2001; Khoa N.X., 2004) and 0.2 - 8.2 ppb (n = 10 in 2005; Thi T.T.V., 2007); the DDTs levels were higher than that specified in the ISQG, Canada (4.48 ppb) and many data in Khoa's study were even higher than the PEL, Canada (16.32 ppb) applied to marine sediment;
- HCHs content in the lagoon sediment (ranging from 5.6 to 92.4 ppb; n = 10 in 2005; Thi T.T.V., 2007) was 6 to 90 times higher than that of the PEL (0.99 ppb); several levels of aldrin and dieldrin observed by Thi's study in

the lagoon (n = 10 in 2005) were 5 to 30 times higher than that of the PEL (4.30 ppb); also, levels of endrin found in several samples (n = 10 in 2005) were about 10 times higher than that of the ISQG (2.67 ppb), but lower than that of the PEL (62.4 ppb).

The high levels of OCPs in the lagoon sediment are likely to be adversely affecting the ecosystem, human health via food chains and the quality and productivity of aquatic biota in the lagoon. High bioaccumulation of DDTs in several benthic species was found in the period 1998 to 2001 in rabbitfish (mean $\pm$ S = 179 $\pm$ 52 ppb; n = 25) and local carp (286 $\pm$ 82 ppb; n = 25) collected in the Tam Giang – Cau Hai lagoon. The levels of DDTs and HCHs found in bivalve (Meretrix meretrix) collected from Thuan An area in the lagoon were in the range of 34 – 53 ppb and 5 – 7 ppb, respectively (Sy H.T., 2008). Although these were preliminary studies only, there should certainly be concerns about the quality of aquatic products as well as the human health implications of consuming them.

As use of OCPs in agriculture has been prohibited in Vietnam since 1995, a decreasing trend in the OCPs level in the lagoon water and sediment can be observed from the above data: decrease in DDTs level from 9.8 - 33.4 ppb in 2001 down to 0.2 – 8.2 ppb in 2005. In spite of that, DDTs are still being used for malaria prevention and illegally for agricultural purposes by many local farmers due to entrenched habits and the high effectiveness of DDTs. As such, concerns about the adverse effects on organisms and human health remain.

Many current studies confirmed that mothers with contact or without contact with OPCs had excess of DDTs and HCHs in their breast milk (Raab U. et al., 2008, Wang Y. R., 2008). DDTs & HCHs are called endocrine disruptors. They and PCBs are accumulated in fatty tissue of human and excreted through breast milk. Their excretion is very slow. Therefore, breast milk is used as bio-monitoring matter to assess the accumulation of DDTs, HCHs and PCBs in human (Ennaceur S. et al., 2008; Minh N. H. et al., 2004). As a result of this, the hazardous chemicals can be estimated in breast-fed babies. High content of OCPs and PCBs in breast milk may negatively affect the development of breast fed babies, disorder endocrine and resist estrogen (Polder A. et al., 2009, Ntow W. J., 2008, Kunisue T., 2006). Therefore, content of OCPs and PCBs in breast milk are one of good indicators to assess human health risk. In term of lactating mothers, the content of OCPs and PCBs in breast milk is the relationship with many factors such as age of mother, the number of their children, food, content of lipid in their body and other environmental factors.

There were many researches on assessing the accumulation of OCPs and PCBs in environment as well as the analyzing and evaluating their presence in breast milk with the aim at defining the origin of morbidity and fatality in human by these chemicals. However, the study on this problem is still a few in Vietnam generally and in Hue city particularly. This study was conducted with the aim of assessing levels of DDTs (p,p'-DDE; o,p'-DDT and p,p' DDT), HCHs (-HCH, -HCH and -HCH) and PCBs (PCB 28, 52, 101, 118, 138, 153 and 180) in breast milk collected from lactating mothers in the suburbs of Hue city.

2. Methods

Human breast milk samples were randomly collected from 30 lactating farmer mothers and from 10 lactating non-farmers mothers (with average age of 26 - 29, range of 18 - 40) living in three suburb communes of Hue city: Thuy Xuan (n 13), Huong Long (n 13) and Thuy duong (n 14) in 2010. Also, breast milk samples were collected from 24 lactating non-farmer mothers in the two wards of Hue city in 2011: Thuy Bieu (n 14) and An Cuu (n 14) (Fig.1). The samples were stored at 4^{0}C at sampling sites and at - 20^{0}C in laboratory prior to analysis.

Sample extraction was conducted as follows (Minh N.H., 2004; Raab U., 2008, Beyer A., 2010): 10 g of the breast milk sample was added onto 10 g pre-cleaned diatomite earth (Merck, Damstadt, Germany) packed in a glass column and then chemicals to be analyzed were extracted with 200 mL diethyl ether at a flow rate of 1 mL/min. The extract was concentrated to 8 mL by vacuum evaporator. One-fifth of the concentrated extract was used for fat content determination by gravimetric method. The remaining extract was purified on chromatogaphic mini-column packed with 2 g activated florisil (Florisil 30 – 60 mesg, Sigma – Aldrich, USA) and 1 cm length of activated anhydrous sodium sulfate (Merck, Germany) top side. The lipid in the purified-extract was removed by concentrated sulfuric acid treatment (Merck, Germany). The lipid-removed extract was evaporated to 1 mL under a gentle stream of nitrogen and was ready for gas chromatographic analysis.

The separation and detection of DDTs, HCHs and PCBs was performed by gas chromatography system (Agilent 7890 A) equipped with an auto-injection system (Agilent 7683B), micro-electron capture detector (μ-ECD) and HP5-MS capillary column (5% phenyl methyl siloxane phase, 30 m x 0,25 mm i.d. x 0,25 μm film thickness). Nitrogen was used as gas carrier at a flow rate of 1.5 mL/min and make-up gas in the detector at 5 mL/min. 1 μL of the final extract was injected into injector operated at 285^{0}C and splitless mode. The temperature of the detector was 300^{0}C. The column oven temperature was programmed from 90^{0}C (held for 2 min) to 150^{0}C at a rate of 30^{0}C/min, to 204^{0}C (held for 3 min) at a rate of 3^{0}C/min, and finally to 280^{0}C (held for 10 min) at a rate of 8^{0}C/min; quantification was made by calibration curve method (Alvarez, M.F., 2008; Burke E.R., 2003; Minh T.B., 2003).

A 1000 ppb standard solution of the three-isomer DDTs (Accustandard, USA), a 1000 ppb solution of the three-isomer HCHs (Accustandard, USA) and a 7-congener standard solution of PCBs (PCB 28, 52, 101, 118, 138, 153 and 180) - PCBs standard "Mix 3" (Dr. Ehrenstorger, Germany) with 10 ppm concentration of each congener were used for preparation of working standard solutions.

Quality control of the gas chromatographic analysis with

micro-electron capture detector (GC/μ-ECD) was done:

- 0.5 - 100 ppb multi-level calibration curves of the OCPs and PCBs were created for quantification with good linearity (r > 0.99);
- The method limit of detection (LOD) for the chemicals to be analyzed (LOD = 3S; S calculated from the data of analysis of the chemicals with concentration of each 0.5 ppb; n = 9) were 0.02 - 0.04 ppb for DDTs, 0.01 – 0.06 ppb for HCHs and 0.01 – 0.05 ppb for PCBs;
- The repeatability of analysis of two levels of each

chemical (5 ppb and 50 ppb standard solutions) was good: RSD = 2.3 - 2.9% and 0.4 - 7.9% (n = 3) for DDTs; RSD = 0.4 - 3.0% and 0.4 - 0.7% (n = 3) for HCHs and RSD = 2.0 – 4.0% and 0.2 – 0.5% (n = 3) for PCBs;

- The method accuracy was good for analysis of a practical breast milk sample spiked with 29.4 ng/g lipid each chemical: recoveries for DDTs, HCHs and PCBs were 115 – 146%, 89 – 132% and 80 – 138%, respectively.

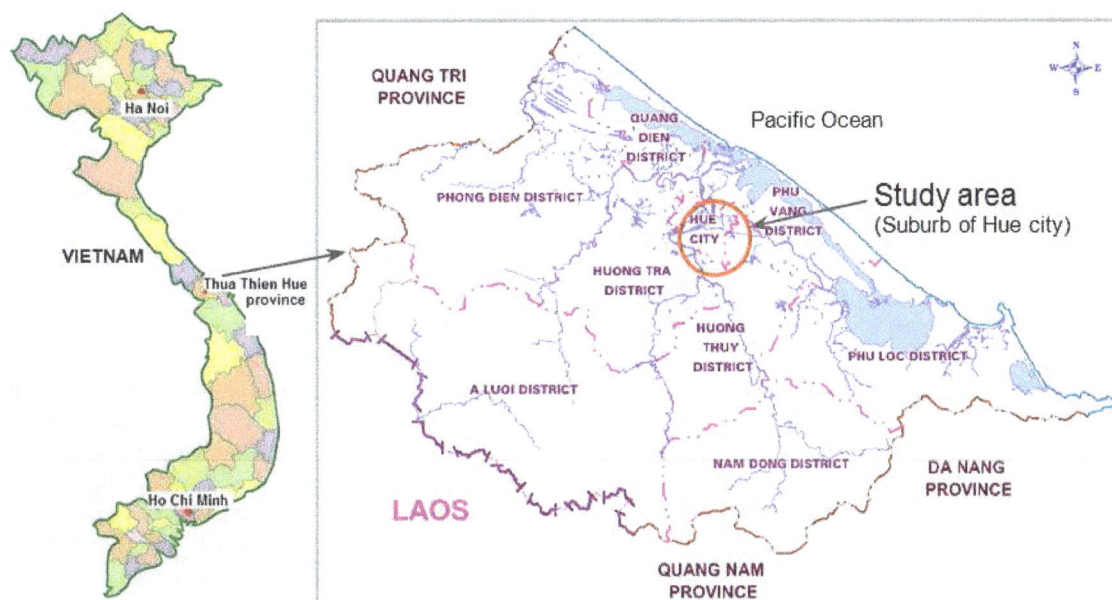

Fig. 1. *Study area in the suburb of Hue city in Thua Thien Hue province.*

3. Results and Discussion

The results of analysis of OCPs and PCBs in the breast milk samples showed in Table 1 and Table 2.

Table 1. *Levels (ng/g lipid) of OCPs and PCBs in the human breast milk samples in three communes in the suburb area of Hue city in 2010[*].*

Chemical	Thuy Xuan (n = 13)		Huong Long (n = 13)		Thuy Duong (n = 14)		Overall (n = 40)	
	Mean	Range	Mean	Range	Mean	Range	Mean	Range
Lipids %	1.6	0.8 - 3.8	1.8	0.4 - 6.3	2.8	0.8 - 6.1	2.1	0.4 - 6.3
p,p'-DDE	52.9	2.3 - 130	42.6	5.4 - 100	25.4	6.3 - 67.7	40.3	2.3 - 130
o,p'-DDT	7.9	1.8 - 15	15	2.0 - 31	11.3	2.3 - 18	11.4	2.0 - 31
p,p'-DDT	28.6	0.8 - 91	48.4	7.7 - 202	28.7	1.3 - 86	35.2	0.8 - 202
DDTs	89.4	4.9 - 171	106	18.1 - 284	65.4	22.7 - 130	86.9	18.1 - 284
α-HCH	21.2	7.0 - 34	32.8	4.2 - 104	15.9	4.4 - 38	23.3	4.2 - 104
-HCH	22.3	3.5 - 68	24.5	3.0 - 103	20.4	4.1 - 97	22.4	3.0 - 103
-HCH	33.0	10.0 - 59	41.4	6.1 - 107	22.6	6.3 - 54	32.3	6.1 - 107
HCHs	76.5	20.5 - 149	98.7	13.3 - 248	58.9	27.0 - 189	78.0	13.3 - 248
PCB 28	<LOD	n.d.	<LOD	n.d.	1.0	<LOD -1.0	0.3	<LOD-1.0
PCB 52	65	<LOD - 65	40	<LOD - 70	24	<LOD - 24	43	<LOD - 70
PCB 101	<LOD	n.d.	<LOD	n.d.	1.2	<LOD - 1.2	0.4	<LOD - 1.2
PCB 118	<LOD	n.d.	0.5	<LOD- 0.5	16	<LOD - 32	5.5	<LOD - 32
PCB 153	<LOD	n.d.	0.2	<LOD- 0.2	1.4	<LOD - 2.2	0.5	<LOD - 2.2
PCB 138	<LOD	n.d.	<LOD	n.d.	0.7	<LOD - 0.8	0.2	<LOD - 0.8
PCB 180	13	0.7 - 27	14	0.7 - 27	19	5.3 - 48	15	0.7 - 48
PCBs	78	0.7 - 92	55	1.0 - 98	63	5.3 - 109	65	0.7 - 108

[*] LOD was 0.02 ppb for PCB 28, 0.02 ppb for PCB 52, 0.01 ppb for PCB 101, 0.04 ppb for PCB 118, 0.03 ppb for PCB 153, 0.03 ppb for PCB 138 and 0.05 ppb for PCB 180; n.d.: no determined

Table 2. Levels (ng/g lipid) of OCPs and PCBs in the human breast milk samples in two wards in the suburb area of Hue city in 2011[*].

Chemical	Thuy Bieu (n = 14)		An Cuu (n = 14)		Overall (n = 28)	
	Mean	Range	Mean	Range	Mean	Range
Lipid (%)	1.9	0.9 - 3.5	1.6	0.4 - 3.9	1.8	0.4 - 3.9
p,p'-DDE	329	38.8 - 1183	655	69.3 - 5547	492	38.8 - 5547
p,p'- DDD	37.3	2.2 - 132	36.4	6.0 - 318	36.9	2.2 - 318
p,p'- DDT	22.3	4.0 - 55.6	22.5	4.7 - 89.9	22.4	4.0 - 89.9
DDTs	389	75 - 1198	714	75 - 5683	553	75 - 5683
α-HCH	9.7	<LOD - 30.2	3.8	<LOD - 18.5	6.8	<LOD . 30.2
-HCH	3.6	1.1 - 6.3	4.1	<LOD - 29.2	3.9	<LOD - 29.2
-HCH	12.0	1.2 - 24.1	3.1	<LOD - 14.8	7.6	<LOD - 24.1
HCHs	25.3	3.5 - 57.8	11.0	<LOD - 62.5	18.3	<LOD - 62.5
PCB 28	0.2	<LOD - 0.9	0.7	<LOD - 3.0	0.5	<LOD - 3.0
PCB 52	0.3	<LOD - 3.3	2.3	<LOD - 11.7	1.3	<LOD - 11.7
PCB 101	<LOD	<LOD	0.5	<LOD - 2.3	0.3	<LOD - 2.3
PCB 118	1.0	<LOD - 3.1	2.3	<LOD - 6.3	1.7	<LOD - 6.3
PCB 153	2.7	1.5 - 4.8	3.7	1.2 - 16.8	3.2	1.2 - 16.8
PCB 138	2.3	0.5 - 5.1	2.6	0.6 - 12.1	2.5	0.5 - 12.1
PCB 180	3.6	1.5 - 7.7	4.5	1.4 - 16.7	4.1	1.4 - 16.7
PCBs	10.1	4.8 - 15.8	16.5	5.6 - 67.9	13.3	4.8 – 67.9

[*] LOD for each chemical and n.d. as in Table 1.

For DDTs: The isomers of DDT were found in all breast milk samples in 2010 and 2011; their levels in the samples in 2011 were in larger range and higher than that in 2010 (p < 0.05); these levels of DDTs were much lower than that in Hanoi city (2100 ng/g lipid, n = 42) and Hochiminh city (Minh N.H., 2004).

For HCHs: The isomers of HCH were detected in all samples in 2010 and in above 50% of total samples in 2011; it was differrent from DDTs, the levels of HCHs in 2010 were in larger range and higher than that in 2011 (p < 0.05).

For PCBs: The levels of PCBs congeners found in 2010 were in lager range and higher than that in 2011 (p < 0.05); while only congener PCB 180 was detected in all samples in 2010, three congeners (PCB 153, PCB 138 and PCB 180) were found in all sample in 2011; the other congeners of PCBs were detected in fewer samples and at lower levels.

The data of OCPs and PCBs found in Hue city (this study) were compared with that in Hanoi and Hochiminh cities (Minh N.H., 2004) as indicated in Figure 1.

For assessing health risk for breast fed babies, estimated daily intake (EDI) of the pollutants (OCPs and PCBs) by babies was estimated and then compared with tolerable daily intake (TDI) according to the guideline of Canada Health Agency (Oostdam, 1999). The EDI was calculated according to fomulas (1), accepting that weight of each baby was 5 kg on average and each baby consumed 700 g breast milk per day (Oostdam, 1999; Minh N. H., 2004; Sudaryanto A., 2006):

$$EDI = \frac{C_{milk} \times 700\,g \times C_{lipid}}{5\,kg} \qquad (1)$$

EDI: estimated daily intake (g/kg wt/day)
C_{milk}: content of pollutans in breast milk (g/g lipid)
C_{lipid}: content of lipid in breast milk (%)

The data of average EDIs of OCPs and PCBs for breast fed babies in the study areas showed in Table 3 were rather lower than the TDIs.

Table 3. EDIs (g/kg wt/day) of DDTs, HCHs and PCBs for breast fed babies in Hue city in 2010 and 2011.

	DDTs		HCHs		PCBs	
	Mean	Range	Mean	Range	Mean	Range
Thuy Xuan (n = 13)	0.20	0.01 - 0.38	0.17	0.05 - 0.33	0.17	0.001- 0.21
Huong Long (n = 13)	0.27	0.05 - 0.72	0.25	0.03 - 0.62	0.14	0.003- 0.25
Thuy Duong (n = 14)	0.26	0.09 - 0.51	0.23	0.11 - 0.74	0.25	0.02 – 0.43
Thuy Bieu (n = 14)	1.01	0.17 - 2.78	0.06	0.01	0.03	0.01 - 0.05
An Cuu (n = 14)	0.89	0.14 - 3.15	0.02	0 - 0.04	0.03	0.01 - 0.06
TDI [*]	20		0.3		1.0	

[*] TDI (g/kg wt/day) according to Canada guideline (Oostdam, 1999).

4. Conclusion

Although OCPs were prohibited for agricultural use since 1995 and transform oils containing PCBs were collected and controled in Vietnam, the pollutants were detected in most of human breast milk samples collected from lactating mothers in the suburbs of Hue city in 2010 and 2011. Generally, the levels of DDTs was higher than the levels of HCHs and PCBs in the samples. In spite of that, the EDIs of the pollutants indicated that there was not much concern about the health risk for breast fed babies in the area under study.

References

[1]   Ennaceur S., Gandoura N., Driss M. R., 2008. Distribution of polychlorinated biphenyls and organochlorine pesticides in human breast milk from various locations in Tunisia: Levels of contamination, influencing factors, and infant risk assessment, Environmental Research, 108, 86 - 93.

[2]   Minh N. H., Someya M., Minh T. B., Kunisue T., Iwata H., Watanabe M.,Tanabe S., Viet P. H., Tuyen B. C., 2004. Persistent organochlorine residues in human breast milk from Hanoi and Hochiminh city, Vietnam: contamination, accumulation kinetics and risk assessment for infants, Environmental Pollution,129, 431 - 441.

[3]   Raab U., Preiss U., Albrecht M., Shahin N., Parlar H., Fromme H., 2008. Concentrations of polybrominated diphenyl ethers, organochlorine compounds and nitro musks in mother's milk from Germany (Bavaria), Chemosphere, 72, 87 - 94.

[4]   Sudaryanto A., Kunisue T., Kajiwara N., Iwata H., Adibroto T. A., Hartono P., Tanabe S., 2006. Specific accumulation of organochlorines in human breast milk from Indonesia: Level, distribution, accumulation kinetics and infant health risk, Environmental Pollution, 139, 107 – 117.

[5]   Polder A., Skaare J. U., Skjerve E., Loken K. B., Eggesbo M., 2009. Levels of chlorine pesticides and polychlorinated biphenyls in Norwegian breast milk (2002-2006), and factors that may predict the level of contamination, Science of the Total Environment, 407, 4584 - 4590.

[6]   Beyer A., Biziuk M. (2010). Comparison of efficiency of different sorbents used during clean-up of extracts for determination of polychlorinated biphenyls and pesticide residues in low-fat food, Food Research International, 43, pp. 831 - 837.

[7]   Devanathan G., Subramanian A., Someya M., Sudaryanto A., Isobe T., Takahashi S., Chakraborty P., Tanabe S. (2009). Persistent organochlorines in human breast milk from major metropolitan cities in India, Environmental Pollution, 157, pp. 148 - 154.

[8]   Ntow W. J., Tagoe L. M., Drechsel P., Klederman P., Gijzen H. J., Nyarko E. (2008). Accumulation of persistent organochlorine contaminants in milk and serum of farmers from Ghana, Environmental Research, 106, pp. 17 - 26.

[9]   Wang Y. R., Zhang M., Wang Q., Yang D. Y., Li C.L., Liu J., Li J. G., Li H., Yang X. Y. (2008). Exposure of mother-child and postpartum woman-infant pairs to DDT and its metabolites in Tianjin, China, Science of the Total Environment, 396, pp. 34 - 41.

[10]   Kunisue T., Muraoka M., Ohtake M., Sudaryanto A., Minh N. H., Ueno D., Higaki Y., Ochi M., Tsydenova O., KamiKawa S., Tonegi T., Nakamura Y., Shimomura H., Nagayama J., Tanabe S. (2006). Contamination status of persistent organochlorines in human breast milk from Japan: Recent levels and temporal trend, Chemosphere, 64, pp. 1601 - 1608.

20

Analyzing the Internet Filtering Policies in KSA and USA

Zohair Malki

Department of Information and Learning Resources, The collage of Computer Science and Engineering, Taibah University, Yanbu Al Bahar, Al Madinah Province, Saudi Arabia

Email address:
zmalki22@gmail.com

Abstract: This paper attempts to define internet filtering and outline the purpose of enforcing the filtering policy, and analyze its effect on the use of the Internet resources. This study further examines and evaluates the regulations regarding Internet access in Saudi Arabia, with an emphasis on the problems that people come across during Internet access with limited access to specific web pages due to the policy of filtering. This paper will also examine the case of the Supreme Court of the United States appellant vs. American library association which is similar to that of the Saudi Arabian case with miniscule differences only in term of ritual and cultural aspects. The paper evaluates the advantages and disadvantages of using the filtering policy and its influence and application to the culture and religion of Saudi Arabia. The study further enlists some of the rules and regulation, developed about this issue with the reference of the case of the Court of the United States appellant's vs. American library association. Finally, this study discusses the impact of different cultural and tradition on the ways of application of the filtering strategy in both sides.

Keywords: Filtering Policy, KSA Internet Filtering, American Library, American Library Association

1. Introduction

In the present day, the World Wide Web has transformed the way of communication, now the entire world seems to be a small village with an ease of access of information than ever before. The recent decade has witnessed a revolution in the popularity and volume of the Internet. Today the Internet is serving millions across the globe as a hub of information and connectivity through remote access. It is also connecting people to information, electronic mail, and tools of delivery for information products. A variety of information that was almost incomprehensible earlier has been made extensively and effortlessly accessible through the web. (Zittrain & Edelman, 2002) However, most of the users of the Internet services are unaware of the rules and regulation of using these services.

Policies affecting the flow of Information and underlining the consequences of the use of technology and accessing available resources through the Internet or databases have received a lot of emphasis in the field of information studies. In the present day while the Internet has become a massive information resource, it must be used wisely and selectively as some of its content is found to be inappropriate for the users and also has experience of dangers of its use in the past.

For example, obtaining illegal or private information on certain people or countries is amongst a commonly practiced vice on the internet. Information Technology is a vital concept for the modern world and is defined as the skill of the new generation. Also, it is imperative to understand the acceptable and unacceptable usage of this new technology in the country as per its traditions and value systems and how the policy related to these services, sometimes protects the users from being affected by illegal practices of its use and at the same time prevents the users from accessing to a particular type of information. It is also essential to consider the access needs of the users and the effect of the rules and regulations on its usage. (Banks, 1998).

In the Gulf countries especially Saudi Arabia, one of the difficult decisions was allowing the access of the Internet to the public, because of its conservative and religious ways of governance. Hence, it took two years by the government to approve Internet access to the public, that too after applying very restrictive policy. As from the illegal and undesirable use and the copyright laws, the access to the Internet services needed to have policies, which were compatible with the rules of the country and consistent with the user's belief. The policy states that citizens should have the right to use the services in an atmosphere that is conducive to the successful completion of Internet services. The Saudi government installed filtering systems to prevent the public access to certain web pages and sites. The question that needs to be asked now is how does this policy come together? It seems that different cultures tend to develop different values, self-concepts, beliefs, lifestyles, and ways of providing services (Burger, 1993). For this reason, there was an obvious need for more comprehensive research to explores and analyzes the problems and the influence of the application of such filtering policy on the culture and tradition of Saudi Arabia.

Allowing Intellectual freedom in the information hungry society with filtering software's is abroad topic requiring deliberations evaluation and analysis. As it can be seen, there are many people who may agree with this concept; however others may not. Although it is now a popular and widespread practice in many places such as companies, libraries, universities, schools, and other providers of Internet access, it remains a controversial topic amongst a section of the population. (Busha, 1972). This paper will explain internet filtering, outline the purpose of applying the filtering policy, and elucidate the filtering policy and its effect on the use of the Internet resources. This paper attempts to investigate and analyze the Internet access policy in Saudi Arabia. It will specifically explore the problems encountered by the people while accessing the Internet with the filtering policy to restrict the access to specific web pages.. The paper also throws light on the advantages and disadvantages of applying this policy and the influence of the culture and religion to the obligation of applying this policy and vice versa. This is followed by the case of Supreme Court of the United States appellant's vs. American library association and some of the rules and regulations which have been developed about this issue. This is quite similar to the Saudi Arabian case where the only differences are in terms of culture and tradition aspects. Lastly, the paper also discusses the details about the effect of the different cultural and tradition on the application of the filtering policy.

The term filtering may be defined as – "Filtering is the process by which particular source or destination addresses can be prevented from crossing a bridge or router onto another portion of the network" (Network Buyer's, 1999). or "The use of a program to screen and exclude from access or availability the webpages or e-mail that is deemed objectionable" (Techtarget, 2003).

Content filtering is widely being used by corporations as part of Internet firewall computers and also by owners of personal computers in their homes, especially by parents to screen the inappropriate content their children have an access from a computer (Tech target, 2003).

2. Internet Filtering in Saudi Arabia

Since 1994, Saudi Arabia (KSA) has been connected with the Internet, but the use was restricted on the medical, state, academic and research purposes (Yam, 2002). The public in Saudi Arabia could subscribe to local intranet found inside

the country such as Al-Naseej. In Al-Naseej network they can use public e-mails and link to local chat rooms. However public can't directly use World Wide Web (Al-Sarami, 1999). Starting Juanuary 1999 Saudi Arabian government starts to allow Saudi users to enter public internet services. Since then Saudi government started to strict the online contents available for internet users and applied filtering policy for public accessed online content. Officials of the Saudi government and the king Abdul-Aziz City for Science and Technology (KACST), the Riyadh-based state institution were made the pivots for coordinating the Internet policy. They targeted to exercise control over Internet content and make a decision about the appropriateness of the contents for the users. Moreover, the KACST was made responsible for issues, operation, regulations maintenance and policies governing the usage of the service, and also for the development of the Internet. (Al-Rasheed, 2001).

Figure 1 shows a list of a number of internet users in KSA year by year expressing the increasing amount of internet users in KSA.

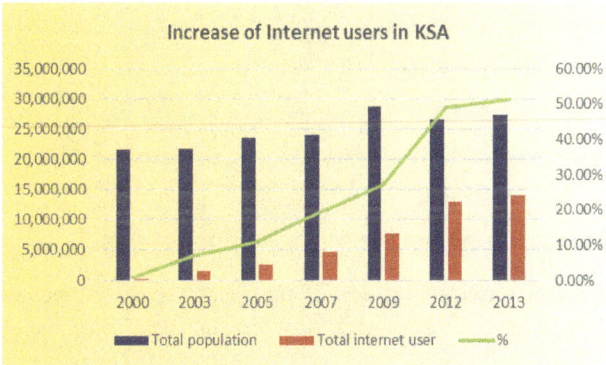

Figure 1. Express the increase of internet users in KSA.

A standing board of trustees framed and affirmed by the government secures the general public by avoiding such material on the Internet that abuses Islam or infringes on the conventions and society of the individuals of the nation (Al-rasheed, 2001). This committee classifies the inappropriate sites as immoral, such as pornographic, sites with sexual content, anti-Islam sites and other. Furthermore, it blocks the subscribers from entering such sites that have any such explicit materials or content.

The Internet is swarming with a lot of explicit an inappropriate content; hence, the government has launched a system to prevent such content from influencing the society. Such system ensures that users subscribed to this service will not receive any inappropriate content in their computer (Yam, 2002). The KACST officials are equipped with software, procedures, and hardware that inhibit internet user from accessing any online content that degrades, or that harms the Islamic Religion values and Muslims traditions. They also installed "firewall" or internet barrier that prevent other parties infringe their websites. This is the reason the government of Saudi Arabia didn't try to quickly provide the service for public, and it is inclined to remove all undesired online contents from being accessed by internet users

(KACST, 1998).

Saudi Arabia's Council of Ministers endorsed this objective early on and called for firewall, administrated by KASCT and prevented users from accessing improper online content. In addition The Council of Minsters created set of rules and policies that control the internet accessing and usage (Zittran & Edelman, 2002). In May 1998, the decision was made public, and it restricted the Internet Services Provider(ISPs) and internet users to surf the internet for prohibited acts such as online gambling, pornography or doing illegitimate activities that violates the social economics, culture, politics and religion. (Al-Adhel, 1998).

(Figure 2 shows a diagram that expresses the niches and blocked websites by KSA. It is not about the pornography websites only but also for various niches.)

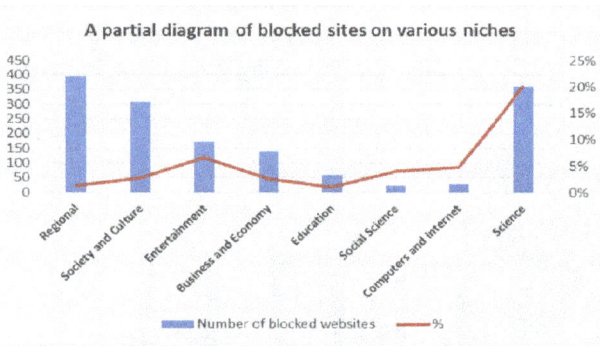

Figure 2. A partial diagram of blocked sites on various niches.

The list is provided by one of the giant search engines named as "Yahoo." Although, this is not the exclusive evidence but still it expresses the picture of the KSA partially.

There is a common thinking that KSA only blocks those sites that contain sexual contents only. But it is not like that. The filtering system of KSA blocks different types of sites on various niches. Even some religious pages are blocked in KSA. Yahoo published list of various websites from different niches in the year of 2012 which are blocked in KSA. Table 1 shows some exclusive part of that list.

Table 1. Examples of few blocked sites on various niches.

Niche/theme of the website	Example from blocked sites	Number of total blocked sites
Religion	al-bushra.org	246
Education and reference	women eb.com	45
Humor sites	createafart.com	81
Entertainment, music, and movies	foxsearchlight.com	251
Pages about Middle Eastern politics, organizations or groups.	hizbollah.org	80
Services allowing circumvention of filtering restrictions	systransoft.com	250
Pornography	Elephanttupe.com	1300

Sometimes internet filter of KSA blocks few sites temporary. Their previous records said that they blocked some famous websites for various reasons at different time. Google and Yahoo published a list jointly in June, 2002 and Table 2 shows few taglines from that list.

Table 2. Example of few temporary blocking sites by internet filter in KSA.

Name of the Website	Blocking Dates
www.amnesty.org/ailib/intcam/saudi	May 18, May 19, May 22, May 24, May 27
www.1-marijuana-seeds.com	May 19, May 22, May 24, May 27
www.al-bushra.org	May 16, May 18, May 19, May 22, May 24, May 27
world.altavista.com	May 18, May 19, May 22, May 24, May 27
www.angelfire.com/ca2/queermuslim	May 18, May 19, May 22, May 24, May 27
www.answering-islam.org	May 18, May 19, May 22, May 24, May 27
www.bahai.com	May 14, May 18, May 19, May 22, May 24, May 27

Although as per the official version of filtering internet content primarily focuses on online materials which are considered offensive to conservative Muslims. In Saudi Arabia the restriction and filtering phenomenon also applied to sites with political content. As an example, in the early 1999, the Committee against Corruption website in Saudi Arabia that is an exiled dissident group faced unauthorized blocking. Users in Saudi Arabia trying to surf forbidden sites get blocked with onscreen message that tells them that the website is forbidden, as shown in Figure 3. The logging into forbidden sites is reported and recorded. Saudi government has left no stone unturned to prevent any effort by the users to evade censorship. Recently, the URL address of common websites for online proxy server facilitating anti-censorship and a web site offering anonymous services was blocked in the Saudi Kingdom. (Zittran & Edelman, 2002).

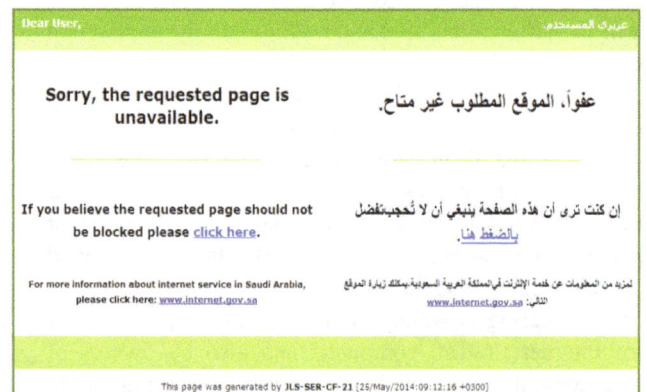

Figure 3. The warning message displayed when trying to access violated sites in Saudi Arabia.

2.1. The Disadvantage of Filtering

The filtering policy imposed by the Government has posed pair of threatening points to the right of publishing information and freedom of expression.

• Firstly it applies prior censorship to the internet content, so, each content has to be subjected to a hard level of blocking.

- Secondly, the filtering technologies developed till date is imprecise tools for content filtering in practical terms.

Even though the motivations for preventing online surfing such as keeping adult content and material away from users have been justifiable, the tools for filtration have intentionally or else, spontaneously blocked the legal surf of cultural, political, and medical content which are available and legal internationally with international standards and laws.

KACST under government orders blocks the selected websites based on their content. This blocking also extends to inappropriate cultural or political materials and content (Yam, 2002). A proxy server like the one used in KACST is also used by authorities to keep an eye the computer terminals accessing web sites and also the duration of such accesses. (Douglas1999).

The concept of filtering the Internet by the government has been successfully implemented in K.S.A despite of the fact that it may seem objectionable to some. Moreover, this policy gives the opportunity of request to the users, for instance any of the users of the internet service can complain request for unblocking using an unblock request form, in case a genuine website is intentionally blocked, stating the cause of why not blocking such site. Similarly, online users can send information about a website that they feel it is offensive and giving reasons for its blocking using a proxy report email address (Al-kdahi, 2003).

2.2. The Filtering Mechanism in Saudi Arabia

Various sources in the computer industry state that the K.S.A. System works in two stages. First, it approves web pages and stores them in a 500-gigabyte storage system in the capital city of Riyadh. Users access these pages from this computer system rather than the on the web. These are the frequently used pages, and they can be accessed quickly without the system having to assess their suitability each time. Requests received from the users to access the pages that are not stored in the cache, are scrutinized for suitability by the second stage of the system supplied by Websense, which is a U.S. based company, which lists and filters thirty categories of potentially unsuitable sites (Whitaker, 2000).

The policy of Internet filtering enables the government to:

- Prevent the public from accessing the online content not respecting Islam traditions or infringes on the people culture in the country
- Blocks any content interfering with the fundamental concepts for example; pornographic, sexual content, anti-Islam sites and other.
- Control on Internet activity or forbidding the distributing or online surfing of content that contains sensitive and valuable information for example, any news damaging anything negating a basic guideline or enactment, or encroaching the sacredness of Islam and its benevolent Shari'ah, or breaking open tolerability.
- Full control over the access to the Internet the ability to control over the service's providers and the connection to

the sources of information.

2.3. Some Practical Implications

It is also a ground reality that the filtering software has been an imprecise tool, and it always blocks loads of material beyond its stated purpose. The most sophisticated products have similar issues even though they are checked by staff members on a regular basis for suitability of the websites. The filtering system becomes disadvantageous in term of blocking a lot of important information especially in the field of Education, health, religion, and entertainment. Users remain dissatisfied about these aspects of the filtering policy.

The Users satisfaction about KASCT (Open Net Initiative 2004) is shown in Figure 4. The filtering system in KSA has continually blocked the websites containing information on education, social, psychology, and health which has hampered the spread of knowledge about latest threats in these fields. For example, the sites having content about AIDS prevention is blocked because it has the words sex, woman, and intercourses in the content of the website. Even the website for pro-filtering was blocked because it shared a commercial server with a pornography site (Richtel, 1999).

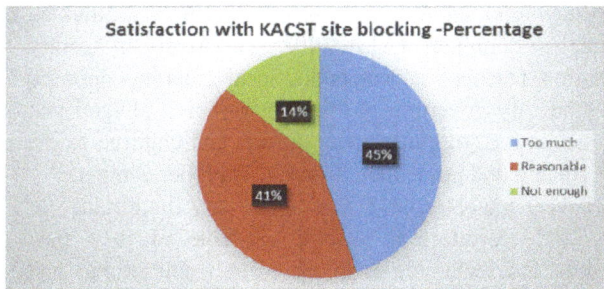

Figure 4. *Users Satisfaction with KACST site blocking.*

3. Children's Internet Protection Act (CIPA): A Case from the USA

The library associations and website publishers challenged constitutionality of children's Internet protection Act (CIPA). Under this proposal, any school or library where software for pornography filter was not installed would be prohibited from federal fund provided with the intention to help in the purchase of link to access the internet.

According to a letter signed by the Computer and Communications Industry Association and Information Technology Association of America, groups that belonging to the advanced technology productions, filtering obligation has become an obstacle for federal regulation of using the internet and accessing its content (CDT, 2000). In addition, Teacher associations and American Library Association have an objection on the blocking and filtering policies (Bruening & Davidson, 2003). The objection of hard filtering policy said that mandatory filtering reflects a political viewpoint. They said that sometimes the filtering is flawed and they fail to block pornography.

Internet filtering including focus on the family is very

important with respect to some conservative groups, and as per the American Family Association. "Children do not have a constitutional right to access and view Internet pornography in our local libraries (AFA, 2003)."

Although the court didn't apply forcing policy on content filtering, in part, The Supreme Court found that the filtering tools is less restrictive to protect users from ethically harmful and undesired content effecting their well-being; it has stated CDA to be unconstitutional. However, despite that COPA is likely vanish as a valid content law, there is still a burning issue among parents, legislators and conservatives which is the explicit sexually content and other hostile content (ALA, 2003). Last year, Congress issued a collection of national legislative proposals in response to that parental concern; it comes after making good survey for the media service by children and online safety. Most of the proposed legislation mentioned the requirement to install blocking and filtering technologies in universities, schools and libraries in which computers are gained by public funds (ALA, 1999)

Moreover, many teachers, scientists and Liberians said that it is very undesirable that some of the considered legislations always connect the use of the internet filtering policies with financial purposes. All education centers and schools, who get "E-rate" fund, are obliged to install filtering and blocking tools on their computers (ALA, 2002). The American Library Association, teachers, educators and local Liberians opposed such mandatory governmental filtering policies, alleged that they must be free to guide their students and children to the proper content and guide them to avoid improper content.

Moreover, many believe that it is very important for society and educators to discuss the role of then new inventions and technologies in the overall educational and public trends (ALA, 2003). It is considered constructive act to discuss such important ideas and exchange them, it will also encourage users to get benefits out of the electronic resources.

Many Libraries and schools have responded to the filtering controversy and the possibility of government regulation in many ways, schools and educators which using commercial filtering tools said that it is very restrictive filtering policies and prohibit in many cases the usual use of internet content.

In addition, in the "Filtering on Home PCs" part, many of the software tools are including some political and cultural trends and bias, that stop the access of internet content that the teacher , in some cases, need his students to get access to such content and evaluate it. As a result the commercial filters became not only filtering "Objectionable" sites but also became an obstacle in the education process. (Langland, 1997)

According to a study by the National Commission conducted in 1998 on Libraries and Information Technology policy, almost all libraries were found to have policies that govern Internet usage. However, only about 15% of these public libraries use filters.

Amendment 1501 passed by the House of Representatives was related to the Child Safety and Protection Act and it required that all libraries and schools have filters installed on

their Internet access computers to block obscenity and child pornography . While the American Library Association (ALA) has no objection to the idea that all libraries install filters on their computers, the ALA believes that filters in libraries are not the only way to protect children and there are many places where children have threats due to access to the internet such as television, movie houses etc. The policy of ALA states that parents should supervise older and younger children when using the Internet and young children should never be left alone in the library.(ALA, 1999).

Conversely, ALA also raised concern in opposition to amendment 1501. The ALA believes this amendment will impose one size fits all federal mandates, which undermines the powers and local decisions made by libraries, schools, and their governing boards responsible to provide a safe and rewarding experience for children in using the Internet (ALA, 1999). It also states that the library users who access the Internet must be responsible for their own searches. Library has also set up guidelines to achieve this goal by offering various resource formats and services to meet the need of the community.

- These guidelines state that parents or guardians are responsible for supervising their children's Internet use and should convey to them the material they should not view and use.
- In addition, all citizens of the community can freely use the public library for Internet access, portioning to recreational purposes, which are limited by the Computer Courtesies list available in the library (ALA, 2003).
- According to library policy on Intellectual Freedom, mentions that anyone accessing sexual explicit or violent materials may lose their library privileges.
- Anyone found using the library for illegal and unacceptable use including harassment; slandering and violation of copyright laws will not only lose their library privileges, but may be subject to prosecution by local, state or federal authorities.
- The policy goes on to state that citizens should have the right to use the library in a conducive atmosphere for successful completion of library business (Langland, 1997).

4. Analysis and Discussion

It is clear that this internet filtering intends to protect the community as a whole and to maintain a certain moral standard throughout the country. On the other hand, the policies of the American government appear to be somewhat different from that of Saudi Arabia. These differences arise mainly from the well-known freedom of speech and knowledge in a democratic society. Intellectual freedom is a core value of the library profession and a basic right in a democratic society. In the United States, everybody is made responsible for their desire of knowledge, and it is deemed to be the government's obligation to provide the public with this freedom. Therefore, the application of the filtration of information over the Internet would surely contradict the core values. The rights of library users, to read, seek information, and speak freely are actively defended by the ALA as guaranteed by the First Amendment. It is quite clear that the ALA assists and promotes libraries that are helping children and adults develop the skills required to understand and effectively utilize information resources essential in the present day global information society. The ALA also affirms that the use of filtering software by libraries clearly violates the library bill of rights by blocking access to consistently protected speech (ALA, 2000). In this case, libraries and schools in the United States have the right to decide wither to apply filtering over the information on the Internet or not.

In USA most of the websites are open to all people. They have few laws and if any websites break those laws then that specific website will be blocked in USA. Here goes the list of laws which may be used to protect a specific group of people from entering few specific sites, the list is shown in Figure 5.

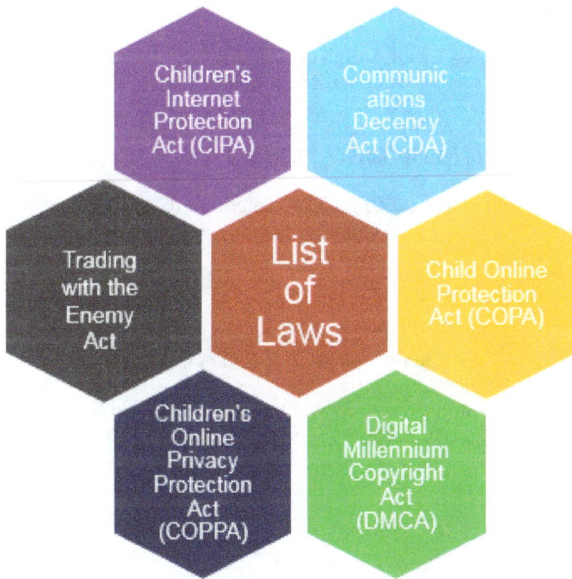

Figure 5. List of laws for preventing group of people of entering specific sites.

Besides these, almost 99.9% websites are open in USA but few of the government and private offices block few websites in offices at office hour so that employer can concentrate on office work perfectly. Figure 6 shows a list of those restricted websites with their percentage of restriction.

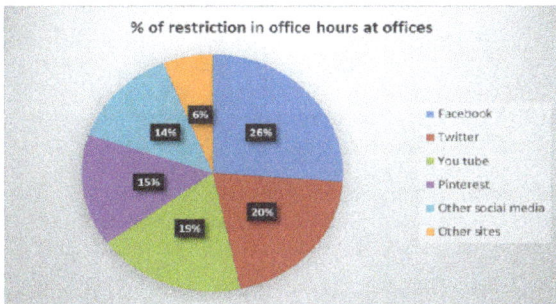

Figure 6. Percentage of restrictions in office hours.

In Saudi Arabia, on the other hand, libraries and schools do not have control over its own rules regarding Internet filtering. The Saudi Arabian government is responsible to manage the content of the web. Thus, the deciding on what type of filters is to be applied becomes a government enterprise ensuring cultural protection. In simple terms, the government like a censor investigates and scrutinizes the suitable content and makes it available to the users.

The handling of Internet filters by the government can be called a natural phenomenon. In the past, when the library being the only source of information in Saudi Arabia, information on politics, and sexual issues were kept in a separate special room of the library; which was studied by people only on special request. Because such kind of information is freely available over the Internet, the Saudi government has attempted to create the necessary filtration systems addressing all such matters (Al-Adhel 1998). In the Saudi Arabia, a standing committee approved by the government, has been formed to protect society from such material on the Internet that violates Islam or encroaches on traditions and culture of the people in the country. This committee including members from the Saudi public citizens will determine which sites are immoral, such as including pornographic, sexual content, anti-Islamic sites and others. (Al-rasheed, 2001)This strategic approach of the government of Saudi Arabia allows the involvement of its citizens to applying the policy of filtering. In term of freedom of expression, the Saudi governments has given its citizens the right to determine whether a piece of information should or should not be blocked because it contains useful information, or there is a need for these kinds of information, for instance research, education, and medical purposes.

This procedure starts with submitting request from the users, for instance, any user of the Internet service can complain in case the site is blocked, giving reasons on why it should be unblocked using an unblock request form. Similarly, users can submit information about any site they consider offensive and reasons explaining why it should be blocked using a proxy report email address (Al-kdahi, 2003).

Although the core questions arises that do young people have the right to explore information and entertainment resources online without divulging who they are and what are the rights of their parents? Library users also access the Internet, and they must be responsible for their own searches. Libraries also have set up guidelines to achieve this goal by offering various resource formats and services meeting the needs of the community. The parents or guardians are responsible to supervise their children during Internet use by explaining to them which materials they should not use or view.

In the United States, there is no restriction on adult to access any kinds of materials on the Internet. However, it may be noted that this freedom may be a source of trouble for those who intentionally use certain information to inflict moral, physical or financial pain to others. Identity theft is one such example where one uses the Internet posing to be someone else causing grave damages to those who are targeted. One must not completely rule out the issues related to filtration in the United States as there are several communities in US that would strongly support the idea of Internet filtration. While at the same time there are many who out rightly reject this matter.

In Saudi Arabia, restriction over the Internet materials applies for both children and adults. The Islamic rules form the basis of the Saudi Arabian culture. Hence in accordance to Islam, information such as pornography, anti-Islam sites, sexual content, and others are prohibited for both children and adults. This would explain the fact that no age exception by the Saudi government for the free, unfiltered use of internet.

5. Conclusion

Internet filters are an important tool to control and maintain the appropriate rating and scrutiny of Web content. However, it is suggested by the arguments on both sides that the application of filters varies across cultures. For instance, in the United States, the public influences the legislation related to such control mechanism. Although certain individuals consider that the application of Internet filters violates the free flow of information, while others argue that a control to a certain extent is needed to keep pornographic materials away from the underage audience.

Nonetheless, other related problems also arise when the cultural context dictates how filters are applied and/or implemented. A closer look at the Internet filters dilemma suggests that there is a censorship regardless of culture. How these filters are being applied is quite interesting. In other words, who are the policy makers? In the US, the Congress and major interest groups impose the policies and legislation regarding filtering lodge. With all fairness, regardless of the source of control, it is a form of censorship. In Saudi Arabia, the government being the only form of censorship is responsible to create and force Internet filtering policies to the society.

References

[1]   U.S. V. American Library Association, 2013. Retrieved at October/24/2013.

[2]   The American Library Association. (2013). Why is Intellectual Freedom. Retrieved on October/27/2013.

[3]   The American Library Association. (2002). ALA applauds federal court ruling on the Children's Internet Protection Act. Retrieved on October/27/2013.

[4]   The American Library Association. (2013). How Do You Guide Children When You Can't Be with Them 24 Hours A Day?. Retrieved on October/27/2013.

[5]   [5] The American Library Association. (2013). What is Censorship, How Does Censorship. Retrieved on October/27/2013.

[6]   The American Family Association. (2013). Your Children and the Internet. Retrieved on November/4/2013.

[7]   Federal Trade Commission.(1999). Children's Online Privacy Protection Rule--Comment P994504(American Library Association Submission). Retrieved on October/20/2013.

[8]   Bruening, P& Davidson, A. (2013). Constitutionality of Internet Filtering Mandate Challenged in Court. Retrieved on November/4/2013.

[9]   The Center for Democracy and Technology. (2000). Free Speech. Retrieved at November/10/2013.

[10]   Banks, M. (1998). Filtering the net library: the case (mostly) in favor. Computer in libraries.18(3), 50.

[11]   Zittrain, J & Edelman, B. (2002). Documentation of the Internet Filtering in Saudi Arabia. Berkman center for Internet & society Harvard School. Re-trieved on October/21/2013.

[12]   Robert H. Burger. (1993). Information Policy: A framework for evaluations and policy research. Ablex Publishing. April, 1993.

[13]   Busha, C. H. (1972.) Intellectual Freedom and censorship: The climate of opinion in midwestern public library. Library Quarterly, 24(3), 283-301.

[14]   Yam, A. (2002) Saudi Arabia censors the web. It Matters. More then the Bits and Bytes. Retrieved on October/21/2003.

[15]   Al-Sarami, N. (1999). Problems and Possibilities; Internet in the Kingdom. Saudi Gazette, March 13.

[16]   Al-Rasheed, A. (2001). The Internet in Saudi Arabia (management view). Communications engineering technical exchange meeting 2001, 30 April- 2may. ARAMCO, Al-dharan.

[17]   Langland, L. (1997). Public libraries, Intellectual freedom, and the internet: To filter or not to filter. PNLA Quarterly 16 (3).14.

[18]   Al-Adhel, A. (1998). The regulations on ISPs forbid them from establishing any linkage to the Internet except via the KACST. The rules were published in the May 6, 1998. Al-Jazira daily, as reported in FBIS.

[19]   Network Buyer's.(1999). Glossary of Industry Terms. Retrieved on October/27/2003.

[20]   TechTarget Corporate Web Site. (2003). Filtering Definition. the guide to the Tech Target network of industry-specific IT Web sites. Retrieved on Octo-ber/29/2003.

[21]   Douglas, J. (1999). The Internet's 'Open Sesame' Is Answered Warily," New York Times. March 18.

[22]   Al-Kdahi, M. (2003).King Abdulaziz City for Science & Technology (KACST). The Internet Services Unit. Retrieved on November10/2013.

[23]   Whitaker, B. (2000). Saudi claim victory in war for control of web. The Guardian, 11 May.

[24]   Richtel, M. (1999). Tables Turn on a Filtering Site As It Is Temporarily Blocked. New York Times, March 11.

[25]   The American Library Association. (1999). American Library Association and ACLU Court Challenge to Children's Internet Protection Act (CHIPA)1999. Retrieved on October/5/2013.

21

Improvement of Granular Subgrade Soil by Using Geotextile and Jute Fiber

Md. Akhtar Hossain, Akib Adnan, Md. Maskurul Alam

Department of Civil Engineering, Rajshahi University of Engineering & Technology, Rajshahi, Bangladesh

Email address:

akhtar412002@yahoo.com (Md. A. Hossain), akib_adnan@yahoo.com (A. Adnan), maskurul090018@gmail.com (Md. M. Alam)

Abstract: Geotextiles and jute fibers both have been successfully used for reinforcement of soils to improve the bearing capacity. In the present study, firstly the geotextile is used as a tensional material for reinforcement of granular soils. Laboratory California Bearing Ratio (CBR) tests were performed to investigate the load-penetration behavior of reinforced and unreinforced granular soils with geotextile. By placing geotextile at certain depth within sample height in one, two or three layers were tested under soaked condition to investigate the effects of the number of geotextile layer on the increase in bearing capacity. Secondly, laboratory tests were performed to investigate the behavior of granular soil reinforced with jute fiber of various aspect ratio mixed with soil at 0.5%, 1.0%, 1.5%, 2.0% by weight of the soil. Finally, the granular soil was reinforced with the combination of geotextile (top and middle layer of the sample) and jute fiber (0.5% and 1% by the weight of soil). The experimental results were then studied and compared to determine the most effective combination of geotextile and jute fiber to reinforce the studied granular soil.

Keywords: CBR Tests, Jute Fiber, Geotextile, Bearing Capacity, Granular Soil

1. Introduction

The uses of geotextile in many engineering applications have become more apparent and have proven to be an effective means of soil improvement. Soil improvement in the broadest sense, is the alteration of any property of a soil to improve its engineering performance. It also comprises any process which increases or maintains the natural strength of soil. In early applications in roads and airfield construction, emphasis was laid on the load bearing function of the geotextile. Resl and Werner (1986) carried out the laboratory tests under an axi-symmetric loading condition using nonwoven, needle-punched geotextiles. The result showed that the geotextile layer placed between subbase and subgrade can significantly increase the bearing capacity of soft subgrades. Fannin and Sigurdsson (1996) carried out a full scale field trial to observe the performance of different geosynthetics in unpaved road construction over the soft ground. Numerous papers have examined the reinforcement of soil (Bergado *et al.*, 2001; Raymond and Ismail, 2003; Park and Tan,2005; Yetimoglu *et al.*, 2005). Current research works are mainly emphasized on improving the strength mechanism and bearing capacity of

the reinforced soil by adding jute fiber and geogrid (Hossain *et al.*, 2015; Allahbakhshi, M. and Sadeghi, H. 2014). In this study, CBR test carried out on nonwoven needle-punched geotextile combines with the granular soils, the geotextile reinforcement placed between three different subgrade layers and the comparison between bearing capacity of soil with and without geotextile reinforcement under axi-symmetric loading condition was investigated. Further tests were carried out of soil reinforced with jute fiber of various aspect ratios and a number of combined reinforcements.

2. Methodology

Samples with various geotextile layers and various percentage of jute fiber were prepared and tested. The test results were compared to determine the combination reinforcement. The combination reinforcement was applied and similar tests were performed.

Samples Tested

1. Sample without any reinforcement.

2. Sample reinforced with 0.5%, 1.0%, 1.5%, 2.0% of 50mm jute fiber.

3. Sample reinforced with 0.5%, 1.0%, 1.5%,2.0% of 100mm jute fiber.

4. Sample reinforced with 0.5%, 1.0%, 1.5%, 2.0% of 150mm jute fiber.

5. Sample reinforced with single layer of geotextile (top, middle, bottom).

6. Sample reinforced with two layer of geotextile (top-middle, top-bottom, middle-bottom).

7. Sample reinforced with three layer of geotextile (top-middle-bottom).

8. Sample reinforced with 0.5% jute fiber (50mm) and single layer (top) of geotextile.

9. Sample reinforced with 0.5% jute fiber (50mm) and single layer (middle) of geotextile.

10. Sample reinforced with 1.0% jute fiber (50mm) and single layer (top) of geotextile.

3. Materials Used

Collection of Soil Sample and It's Geotechnical Properties
Soil sample obtained locally is used for the present experimental investigations. Sample used in this research was collected from the bank of river Padma at Talaimari, Rajshahi. The required properties of the soil were determined and are presented on Table.3.1.

Table 3.1. Basic Geotechnical Properties of Soil Sample.

Optimum Moisture Content	13
Specific Gravity	2.635
Angle of Internal Friction	34.14
Finness modulus	2.611

Table 3.2. Basic Engineering properties of geotextile.

Weight(g/m²)	175
Thickness(mm)	0.9
Static Puncture(N)	170
Tensile Strength(KN/m)	13
Elongation at Peak Stress	45-50%

Table 3.3. Basic Properties of Jute Fiber.

Weight per Unit Length (gm/cm)	0.36
Diameter (mm)	4
Length (mm)	50(A.R. = 12.5) 100(A.R. = 25) 150(A.R. = 37.5)

4. Test Results

Soil sample was mixed with 0.5%,1.0%,1.5%,2.0% jute fiber of various length(10mm,20mm,40mm) and reinforced with 1 layer, 2 layers,3 layers of geotextiles.CBR tests were performed for all test samples. Optimum moisture content

of sample with and without jute fibers (differently for each soil-fiber ratio) was experimentally achieved by Modified Proctor test. Later on various blends on jute fibers and geotextile layers were used to achieve similar or improved CBR values with less layers of geotextiles to reduce the overall cost of a project.

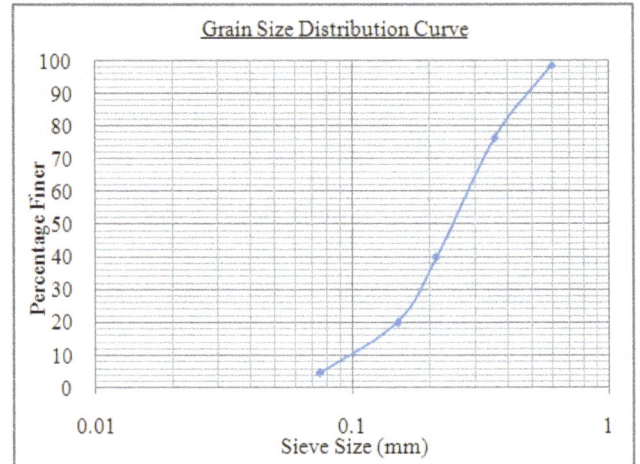

Fig. 4.1. Grain size distribution of Sample.

Table 4.1. Optimum moisture content of sample with and without jute fiber.

Jute Fiber Content(% by weight)	Optimum Moisture Content (% from Modified Proctor test)
0.0	13.5
0.5	14.8
1.0	15.6
1.5	17.1
2.0	18.6

4.1. CBR Test Results for Various Lengths and Amount of Jute Fiber

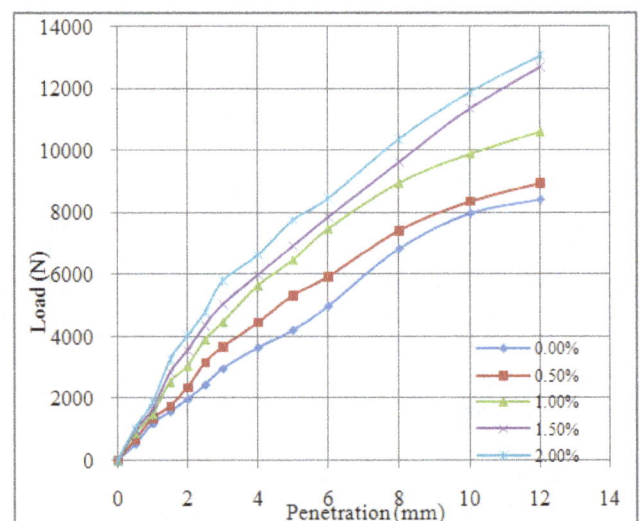

Fig. 4.2. Load vs Penetration Plot for various % of jute fiber (50 mm) mixed subgrade soil in unsoaked condition.

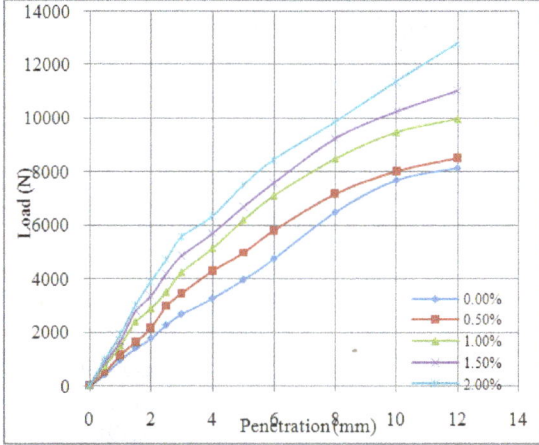

Fig. 4.3. *Load vs Penetration Plot for various % of jute fiber (50 mm) mixed subgrade soil in soaked condition.*

Table 4.2. *CBR Test Results for jute fiber (50 mm) mixed sample.*

% of jute fiber by weight	Unsoaked		Soaked	
	CBR Value	% Increase in CBR	CBR Value	% Increase in CBR
0.0%	17	-	15	-
0.5%	25	47.1	23	53.2
1.0%	29	70.25	27	80.1
1.5%	34	100.1	29	93.33
2.0%	36	111.5	32	113.6

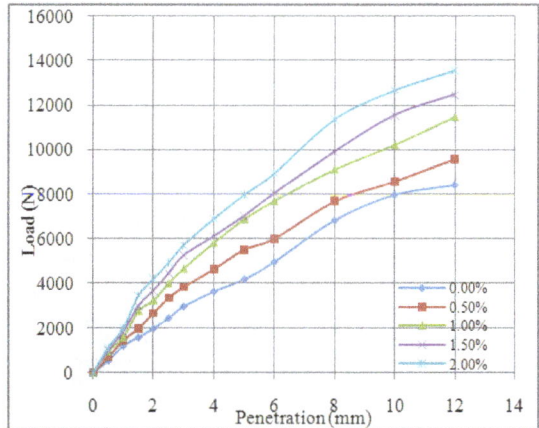

Fig. 4.4. *Load vs Penetration Plot for various % of jute fiber (100 mm) mixed subgrade soil in unsoaked condition.*

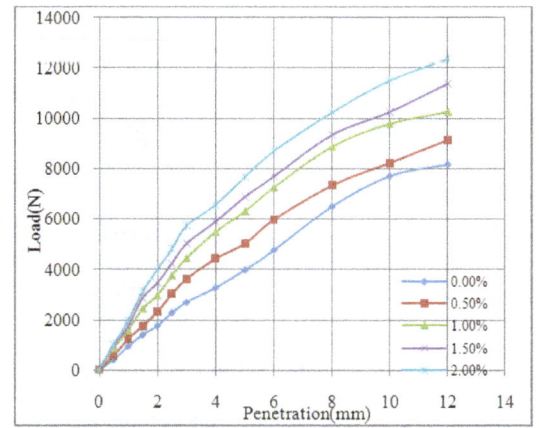

Fig. 4.5. *Load vs Penetration Plot for % of jute fiber (100 mm) mixed subgrade soil in soaked condition.*

Table 4.3. *CBR Test Results for jute fiber(100mm)mixed sample.*

% of jute fiber by weight	Unsoaked		Soaked	
	CBR Value	% Increase in CBR	CBR Value	% Increase in CBR
0.0%	17	-	15	-
0.5%	29	70.5	24	60
1.0%	33	94.1	28	86.6
1.5%	35	105.4	30	100
2.0%	37	117.7	31	106.6

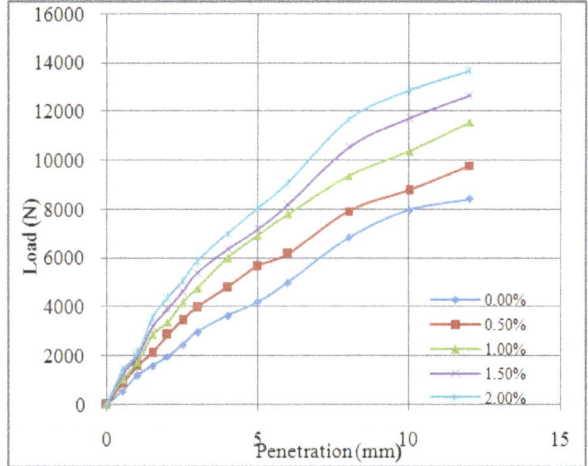

Fig. 4.6. *Load vs Penetration Plot for % of jute fiber (150 mm) mixed subgrade soil in unsoaked condition.*

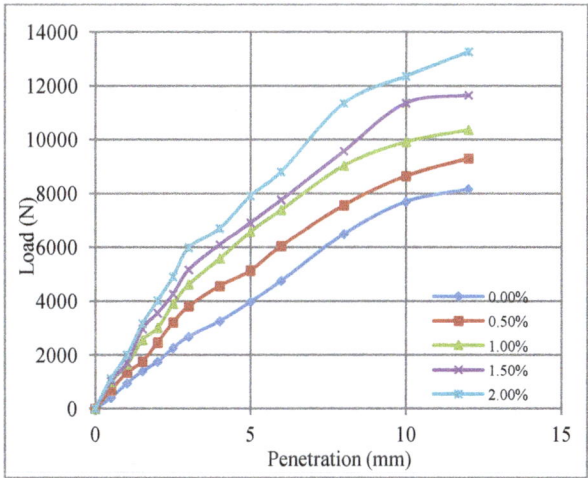

Fig. 4.7. *Load vs Penetration Plot for % of jute fiber (150 mm) mixed subgrade soil in soaked condition.*

Table 4.4. *CBR Test Results for jute fiber (150 mm) mixed sample.*

% of jute fiber by weight	Unsoaked		Soaked	
	CBR Value	% Increase in CBR	CBR Value	% Increase in CBR
0.0%	17	-	15	-
0.5%	30	76.4	25	66.6
1.0%	35	105.8	30	100
1.5%	36	111.7	31	106.6
2.0%	38	123.5	32	113.3

4.2. Comparison of CBR Values due to Mixing of Jute Fiber

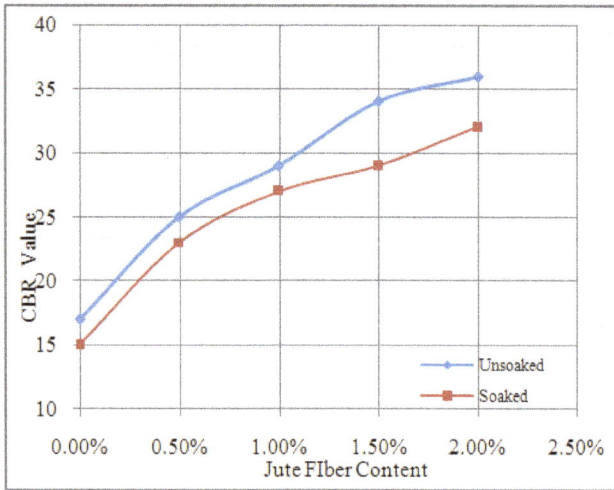

Fig. 4.8. *Increased CBR value vs jute fiber (50 mm) content plot.*

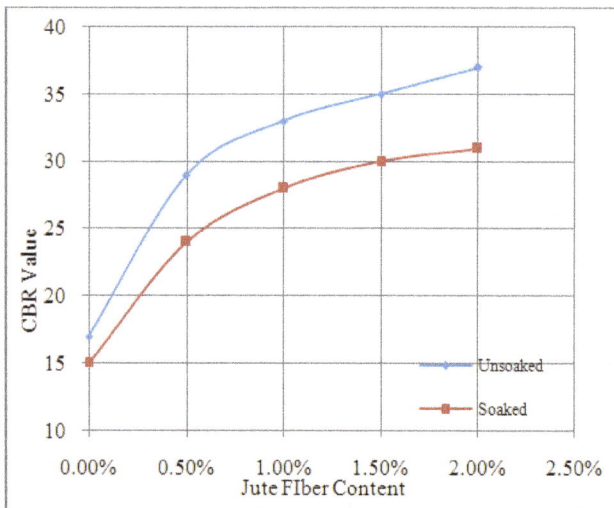

Fig. 4.9. *Increased of CBR value vs jute fiber (100 mm) content plot.*

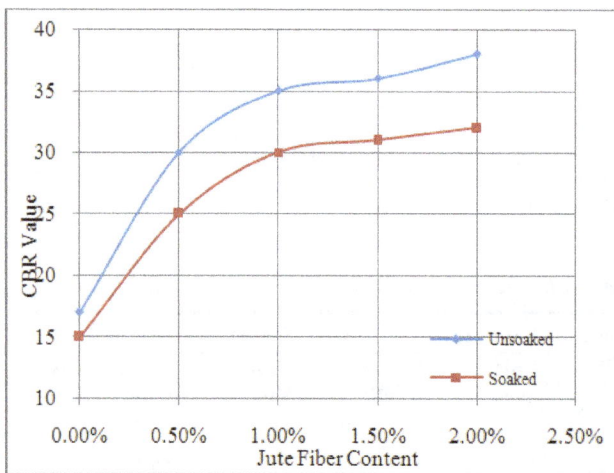

Fig. 4.10. *Increased of CBR value vs jute fiber (150 mm) content plot.*

4.3. CBR Test Results for Various Geotextile Layers

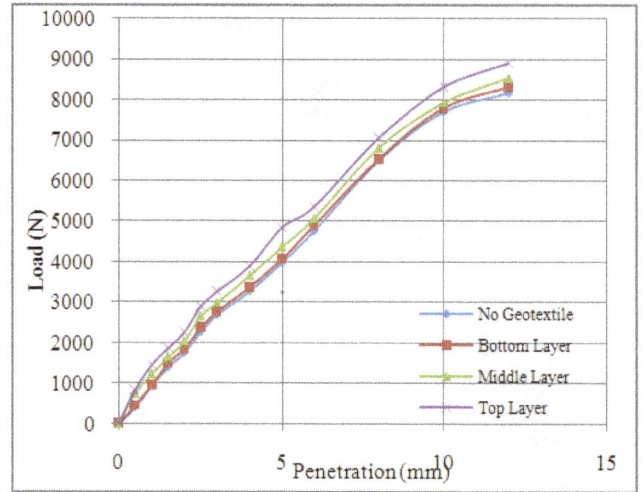

Fig. 4.11. *Load vs penetration plot for soil reinforced with single layer of geotextile under soaked condition.*

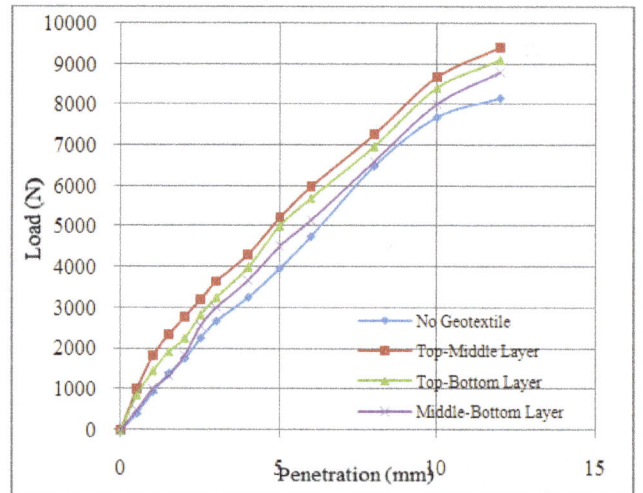

Fig. 4.12. *Load vs penetration plot for soil reinforced with double layer of geotextile under soaked condition.*

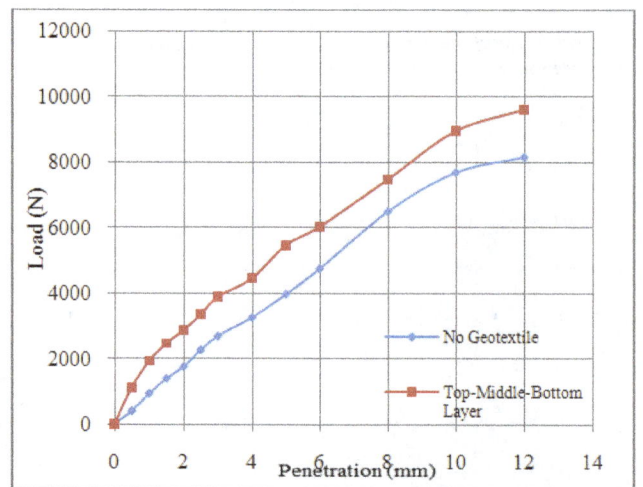

Fig. 4.13. *Load vs penetration plot for soil reinforced with triple layer of geotextile under soaked condition.*

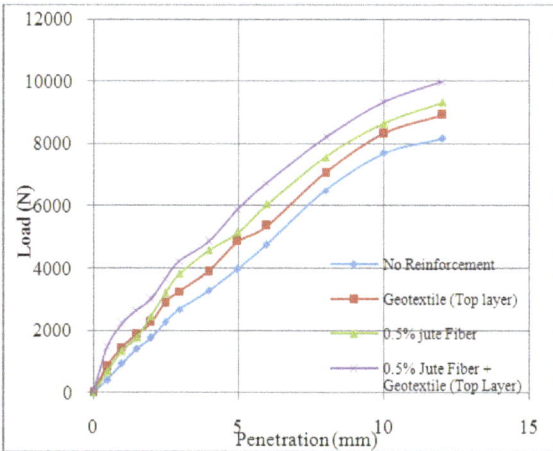

Fig. 4.14. *Load vs penetration plot for soil reinforced with 1 layer of geotextile at top and 0.5% jute fiber (150 mm) under soaked condition.*

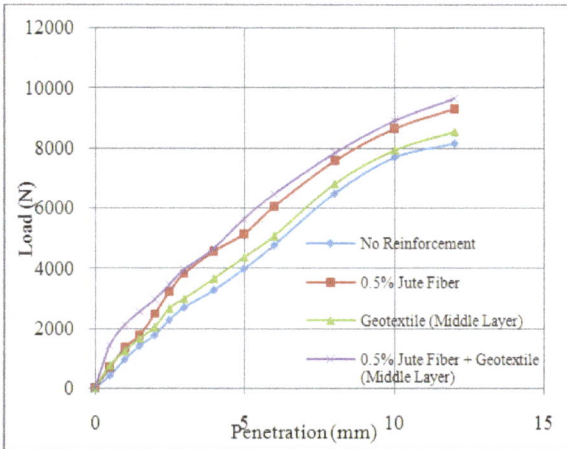

Fig. 4.15. *Load vs penetration plot for soil reinforced with 1 layer of geotextile at middle and 0.5% jute fiber (150 mm) under soaked condition.*

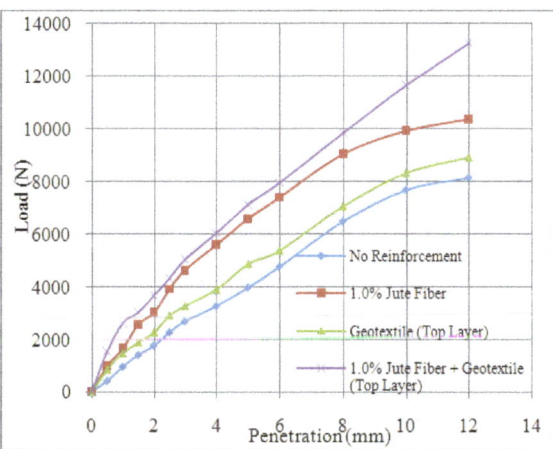

Fig. 4.16. *Load vs penetration plot for soil reinforced with 1 layer of geotextile at top and 1.0% jute fiber (150 mm) under soaked condition.*

Table 4.5. *CBR values for soil reinforced with combination of geotextile and jute fiber.*

Jute fiber(%)	Without Geotextile	One layer at top		One layer at middle	
		CBR	% Increase	CBR	% Increase
0.0	15	19	26.6	18	20
0.5	25	32	113.3	31	106.6
1.0	30	51	240	-	-

5. Results and Discussion

By studying Fig. 4.2, Fig. 4.3, Fig. 4.4, Fig. 4.5, Fig. 4.6, Fig. 4.7 which show the load vs penetration plots of the soil sample with jute fiber reinforcement at various percentages, it can be said that increasing both the percentage of jute fiber and aspect ratio increases the bearing capacity of the soil. Fig. 4.8, Fig. 4.9, Fig. 4.10 shows the increase of CBR value due to adding jute fiber. Fig. 4.11, Fig. 4.12, Fig.4.13 shows the load vs penetration plot for soil reinforced with geotextile. From the Fig. 4.11, Fig. 4.12, Fig.4.13 we can conclude that placing geotextile at top layer is more effective than placing it anywhere else in the soil sample. Fig. 4.14, Fig. 4.15, Fig 4.16 shows the load vs penetration plot for soil reinforced with combination reinforcements discussed above. From Fig. 4.14, Fig. 4.15, Fig 4.16 it can be said that placing the geotextile at top and increasing jute fiber content increases the CBR value most effectively. But from economic point of view increasing the aspect ratio of jute fiber is more effective than increasing its percentage in the mixture.

6. Conclusion

The following observations were made on the behavior of unreinforced soil, soil reinforced with geotextile, jute fiber mixed soil and soil reinforced by both geotextile and jute fiber. From the tests performed it is clearly evident that increasing the percentage of geotextile will increase the load bearing capacity of the soil. Though bearing capacity of sample having the same percentage of jute fiber varies with the length of jute fiber used and CBR value increases with the increase of jute fiber which can be observed by studying Table 4.2, Table 4.3, Table 4.4, as a result 50mm and 100mm jute fibers were rule out in case of combined reinforcement. The effectiveness of geotextiles is governed by the no. of layers used and the placement of the layer. Tests show that placing the layer of geotextile at the top and middle of the soil sample gives the most effective results. The most effective result is achieved by using the geotextile and jute fiber simultaneously. Tests show that a single layer of geotextile coupled with 0.5% and 1.0% jute fiber gives almost the same CBR values(Table 4.5) as a sample reinforced with 2/3 layers of geotextile or 1.5/2.0% jute fiber which will effectively reduce the cost of the project. Through the tests performed earlier it is evident that:

1. For jute fiber, CBR value increases with the increase of both fiber content and aspect ratio. It is worth mentioning that the highest amount of change is recorded for 0.5% and 1.0% fiber content for all aspect ratios.

2. Tests show that placing of geotextile at top is more effective than placing it at middle or at bottom.

3. Increased layer of both geotextile and length and percentage of jute imparts increasing bearing capacity to the granular subgrade soil.

4. Combination reinforcement with single layer geotextile and only 0.5% and 1.0% jute fiber is as effective as two or three layers of geotextile.

References

[1]   Allahbakhshi, M. and Sadeghi, H. (2014) "Behaviour of industrial machinery foundation on prestressed geogrid-reinforced embankment over soft soil under static load", *Journal of Science Publishing Group*, Vol. 2, Issue 6, pp. 65-73.

[2]   Al-Moussawi, H.M. and Andersland, O.B. (1987) Discussion on the paper "Behavior of Fabric- versus Fiber- Reinforced Sand", *J. Geotech. Engg., ASCE*, Vol. 113(7), pp.381-387.

[3]   Datta, M. (1999) "Engineering Properties of Coal Ash", *Proceedings of Indian Geotechnical Conference*, New Delhi, pp.41-45.

[4]   Bergado, D.T., Youwai, S., Hai, C.N. and Voottipruex, P. (2001) "Interaction of nonwoven needle-punched geotextiles under axisymmetric loading conditions". *Geotextiles and Geomembranes*, Vol. 19, pp.299-328

[5]   Fannin, R.J., O. Sigurdsson (1996) "Field observations on stabilization of unpaved roads with geosynthetics". *Journal of Geotechnical Engineering, ASCE*, 122 (7), pp.544–553.

[6]   Haeri, S.M., R. Nourzad, A.M. Oskrouch (2000) "Effect of geotextile reinforcement on the mechanical behavior of sands". *Geotextiles and Geomembranes* 18 (6), pp.385–402.

[7]   Hossain, M.A., Hossain, M.S. and Hasan, M.K. (2015) "Application of Jute Fiber for the Improvement of Subgrade Characteristics" *Journal of Science Publishing Group*, Vol. 2, Issue 3, pp. 26-30.

[8]   Kaniraj, S.R., and Rao, G.V. (1994) "Trends in the use of geotextiles in India", *Journal of geotextiles and geomembranes, pp.*13389 - 402.

[9]   Latha, G.M, Murthy, V.S., (2007) "Effects of reinforcement form on the behavior of geosynthetic reinforced sand". *Geotextiles and Geomembrane*, Vol. 25, pp. 23–32.

[10] Park, T., S.A. Tan (2005) Enhanced performance of reinforced soil walls by the inclusion of short fiber. *Geotextiles and Geomembranes*, Vol 2 (4), pp. 348–361.

[11] Raymond, G., Ismail, I.,(2003) "The effect of geogrid reinforcement on unbound aggregates". *Geotextiles and Geomembranes*, Vol. 21, *pp.*355–380.

[12] Resl, S., Werner, G., (1986) "The influence of nonwoven needle- punched geotextiles on the ultimate bearing capacity of the subgrade". *Proceedings of the Third International Conference on Geotextiles*, Vienna, Vol. 4, *pp.* 1009– 1013.

[13] Yetimoglu, T., Salbas, O., (2003) "A study on shear strength of sands reinforced with randomly distributed discrete fibers". *Geotextiles and Geomembranes*, Vol. 21 (2), *pp.*103–110.

22

Exploration of Pertinent Apparel Production Systems for Bangladeshi Clothing Industries

Nasrin Ferdous, Reashad Bin Kabir

Department of Apparel Manufacturing Management & Technology, Shanto-Mariam University of Creative Technology, Dhaka, Bangladesh

Email address:

nasrinferdous@gmail.com (N. Ferdous), robinreashad@yahoo.com (R. B. Kabir)

Abstract: Like most industries that adopt a reactive manufacturing strategy, the clothing industry changes only when external forces dictate that the current approach or strategy will no longer satisfy the prevailing business environment. This paper presents the Apparel Manufacturing Systems and discusses the underlying premises that support the success of Apparel Production Systems both in the formative stages and during sustained operation. A review of the known origins of manufacturing processes illustrate how the production system can be used to advantage for clothing manufacture. No single solution fits all products/tasks in fully optimizing Socio technical System (STS), given different technologies, environment and people, etc. An attempt was made in this paper to show the handy uses of diverse production system in Bangladeshi Garments Industries.

Keywords: Clothing Industry (CLI), Production (PRN), Manufacturing (MAN), Production Management System (PMS), Product Line (PRL)

1. Introduction

An apparel Production system is an integration of materials handling, production processes, personnel, and equipment that directs workflow and generates finished products. Three types of production systems commonly used to produce apparel are progressive bundle, unit Production, and modular production. Each system requires an appropriate management philosophy, materials handling methods, floor layout, and employee training. Firms may combine or adapt these systems to meet their specific production needs. Firms may use only one systems a combination of systems for one product line, or different systems for different product lines in the same plant [1].

Since the mid-1970s, Apparel Manufacturers and other Apparel companies that contract for the production of apparel, have searched for strategies suitable to capture the increasingly elusive apparel customer while searching for ways to cut costs and deliver more product variety at a faster pace [2,3,4]. During these decades, low-cost labor from a number of global sources was a lure that drew U.S. apparel companies to use off-shore manufacturing facilities however the Rising costs of fuel for transportation and the demand for higher wages in many countries are causing apparel industries to re-examine their current off-shore strategies [5, 6].

As apparel companies consider a return to domestic production, they will be making a choice in production facilities, including production systems. These companies must hire or acquire production systems that will accommodate variations in style changes, shorter lead times and smaller size, in other words, companies need production system that will provide flexibility, speed and cost reductions. Selection of the right production system is considered critical to market success [5, 7, 8].

The ever improving living standards and rapidly changing fashion trends are pushing apparel manufacturers to respond as fast as possible to model and quantity changes and to produce high-quality, low-cost products. Given this challenge, new production systems were adopted and experimented in the apparel industries [9].

2. Literature Review

2.1. Classification Production System

Production System can be classified in different ways. A

number of writers and experts segmented this in different ways. They are as follows:

Types of production system based on order quantity

a. Individual Production: It is also known as making through. It is the traditional method in which entire garment is assembled by one operator. Each product is made only once or in very small quantities. This system requires highly experienced operators and versatile machines.

b. Batch Production: It is used for larger, though fixed quantities of large identical articles either for stock or to order.

c. Mass Production: In this production, large quantities of identical products are made continuously [10].

Types of production system based on customer or consumer nature

a. Bespoke Production: This type of production is on behalf of individual clients, according to individual size and requirements.

b. Industrial Production: Production is geared to an anonymous, statistically and/ or demographically and culturally defined target consumer group [11].

2.2. Commonly Used Production Systems in Bangladeshi Apparel Industries

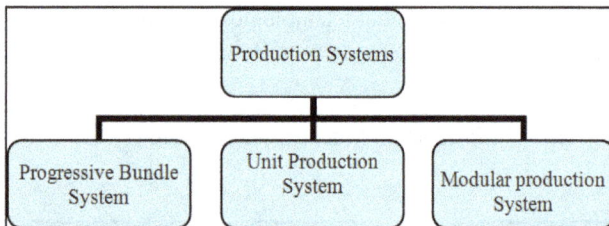

Figure 1. Commonly used to Production Systems in Apparel Industries.

Progressive bundle system (PBS)

The progressive bundle system (PBS) gets its name from the bundles of garment parts that are moved sequentially from operation to operation, This system often referred to as the traditional production system has been widely used by apparel manufacturers for several decades and sill in use today. The Technical Advisory Committee of AA.MA (1993) reports that 80% of apparel manufacturers used the bundle system. The committee also predicts that use of bundle system would decrease as firms seek more flexibility in their production systems.

Bundles consist of garment parts needed to complete a specific operation or garment component. For example, an operation bundle for pocket setting might include shirtfronts and pockets that are to be attached. Bundle sizes may range from 2 to 100 parts. Some firms operate with a standard bundle size, while other firms vary bundle sizes according to cutting orders, fabric shading, size of the pieces in the bundle, and the operation that is to be completed. Some firms use a dozen or multiples of a dozen because their sales are in dozens.

Bundles are assembled in the cutting room where cut parts are matched up with corresponding parts and bundle tickets.

Bundle tickets consist of a master list of operations and corresponding coupons for each operation. Each bundle receives a ticket that identifies the style number, size, shade number, list of operations for routing, and the piece rate for each operation. Operators retain a corresponding segment of the bundle coupon for each bundle they complete. At the end of the workday, bundle coupons are turned in, and the earned time from completed bundle tickets is totaled to determine the operator's compensation. Firms may use electronic bundle tickets or smart cards that accompany each bundle and that are swiped at each workstation along with identification cards. This reduces paper-work, facilitates access to information, and eliminates lost bundle tickets.

Bundles of cut parts are transported to the sewing room and are given to the operator scheduled to complete the operation. One operator is expected to per- form the same operation on all the pieces in the bundle, to retie the bundle, to process the coupon, and to set it aside until it is picked up and moved to the next operation. A progressive bundle system may require a high volume of work in process because of the number of units in the bundles and the large buffer of backup work that is needed to ensure a continuous workflow for all operators (Figure 2). The firm's materials handling system facilitates bundle movement between operations [1].

A [12]

B [13]

Figure 2. Progressive Bundle System.

However, the progressive bundle system may be used with a skill center or line layout depending on the order that bundles are advanced through production. Each style may have different processing requirements and thus different routing. Routing identifies the basic operations, sequence of production, and the work centers where those operations are to be performed, some operations are common to many styles and at those operations, and work may build up waiting to be processed.

The progressive bundle system is driven by cost-efficiency for individual operations. Operators perform the same operation on a continuing basis, which allows them to increase their speed and productivity. Operators who are compensated by piece rates become extremely efficient at one operation and may not be willing to learn a new operation because it reduces their efficiency and earnings. Individual operators that work in a progressive bundle system are independent of other operators and the final product.

Slow processing, absenteeism, and equipment failure may also cause major bottlenecks within the system. Large quantities of work in process are often characteristic of this type of production system. This may lead to longer throughput time, poor quality concealed within bundles, large inventory, extra handling, and difficulty in controlling inventory

The success of a bundle system may depend on how the system is set up and used in a plant. This system may allow better utilization of specialized machines, because output from one special-purpose automated machine may be able to supply several operators for the next operation. Small bundles allow faster throughput unless there are bottlenecks and extensive waiting between operations. [11]

Positive sides of progressive bundle system:
- This system may allow better utilization of specialized machines, as output from one special purpose automated ma-chine may be able to supply several operators for the next operation.
- Small bundles allow faster throughput unless there are bottlenecks and extensive waiting between operations.
- Weakness of progressive bundle system:
- Slow processing, absenteeism, and equipment failure may also cause major bottlenecks within the system
- Large quantities of work in process. This may lead to longer throughput time, poor quality concealed by bundles
- Large inventory, extra handling and difficulty in controlling inventory. [14]

Unit production system (UPS)

As a mechanical system this has been in use for many years, but a major advance was made in 1983 when computers were first used to plan, control and direct the flow of work through the system.

The operational principles are as follows:

All the components for one garment are loaded into a carrier at a workstation specially designed for this purpose. The carrier itself is divided into sections, with each section having a quick-release clamp, which prevents the components from falling out during movement through the system. When a batch of garments has been loaded into carriers they are fed past a mechanical or electronic device, which records the number of the carrier and addresses it to its first destination. Some of the more intelligent systems address the carriers with all the destinations they will have to pass through to completion.

The loaded carriers are then fed onto the main powered line, which continually circulates between the rows of machines. This main, or head, line is connected to each workstation by junctions, which open automatically if the work on a carrier is addressed to that particular station. The carrier is directed to the left side of the operator and waits its turn along with the other carriers in the station.

When the operator has completed work on one carrier, a push button at the side of the sewing machine is pressed and this actuates a mechanism, which transports the carrier back to the main line. As one carrier leaves the station, another is automatically fed in to take its place. When the carrier leaves the station it is recorded on the data collection system, and then addressed to its next destination.

Unit Production System requires substantial investments, which are not always justified by conventional payback calculations. Apart from the measurable tangible benefits, UPS also have many intangible benefits such as a more orderly and controlled flow of work, and the ability via the control computer of simulating the production situation some time in advance. These intangibles are difficult to measure, but in themselves make a very positive contribution to the overall viability of the unit.

Overall, unit production systems have major advantages over the other entire manual and the mechanical systems used for the mass production of clothing. Most importantly, they provide a clothing factory with the capability to respond quickly to any changes, which might occur. In the fast moving fashion business, this is essential.

Positive sides of unit production system:
- Bundle handling completely eliminated.
- The time involved in the pick-up and disposal is reduced to minimum.
- Output is automatically recorded, eliminates the operator to register the work.
- The computerized systems automatically balance the work between stations.
- Up to 40 styles can be produced simultaneously on one system.

Weakness of unit production system:
- Unit production system requires high investments.
- The payback period of the investment takes long time.
- Proper planning is required to be effective. [15]

Table 1. Comparison between PBS and UPS [16].

Parameters	Progressive Bundle System (PBS)	Unit Production System (UPS)
Transportation	-Manual transportation, many times helper are hired for this bundle transportation job. -Operators stop their work to fetch bundles. -Less effective in terms of production management. Resulted long response time.	-In this system an automated mechanical system carries pieces to each work stations. -Easier pick up and dispose at each work station. Resulted quick response time
Through put time	-Compare to UPS, through put time longer in PBS. How much long will depend on the bundle size and no. of bundles kept in between two operators.	-Through put time in UPS is less compare to PBS. But it is not the minimum time as in this system there is WIP in between two operators.
Direct Labor content	- Direct labor content is high because usually operator does tying and untying of bundles, positioning components, pulling the bundle ticket and handling of work pieces.	-Direct labor content is less than PBS because an operator only sews the garment part rather than other tasks. In this system garment parts are held by the over head hanger, so less handling of garment components.
WIP level	-In PBS generally operators are asked to sew as much pieces as they can without considering back and front operators. This resulted piling up of work in the operations with higher work content.	-Less WIP in between operators. As workstation has limit of holding no. of hangers. Also after completion of operation hangers are transported to the next operation automatically.
Cutting work requirement	-As a result of High Work In Process (WIP) is required by sewing section, cutting sections are required to perform 60-70% more than actual production can handle.	-Lower WIP results in less cutting works. A balanced flow of material established in between cutting and sewing line.
Inventory Level	-Due high WIP and higher cutting, fabrics and trims need to stock in advance	-Less inventory for fabric and trims.
Excess labor requirement	- Usually in PBS needs more overtime works, repair work due to some unfinished operations.	-Plant with UPS system needs less overtime as planning is easy in this manufacturing system.

Modular Manufacturing Systems (MDS)

A modular garments production system is a contained, manageable work unit that includes an empowered work team, equipment, and work to be executed. Modules frequently operate as minuscule factories with teams responsible for group goals and self-management. The number of teams in a plant varies with the size and needs of the firm and product line in garments. Teams can have a niche function as long as there are orders for that type of garments product, but the success of this type of garments operation is in the flexibility of being able to produce a wide variety of products in small quantities in garments.

Positive sides of modular manufacturing systems:
- High flexibility
- Fast throughput times
- Low wastages
- Improved Quality

Weakness of modular manufacturing systems:
- A high capital investment in equipment.
- High investment in initial training.
- High cost incurred in continued training [15]

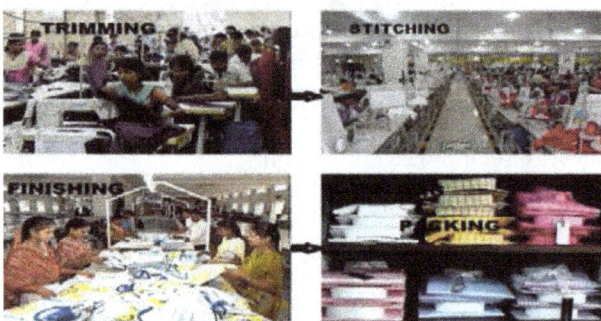

Figure 3. Modular Garments Production System. [14].

It seems that MPS is the perfect solution for the apparel manufacturer to respond to the quick replenishment requirement by the retailers. [17]. Although MPS can provide obvious benefits for the apparel manufactures, the diffusion of MPS in the apparel industry is slow. The possible reasons for this phenomenon are studied by the researchers, most from the perspective of human resource practices [18, 19].

Table 2. Predicted Attributes of the Three Commonly Adopted Production Systems [20].

System Attributes	Production Systems		
	Bundle	Progressive Bundle	Modular
Workflow	Push	Push	Pull
Method of retrieval to workstations	Brought to operator or self-retrieved from general storage	Brought to operator from operator by cart or conveyor	Hand off
Work-in-process(WPI) inventory	High levels (racks or carts of bundles)	Moderate (enough to balance the lines)	Zero to minimal
Number of task per operator	Single task or whole garment	Single task	Single to multiple tasks
Interaction between operators	No teamwork	No teamwork	No teamwork

3. Conclusion and Implications

A vital thing is pointed out by manufacturing connoisseur and researchers that bewilderment subsist with the terminology concerning apparel production systems, and contemporary apparel literature does not afford industry-based and empirically scrutinized definitions for the three most widespread production systems (i.e., bundle, progression bundle and modular).

In Bangladesh primarily Progressive Bundle System (PBS) is used as 80 percent of the Exporting products are Knitted Products. For knitted products, all items bundle together. Example: Esquire Knit Composite, etc.

For woven garments, a group of item bundle tighter, Example: Ananta Apparels, etc.

But in some industries Unit Production System (UPS) is used. Example: Pacific Jeans.

Modular Manufacturing Systems implemented by all industries in sample section.

In a nutshell, it can be said that, all sorts of production systems are available in Bangladeshi apparel industry; nonetheless the use of Progressive Bundle System (PBS) is significantly higher than others.

References

[1]   Ruth E. Glock, Grace I. Kunz. ; (2005), *Apparel Manufacturing: Sewn Product Analysis* (4th ed.), P. 345.

[2]   Doeringer, P., & Crean, S. (2006). Can fast fashion save the U.S. apparel industry? *Socio-Economic Review*, 4, 353-377.

[3]   Monacarz, H.T. (1992), *Information technology vision for the U.S. fibre/textile / apparel industry* (NISTIR No. 4986). Gaithersburg, MD: U.S. Department of Commerce.

[4]   Park, H., & Kincade, D.H. (2011). A historic review of environmental factors and business strategies for U.S. apparel manufacturing industry, 1973-2005. *Research Journal of Textile and Apparel*, 15(4), 102-114.

[5]   Anonymous (May 14, 2011). Moving back to America: Multinational manufacturers. *The Economist*.

[6]   Friedman, A. (2012, September 5). Back in the U.S. they are WWD, Section II. Retrieved from http://www.wwd.com

[7]   Kim, Y., & Rucker, M. (2005). Production sourcing strategies in the U.S. apparel industry: A modified transaction cost approach. *Clothing and Textile Research Journal*, 23 (1), Pg.1-12.

[8]   Su, J., Dyer, C.L., & Gargeya, V.B. (2009), Strategic sourcing

and supplier selection in the U.S. textile-apparel-retail supply network. *Clothing and Textile Research Journal*, 27 (2), Pg.83-91.

[9]   G. Pan, (2014), A Quantitative Analysis of Cellular Manufacturing in Apparel Industry by Using Simulation, Journal of Industrial Engineering and Management, Vol :7 Iss: 5, Pg.1385-1396.

[10]   H. Eberle, et al.; (2008), *Clothing Technology: From fibre to fashion*, (5th ed); Pg. 128.

[11]   H. Eberle, et al.; (2008), *Clothing Technology: From fibre to fashion*, (5th ed) ; Pg. 128.

[12]   Ruth E. Glock, Grace I. Kunz.; (2005) *Apparel Manufacturing: Sewn Product Analysis* (4th ed.), Pg. 346.

[13]   The Indian Textile Journal (http://www.indiantextilejournal.com/articles/FAdetails.asp?id=1988)

[14]   B.Sudarshan, D. Nageswara Rao ; (2013) "Application of Modular manufacturing System in Garment Industries" ; *International Journal of Scientific & Engineering Research* ; Vol 4 ; Is 12; Pg: 2083-2089.

[15]   V. R. Babu; "Garment production systems: An overview"; *The Indian Textile Journal*.

[16]   *Online Clothing Industries* (http://www.onlineclothingstudy.com/2011/02/comparison-between-progressive-bundle.html).

[17]   X. Wang, CH. Chiu, and W. Guo (2014), "Improving the Performance of Modular Production in the Apparel Assembly: A Mathematical Programming Approach", *Mathematical Problems in Engineering*, Vol 2014, ID 472781, Pg.7 .

[18]   J. T. Dunlop and D. Weil (1996), "Diffusion and performance of modular production in the U.S. apparel industry," *Industrial Relations*, vol 35, no 3, Pg. 334–355.

[19]   P. Berg, E. Appelbaum, T. Bailey, and A. L. Kalleberg (1996), "The performance effects of modular production in the apparel industry," *Industrial Relations*, vol. 35, no. 3, Pg. 356–373, 1996.

[20]   D. Kincade, J. Kim, & k. Kanakadurga (2013). An empirical investigation of apparel production systems and product line groups through the use of collar designs. *Journal of Textile and Apparel, Technology and Management*, vol 8, iss 1, Pg. 1-15.

23

A Review of Distributed Generation Resource Types and Their Mathematical Models for Power Flow Analysis

Haruna Musa

Department of Electrical Engineering, Bayero University, Kano, Nigeria

Email address:

harunamusa2@yahoo.co.uk (H. Musa), hmusa.ele@buk.edu.ng (H. Musa)

Abstract: The emergence of Distributed Generation (DG) in distribution network has changed the configuration of this century's power system in terms of power flow. The reason for this is that DG affects the power flow and voltage conditions in the distribution system; contrary to its traditional unidirectional nature in radial configuration. It is worth mentioning that the change in the direction of power flow is not limited to the distribution network, but can as well extend to the transmission or sub-transmission systems, especially when DG penetration is high. This paper gives an overview of DG types and modeling techniques of the DG for power flow analysis during planning and operations. Various DG technologies are highlighted, different models of DGs are presented and some key challenges ahead with current drive towards smart grid networks is also discussed.

Keywords: Distribution System, Modeling, Distributed Generation, Distributed Generation Resources, Power Flow, Power Converters

1. Introduction

The increasing demand for green energy sources to replace fossil fuel sources has brought some rapid increase in penetration of DG in power systems. The penetration is certainly going to increase to a level that would have an impact on the entire system operation and performance [1]. DG penetration has increased more than 5 times during the last 8 years in some developing countries of Asia and this shows an increase that exceeds that of EU and US [2]. Therefore, a study on cumulative effect of high penetration on the entire network at all levels is necessary. Even though DG is associated with benefits, there are a lot more technical issues that are still not well understood and addressed [3]. For example, the studies conducted that relate interactions between transmission and distribution systems are based on simulations only. Unlike the simulations conducted on systems, real applications requires more adequate models of the DG units that can give better indication of the interactions in terms of power flows in the network. The models should meet certain basic requirements that can allow investigations at both local and global levels to be conducted. DG's interaction has been studied from the local/micro-grid point of view to the overall/global system point of view level all

with the aim of assessing its impacts on the overall network as the level of penetration increases. It was established that the impact of DG is no longer restricted to the local load or distribution network where these units are connected, but have impact on the entire transmission system [4]. Alot of studies were conducted on large DG penetration that focuses on system control and stability. However, these studies did not fully consider the various kinds of DG units, but mainly concentrated on stability aspect and control issues from the transmission system point of view. Thus, in [4] and [6], the impacts of DG penetration levels on power system transient stability are studied for different scenarios, and in [5], the impact of selected DG units, i.e. fuel cells (FCs) and micro turbines (MTs), on power system stability for various penetration levels were investigated. On the other hand, some studies have concentrated on the effect of DG units on the distribution network [7-10]. For example, in [10], the stability analysis of a distribution network with selected DG units, i.e. wind generators, and micro-turbines, was presented. Mathematical models of the DGs were integrated in three phase load flow algorithm and the special topological characteristics of distribution networks were fully utilized. The same algorithm was further improved to handle unbalanced three phase network based on ladder iterative method [11]. The interesting aspect of the algorithm is its capability of handling more than one source without system constraints violation.

This paper presents an overview of DG types and models developed so far for the purpose of power flow analysis. Although there have been numerous publications on the topic, a specific review work on the methods and models has not taken place. The paper has highlighted the work done by researchers to date with contributions made and outlined the key challenges ahead that are yet to be addressed with current drive towards smart grid networks.

2. DG Resource Types

Generally Distributed generation resources are defined as those resources that are directly used in the generation of electric power for connection to distribution system. These sources include traditional and non-traditional such as renewable, non-renewable and energy storage technologies like batteries, flywheels, superconducting magnetic energy storage, to mention but a few as categorized in figure 1.0 with illustrations of various technologies.

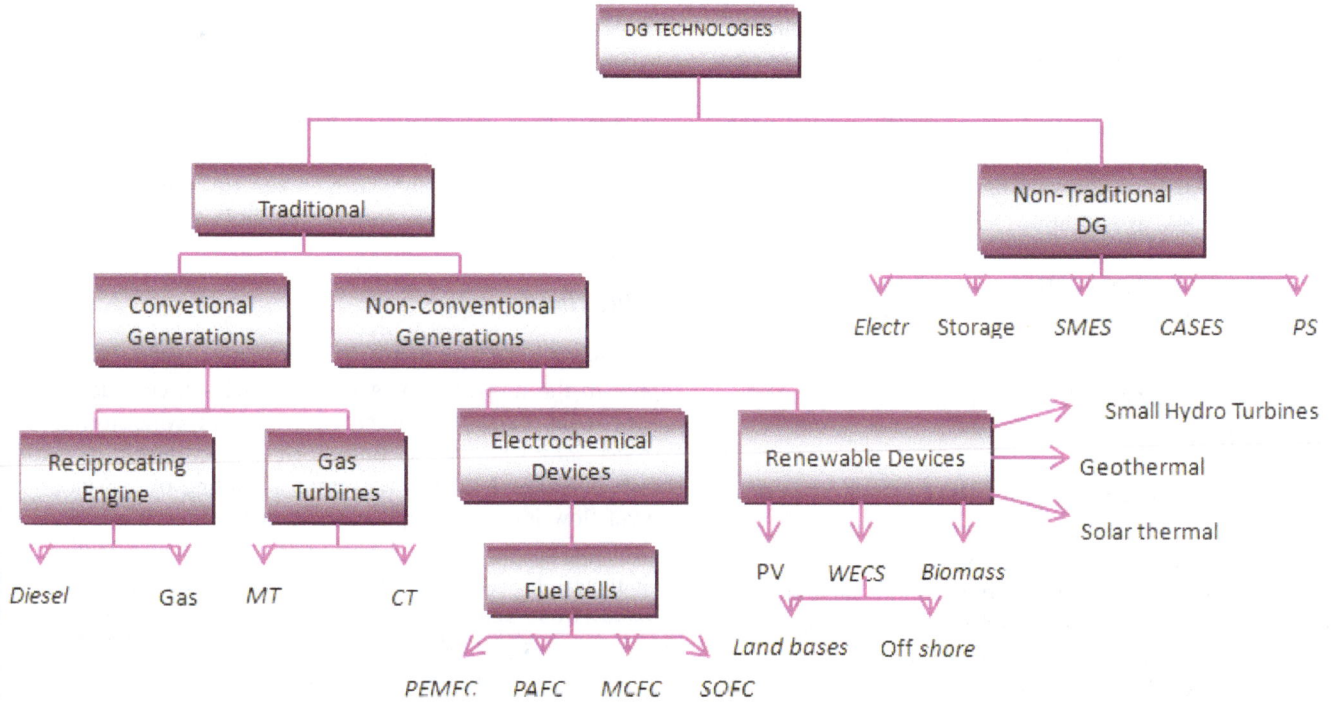

Fig. 1. *Distributed Generation Technologies [43]*

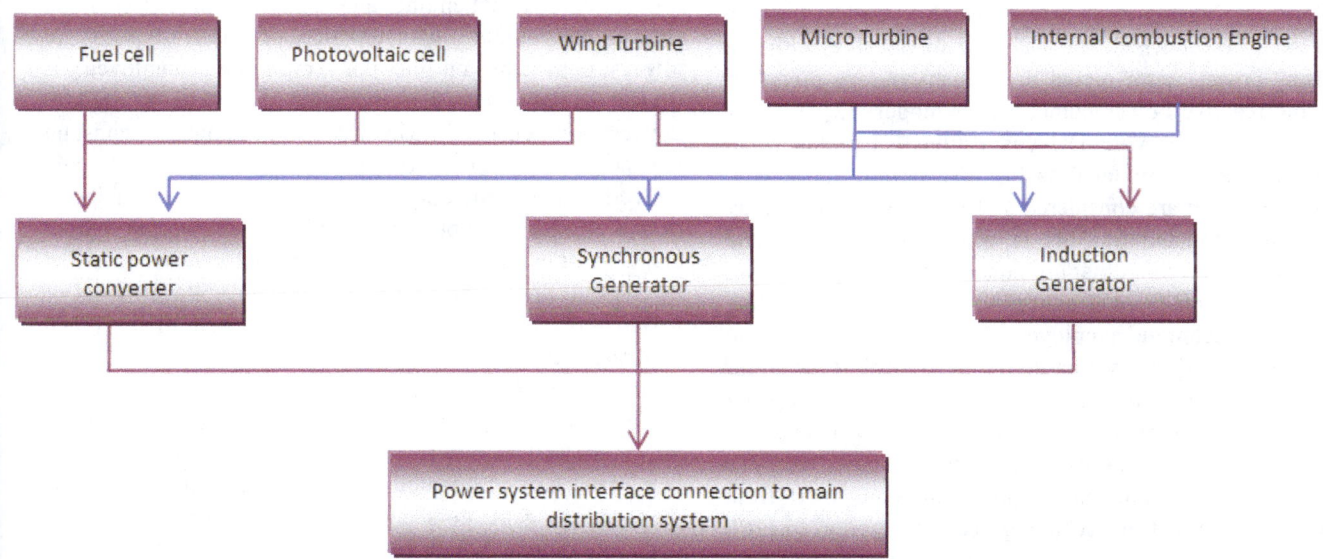

Fig. 2. *DGs and converter circuits connections for grid interfacing [13]*

The traditional DGs are those generators that utilize combustion engines such as low speed turbines, reciprocating engine and gas micro-turbine. These resources even through are small in size but have wide spread geographically. On the other hand the Non-traditional DGs are those sources that produce power with zero emissions and are very friendly to environment. Most of these sources usually outputs DC power, therefore conversion to AC power is necessary before integration into an existing AC distribution network. For this reason sources such as photovoltaic and fuel cells uses power electronic converters (inverters) for grid interfacing as shown in figure 2.0.

The traditional internal combustion engines (rotary machines) are mainly synchronous generators and are interfaced directly to the grid. Wind turbines are also considered as rotary machines and are mainly induction type generators which can be interfaced directly with the grid. In some wind applications as well as some combustion engines like micro-turbines, power converter/inverter devices are employed for grid interface as the benefits associated with the electronic interface justifies the additional cost and complexity involved. Therefore, the energy generated from the various DG sources is injected into the grid via synchronous Generators, static power converters or induction generators. The nature of operation of these generators or converters determines the models of DG to be employed in the power flow solution.

3. DG Resources and Conversion Devices Modeling

In general modelling requires system representation mathematically such that the mathematical model gives sufficient information about actual systems that covers all the necessary system behaviour within certain constraints. A proper model of the DG that can adequately represent a DG type with a view of assessing its impact on the network is of great importance. The model should be represented in such a way that the impact evaluation approach on the network due to the DG presence can easily be conducted. Many models have been developed by researchers on DG for load flow analysis in which the DGs are modeled as either a constant power factor model or constant voltage model or variable reactive power output power model [12]. In their analysis the buses with DG connection that yields small output power are modeled as PQ nodes while those with large DG output are considered as PV nodes. Similar study was also conducted by [14] in which asynchronous and synchronous DG units that are connected to grid via power converters are modeled as either PV or PQ nodes depending on control technique employed. Other models were also discussed in [15] that are based on the control of generator excitation and in this case the synchronous DGs together with static voltage regulators are modeled as PV nodes.

Generally, DG classification can be based on its construction, size, and output power duration apart from classification based on technology earlier reported in [16]. Also for the purpose of modelling DGs have been categorized into four classes be it traditional or non-traditional as suggested by authors in [17]. The authors have classified the DGs into four major groups based on combined power transfer capability as well as terminal characteristics as presented;

1 Those DGs that can provide only active power (P) and can be integrated to the main grid by employing power converters are classified as one type. Typical examples of such are the fuel cell, photovoltaic, micro turbine and wind turbine.
2 The second classification of DG are those that are capable of providing both active (P) and reactive (Q) power. Such DG units type are based on synchronous machines which are commonly found in gas turbine and cogeneration.
3 The third are those that are only capable of supplying only reactive power. An example of such is the Synchronous compensators.
4 Some DGs can provide active power (P) and at the same time consumes reactive power (Q). The well known examples of such type are the induction generators and the doubly fed induction generator (DFIG) systems.

3.1. DG Models

Presence of DG in network means more active power supply to the network. Integration of DG sources such as photovoltaic, fuel cells, micro turbine and wind turbine systems into grid is via power electronic interfaces as shown in figure 2.1. For this reason modeling of DG units in load flow is always dependant on control technique that is been used in the converter circuit. The distribution system has some distinct characteristics which are always different from transmission systems in the sense that it is radial or weekly meshed with the lines having high R/X ratio and may be significantly unbalanced [18].

The sources of energy from the DGs can be categorized into stable energy sources (fuel cell, micro turbine and internal combustion engine) and unstable sources (wind and solar). The output characteristics of these sources are always dependant on the conversion unit employed. A typical example is when induction generator is used to convert wind energy in which case the output is a constant real power (P) with reactive power (Q). However, if static converter is used the output will be a constant power factor output in normal operation condition. Based on output characteristics of the sources, DG model can be classified as constant power factor model, constant voltage model or variable reactive power model as proposed by [19] and explained herein as follows;

A Constant Power Factor Model

This type of model has a specified real and reactive power as well as power factor. The commonly used *DGs* that can be represented by such a model are the synchronous generators and power electronic based units that have outputs which can be adjusted by controlling the excitation current and trigger angles of the units for synchronous generator and power electronic converters respectively. The reactive power of such *DGs* can be calculated as presented in [20] based on;

$$Q_{iDG} = P_{iDG} \tan\left(\cos^{-1}(\text{PF}_{iDG})\right) \qquad (1)$$

while the injected DG equivalent currents is obtained as;

$$I_{iDG} = I^i_{iDG}(V_{iDG}) + jI^i_{iDG}(V_{iDG}) = \left(\frac{P_{iDG} + jQ_{iDG}}{V_{iDG}}\right) \qquad (2)$$

where P_{iDG}, and V_{iDG} are the real output power and voltage at the terminal of the DG respectively, PF_{iDG} is the power factor for the DG installed at bus i while Q_{iDG} is the calculated reactive power output.

B Variable Reactive Power Model

Typical examples of such model are the induction generators that have variable reactive power generation. The real power output is dependent on the wind speed which is calculated based on wind turbine power curve, while the reactive power is dependent on real power output, and generator impedance. Even though the parameters are available, the calculation is usually cumbersome and difficult to execute efficiently. Hence the calculation is based on steady-state which can be represented as a function of the DG`s real power as in [21] by using;

$$Q_{iDG} = -Q_0 - Q_1 P_{iDG} - Q_2 P^2_{iDG} \qquad (3)$$

where Q'_{iDG} is the reactive power function consumed by the wind turbine while the Q_0, Q_1 and Q_2 are obtained experimentally. In situations when the reactive power required by the load cannot be supplied by the distribution network, capacitor banks are used to correct the power factor of the system. The reactive power output of the induction generator is expressed as;

$$Q^i_{i,g} = Q^1_{i,g} + Q^c_{i,g} \qquad (4)$$

where $Q^c_{i,g}$ is the reactive power supplied by capacitor bank.

C Constant voltage model

The constant voltage model is meant for controllable *DGs* of large-scale systems where the specified parameters of this *DG* model are the real power output and bus voltage magnitude. The equivalent current to be injected is integrated into the power flow analysis after the reactive power output of the *DG* which is necessary to keep the bus voltage magnitude on the specified value are evaluated by using two-loop algorithm developed in [22]. The generated output power after evaluating the required reactive power is given by;

$$P^{k,m+1}_{i,g} + jQ^{k,m+1}_{i,g} = P^{k,m}_{i,g} + j\left(Q^{k,m}_{i,g} + \Delta Q^{k,m}_{i,g}\right) \qquad (5)$$

where $\Delta Q^{k,m}_{i,g}$ is the required reactive power variation for the *mth* inner and *kth* outer iterations for the two-loops algorithm.

3.2. Power Flow with Distributed Generator

Many algorithms for power flow of distribution systems with high penetration of DG have been proposed by many researchers. Based on findings, it can be categorized into node based and branch based methods [23]. In the node based methods, node voltage or current injections are employed as state variables for solving the power flow problems [24-26]. This method includes methods such as network equivalence method, Z-bus method, Newton-Raphson algorithm and fast decouple algorithm. On the other hand, the branch based involves use of branch currents or powers as the state variable for the power flow solution [27 – 29]. The technique is applicable to sweep based and loop impedance methods.

Since the DG units are modeled as either PQ or PV, their incorporation into power flow will involve use of active and reactive powers as flow variables instead of complex currents as variables. In the case of PV model the reactive power limits of the generator is constantly checked so that it is kept within limits otherwise the model will become a PQ model if demand for reactive goes beyond its limit. The root bus is considered as the slack bus with known voltage magnitude and angle. It most also be assumed that the initial voltage for all the other nodes are equal to the root node voltage and the initial power loss of all the branches be equal to zero. The integration of the mathematical models of DGs into the three-phase load flow can then be effectively conducted for the purpose of analysis based on the following models as in [14];

3.2.1. Induction Generator Model

The Induction generator output power is always a function of two parameters which are slip and voltage and is expressed as;

$$P = P(V, s) \qquad (6)$$

$$Q = Q(V, s) \qquad (7)$$

where P and Q are the active and reactive powers produced respectively, s is the slip of the generator speed, and V is the bus voltage. By making an assumption that P is constant and neglecting the dependence of reactive power on the slip, equations (6) and (7) are now reduced to;

$$P = P_s = constant \qquad (8)$$

$$Q = f(V) \qquad (9)$$

The model of equations (8) and (9) are appropriate model for squirrel cage induction generator for power flow studies.

3.2.2. Synchronous Generator Model

Synchronous generator model is categorized into two types depending on the excitation system type. As shown Fig.3.0 the regulated excitation is modeled as either constant voltage (PV) or constant power factor model (PQ).

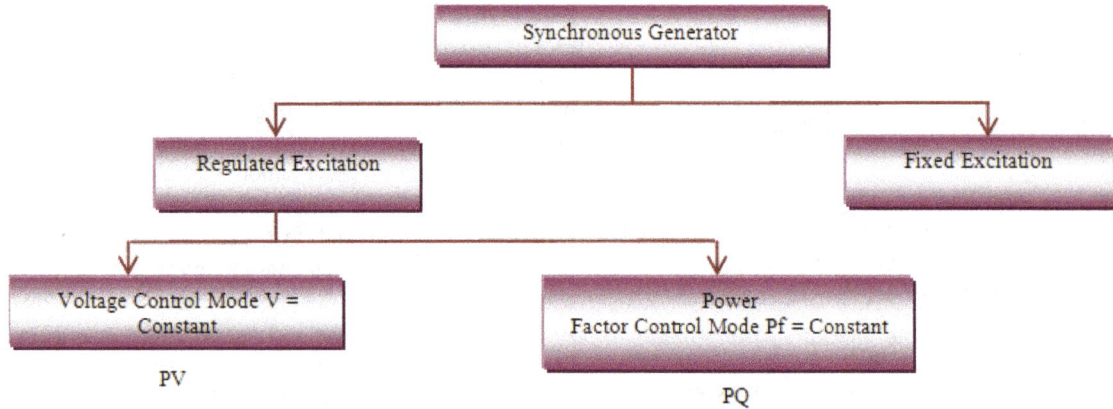

Fig. 3. Synchronous generator model classification

The fixed excitation model can be described by equation (10);

$$Q = \sqrt{\left(\frac{E_d}{X_d}\right)^2 - P^2 - \frac{V^2}{X_d}} \qquad (10)$$

where P and Q are the active and reactive power of DG respectively, E_d is the no-load voltage and is maintained constant, X_d is the synchronous reactance, V is the generator terminal voltage. Assuming that P is constant;

$$P = Constant \qquad (11)$$

$$Q = f(V) \qquad (12)$$

This model is the same as that in equation (8) and (9) except Q is positive in equation (12), which means the synchronous generator without excitation voltage regulation may inject reactive power to the grid.

Table 1. Methods, models and contributions on DG modelling.

S/NO	REF.	Methods	Models	Contributions
1.	Syafii[6]	Unbalanced three-phase distribution power-flow	DG modeled as a PV node or as a PQ node	Improvements in voltage profile as well as reduction in the total system losses
2.	Liu Qingzhen[7]	Backward/Forward Sweeps technique power flow with multiple node styles	PV node, static voltage control node and PI node models	DGs influence on voltage profile, power losses and voltage regulator operating status established
3.	Kamh, M.Z[8]	New power-flow algorithm in the sequence-component frame for voltage-source converter (VSC)	Unified model for various VSC configurations and its host DER unit	Significant improvement in computational efficiency achieved
4.	Elsaiah, S. [9]	Backward/forward power flow method	Induction machine modeled as an admittance load	Maximum torque achieved at less computational efforts.
5.	Khushalani, S. [10]	Unbalanced three-phase load flow algorithm that can handle multiple sources	Constant voltage and Constant power factor DG models	A Software is developed for switching of DG operating modes
6.	Khushalani,S. [11]	Three phase unbalanced power flow algorithm	DG modeled as PQ or PV node.	Effect DG has on voltage profile and currents demonstrated
7.	Moghaddas-Tafreshi S.M.[14]	Backward/forward sweep technique for three phase unbalanced power flow	Both conventional and renewable DG modeled as PQ or PV node.	The algorithm proposed is fast and the operation of machine determines the model of DG for Power Flow solutions
8.	Zhu, Y[18]	Adaptive power flow based on compensation method	DG modeled as PV/PQ and load as linear load model with other network components	Faster and more reliable than conventional
9.	Harrison, G.P[33]	Optimal power flow based on reverse load-ability approach	Model based on fixed-power factor DG handled as negative	Maximization of DG capacity and identification of available

S/NO	REF.	Methods	Models	Contributions
			loads	headroom within a network
10.	Hussein, D.N[34]	Efficient load flow technique which is based on compensation method	Different models of DG types for integration in distribution network	Improvement of system reliability and performance which resulted in deferral of network up grade
11.	Gayme, D[35]	Optimal power flow with storage devices	Model formulated based on simple charge/discharge dynamics for energy storage devices	Significant reduction in generation cost achieved based on demand-cost function
12.	Kamh, M.Z[36]	Sequence frame-based, power-flow algorithm	Type-3 wind driven DG unit model based on steady-state fundamental-frequency	Operating limits of rotor-side and the grid-side converters under balanced and unbalanced conditions were established
13.	Hany E. Farag [37]	Three-phase distribution system power flow for radial topology by extending elements of bus incidence matrix	Models of feeders, voltage regulator (VR), Exact load and DG in its different operation modes developed.	Impact of VR and high penetration of DG on voltage profile and system power losses established.
14.	Kamh, M.Z.[38]	Sequential sequence-frame power-flow solver (sequential-SFPS) algorithm	Sequence-frame model of Type-3 wind generation	Shows the computational efficiency of the sequential-SFPS algorithm
15.	Nayak, S.K[39].	Simulations in Matlab/Simulink environment	Dynamic model for performance analysis of micro-turbine generator (MTG) system in grid connected and islanding mode	Micro-turbine generator system performance indicators as load varies developed
16.	Sexauer, J.M[40]	Probabilistic load-flow-based approach	Statistical models for load, wind speed, and solar irradiance.	Actual assessment of the impact of DG units on the power system obtained using probabilistic approaches
17.	Antonios G [41]	Stochastic power flow in a distribution line with dispersed photo-voltaic (PV) penetration	DG model based on extensive stochastic modeling for probabilistic load model	A cost index for losses is determined which is useful for DG placement and sizing.
18.	Ruiz-Rodriguez F.J. [42]	Analytical method which combines cumulant method with the Cornish–Fisher expansion employed to solve probabilistic load flow	A probabilistic DG model that takes into account the random nature of solar irradiance developed	The proposed method is found to be effective for assessment of the impact of PV-DG on the voltage profiles of distribution networks.

3.2.3. Static Power Converter/Inverters Model

The static power converters are sometimes referred to as power electronic interface and are required for direct connection to grid. These power electronic devices are basically made of high power transistors and employs transistor switching techniques to control the flow of electric power. In this mode of operation at any instant of time, the transistor is either fully on or it is fully off. The transistors are configured into different topologies for the purpose of achieving the desired power conversion function such as DC-to-DC, AC-to-DC, or DC-to-AC. For such functions to be achieved pulse width modulation (PWM) is used for varying the on and off times of the transistors for a desired voltage and frequency. The modern transistors used for this purpose are made from insulated gate bipolar transistors (IGBT). As micro-sources like micro-turbines, wind turbines,

photovoltaic array and even fuel cells output powers are small DC (<100KW), power electronic interfaces are required for conversion DC/AC or AC/DC/AC as shown in fig. 2.

The rule of thumb employed for the modeling of static power converters/inverters for power flow analysis by most researchers as proposed in [30, 31] is the converter controls which will determine the type of model to be employed. If the converter is designed to control active power and Voltage independently then the model is *PV* and when it is designed to control active power and reactive power independently the model is a *PQ* node [32].

4. Summary of Current Works

In this section a summary of previous work done on DG modeling is presented together with methods adapted and models formulated as well as the contribution made from the studies conducted. The papers reviewed in Table 1.0 are not only limited to IEEE Xplore digital library but also to scienceDirect web site.

The reason for the summarized overview is to provide researchers at a glance the models already formulated and achievements made based on the methods already proposed. It can be observed that the models are formulated based on output characteristic of the DG, while the methods have touched on both balanced and unbalanced three phase power systems. The unbalanced three phase system is a typical scenario of distribution networks.

5. Future Research and Challenges Ahead

The emerging smart grid concept that is likely going to affair in tomorrow's distribution network requires repeated and fast load flow solutions for efficient planning, automation and optimization of power system. This therefore necessitates continuous development of appropriate models of DG sources involved in distribution networks power flow. Due to complex nature of some of the DG sources, researchers have employed the use of some assumptions in order to reduce the number of state variables for existing models developed. This situation suggests further work towards development of newer models that are based on state of art modeling techniques with more number of state variables for proper dynamic analysis of systems comes necessary. Hence, mathematical models are required that carries adequate information about the system for proper investigation of system performance depending on interest.

- In general increase in penetration of DG requires formulation of new models of the power system that needs creation of decoupling power system analysis tool which can analyze the system using hybrid load flow.
- Due to the unbalance nature of distribution networks, the concept of impedance matrix and nodal current injections used in generalized single equation load flow method cannot be employed as maintaining them constant during the analysis is impossible. This hinders the modeling of many components in distribution network. A new method that is more robust than commonly used forward/back sweep method is required which can allow inclusion of components such as transformers, voltage regulators, shunt capacitors and various types of loads.
- It is evident that more challenges are likely to come up as network modeling has not been done with full details from network operator's point of view. This means that it is only when these details are obtained and included in the models, that development of optimization and integration methods of DG resources can become an issue.

The techniques developed so far are only pointers towards finding answers to questions about the detailed characteristics of the existing models already developed along with the systematic approach to their allocation in distribution network during planning and operation. Newer modeling efforts are therefore required especially with the introduction of smart energy meters, electric vehicles and other demand side resources.

6. Conclusions

Review of the various DG modeling techniques for power flow analysis has been presented in this paper. Various types of DGs are considered for different model types depending on grid interfacing device employed. Also the power generated by the DG type to be injected into the grid and the characteristic of the interface device determines the type of DG model for power flow studies. Based on this, models are categorized into constant power factor model, constant voltage model and variable reactive power model. These models are developed for the purpose of integration into three- phase load flow analysis during planning and operations of distribution system.

On the whole the review has identified fruitful key areas of active research work for researcher's reference and at the end the challenges ahead with current drive towards smart grid networks are outlined.

Acknowledgements

The author H. Musa acknowledges with gratitude the financial support in form of Research fellowship offered by Bayero University Kano Nigeria and the provision of suitable research facilities.

References

[1] A. Narang, "Impact of large scale distributed generation penetration on power system stability," Natural Resources Canada, CETC, March 9, 2006.

[2] K. K. Sharma, B. Singh "Distributed Generation- A New Approach" International Journal of Advanced Research in Computer Engineering & Technology (IJARCET) Volume 1, Issue 8, October 2012

[3] T. Ackermann, G. Anderson, and L. Soder, —Electricity market regulations and their impact on distributed generation,‖ in Proc. Conf,Electric Utility Deregulation and Restructuring and Power Technologies 2000, London, U.K., Apr. 4–7, 2000, pp. 608–613.

[4] J.G. Slootweg, S. de Haan, H. Polinder, W. Kling, Modeling new generation and storage technologies in power system dynamics simulations, in: Proceedings IEEE Summer Meeting, Chicago, July 2002.

[5] A. M. Azmy and I. Erlich, "Impact of distributed generation on the stability of electrical power system," in Proc. IEEE Power Engineering Society General Meeting, vol. 2, pp. 1056–1063, June 2005.

[6]   Syafii, K.M. Nor, M. Abdel-Akher, "Analysis of three phase distribution networks with distributed generation" IEEE 2nd International on Power and Energy Conference (PEC), pp.1563 – 1568, 2008.

[7]   Liu Qingzhen, Cai Jinding, "A Integrated Power Flow Algorithm for Radial Distribution System with DGs Based on Voltage Regulating" *Asia-Pacific Power and Energy Engineering Conference (APPEEC)* pp.1-4, 2010

[8]   M.Z. Kamh, R. Iravani, "A Unified Three-Phase Power-Flow Analysis Model For Electronically Coupled Distributed Energy Resources" , IEEE Trans. on Power Delivery vol. 26 , no. 2 , pp. 899 – 909, 2011.

[9]   S. Elsaiah, M. Benidris, J.Mitra, "Power flow analysis of distribution systems with embedded induction generators" *North American Power Symposium (NAPS)*, pp.1 – 6, 2012

[10] S. Khushalani, N. Schulz, "Unbalanced Distribution Power Flow with Distributed Generation" IEEE PES Transmission and Distribution Conference and Exhibition, 2005/2006 PP. 301 – 306, 2006

[11] Khushalani, J.M. Solanki, N.N. Schulz, "Development of Three-Phase Unbalanced Power Flow Using PV and PQ Models for Distributed Generation and Study of the Impact of DG Models" IEEE Trans on Power Systems, vol. 22 , no. 3 , pp. 1019 – 1025, 2007

[12] Engineering guide for integration of distributed generation and storage into power distribution systems', EPRI Technical Report TR-100419 Report, December 2000

[13] Teng J.H., ─Modeling distributed generations in three phase distribution load flow, *IEE Proceeding Generation Transmission Distribution.*, vol. 2, no. 3, pp 330–340, 2008.

[14] Moghaddas-Tafreshi S.M. and Mashhour E., ─Distributed generation modeling for power flow studies and a three-phase unbalanced power flow solution for radial distribution systems considering distributed generation, *Electrical Power Systems Research* 79 (2009) pp 680–686.

[15] Chen H., Chen J., Shi D. and Duan X., "Power flow study and voltage stability analysis for distribution systems with Distributed Generation" IEEE Power Engineering Society Meeting 10-22 June, 2006, pp 8.

[16] El-Khattam W., and Salama M. M. A, "Distributed Generation Technologies, Definitions and Benefits" *Electric Power System Research* Vol. 71, no. 2, pp. 119-128 2004

[17] Hung D. Q., Mithulananthan N., and Bansal R. C., "Analytical expression for DG allocation in primary distribution network," *IEEE Trans. Energy Convers.*, vol. 25, no. 3, pp. 814-820, 2010.

[18] Y. Zhu, K. Tomsovic, Adaptive power flow method for distribution systems with dispersed generation, IEEE Trans. Power Deliv. 17 (3) (2002) 822–827.

[19] J.-H. Teng, "Modelling distributed generations in three-phase distribution load flow" IET Gener. Transm. DistribVol. 2, No. 3, pp. 330–340, 2008

[20] Chen Th, Chen Ms, Inoue T. 'Three-phase cogenerator and transformer models for distribution system analysis', IEEE Trans. Power Deliv., 1991, 6, (4), pp. 1671–1681

[21] Feijoo Ae, Cidras J 'Modeling of wind farms in the load flow analysis', IEEE Trans. Power Syst., 2000, 15, (1), pp. 110–115

[22] Teng Jh: 'A direct approach for distribution system load flow solutions', IEEE Trans. Power Deliv., 2003, 18, (3), pp. 882–887

[23] W.C. Wu, B.M. Zhang, A three-phase power flow algorithm for distribution system power flow based on loop-analysis method, Elect. Power Energy Syst. 30 (2008) 8–15.

[24] A.V. Garcia, M.G. Zago, Three phase fast decoupled load flow for distribution networks, IEEE Proc – Gener. Transm. Distrib. 143 (2) (1996) 188–192.

[25] P.A.N. Garcia, J.L.R. Pereira, "Three-phase power flow calculations using the current injection method", IEEE Trans. Power Syst. 15 (2) (2000) 508–514.

[26] J.-H. Teng, C.-Y. Chang, A novel and fast three-phase load flow for unbalanced radial distribution systems, IEEE Trans. Power Syst. 17 (4) (2002) 1238–1244.

[27] A.G. Bhutad, S.V. Kulkarni, S.A. Khaparde, Three-phase load flow methods for radial distribution networks, in: Conf. on Convergent Technologies for Asia- Pacific Region, TENCON 2, 2003, pp. 781–785.

[28] D. Thukaram, H.M.W. Banda, I. Jerome, "A robust three-phase power flow algorithm for radial distribution systems", Elect. Power Syst. Res. 55 (3) (2000) 191–200.

[29] R. Ranjan, B. Venkatesh, A. Chaturvedi, D. Das, "Power flow solution of three phase unbalanced radial distribution networks", Elect. Power Comp. Syst. 32 (4) (2004) 421–433.

[30] S. Naka, T. Genji, Y. Fukuyama, "Practical equipment models for fast distribution power flow considering interconnection for distributed generators", IEEE Power Engineering Society Summer Meeting, vol. 2, 2001, pp. 1007–1012.

[31] Pecas Lopes J.A., Moreira C.L, Madureira A.G., " Defining control strategies for MicroGrids islanded operation" IEEE Trans Power System, vol. 21no.2, pp.916–24, 2006.

[32] H. Chen, J. Chen, D. Shi, X. Duan, "Power flow study and voltage stability analysis for distribution systems with distributed generation", IEEE Power Engineering Society General Meeting, pp.8-12, 18–22 June, 2006

[33] G.P. Harrison, A.R.Wallace, "Optimal power flow evaluation of distribution network capacity for the connection of distributed generation" *IEE Proceedings- Generation, Transmission and Distribution* vol. 152 , no. 1 pp. 115 – 122, 2005

[34] D.N. Hussein, M. El-Syed, H.A. Attia, "Modeling and simulation of distributed generation (DG) for distribution systems load flow analysis" *Eleventh International Middle East Power Systems Conference (MEPCON 2006)* vol. 1 pp.285 – 291, 2006

[35] D. Gayme, Ufuk Topcu, "Optimal power flow with distributed energy storage dynamics" American Control Conference (ACC), pp.1536 – 1542, 2011

[36] M.Z. Kamh, R. Iravani, "Three-Phase Steady-State Model of Type-3 Wind Generation Unit—Part I: Mathematical Models" IEEE Trans. Sustainable Energy, vol. 2, no. 4, pp. 477-486, 2011

[37] Hany E. Farag, E.F. El-Saadany, Ramadan El Shatshat, Aboelsood Zidan, "A generalized power flow analysis for distribution systems with high penetration of distributed generation" Electric Power Systems Research vol. 81, pp.1499–1506, 2011

[38] M.Z. Kamh, R. Iravani, "Three-Phase Steady-State Model of Type-3 Wind Generation Unit—Part II: Model Validation and Applications" IEEE Trans. Sustainable Energy, vol. 3, no. 1, pp. 41-48, 2012.

[39] S.K. Nayak, D.N. Gaonkar, "Modeling and performance analysis of micro-turbine generation system in grid connected/islanding mode" *IEEE International Conference on Power Electronics, Drives and Energy Systems (PEDES)*, Page(s): 1 – 6, 2012

[40] J.M. Sexauer, ; S. Mohagheghi, "Voltage Quality Assessment in a Distribution System With Distributed Generation—A Probabilistic Load Flow Approach" *IEEE Trans. Power Delivery* vol. 28 , no. 3 pp.1652 – 1662, 2013

[41] Antonios G. Marinopoulos, Minas C. Alexiadis, Petros S. DokopouloN "Energy Losses in a distribution line with distributed generation based on stochastic power flow" Electric Power Systems Research vol. 81, pp. 1986– 1994, 2011

[42] F.J. Ruiz-Rodriguez, J.C. Hernández, F. Jurado, "Probabilistic load flow for photovoltaic distributed generation using the Cornish–Fisher expansion" Electric Power Systems Research vol. 89, pp.129– 138, 2012

[43] M. F. Akorede, H. Hizam, and E. Pouresmaeil, "Distributed energy resources and benefits to the environment," *Renewable and Sustainable Energy Reviews,* vol.14, pp. 724-734, 2010.

24

Vertical Profiling and Contamination Risk Assessment of Some Trace Metals in Lagos Lagoon Axis

Popoola Samuel Olatunde, Nubi Olubunmi Ayoola, Oyatola Opeyemi Otolorin, Adekunbi Falilu Olaiwola, Fabunmi Gaffar Idera, Nwoko Chidinma Jecinta

Department of Physical and Chemical Oceanography, Nigerian Institute for Oceanography and Marine Research, Lagos, Nigeria

Email address:

popoolaos@niomr.gov.ng (P. S. Olatunde)

Abstract: Lagos Harbour (a Lagos lagoon axis) serves as a route for goods transportation, coupled with proliferation of urban and industrial establishments. The socioeconomic activities within has often led to the introduction of substantial wastes, marine debris and spills into the harbor. In an attempt to monitor the pollution status of aquatic organisms in the Lagos Lagoon axis, vertical profiling of the trace metal contents was carried out to reveal, the vertical variations in the monthly trends of; Cu, Zn, Pb, Cr, Mn, Ni, and Fe between August and October,2014.Chemical analysis was carried out in the Nigerian institute for oceanography and marine research (NIOMR) wet and instrumentation laboratory, using aqua regia digestion methods and 200AA series Atomic Absorption Spectrophotometer. The observed levels of the trace metals shows a decreasing order of abundance; Fe > Cr > Pb > Mn> Ni > Zn > Cu; Fe > Pb > Cr > Ni > Mn > Zn >Cu and Fe > Cr > Pb > Ni > Mn > Zn > Cu for 0.2meters, 2.6meters and 5.2meters respective depth profile. This study reveal a general contamination trends; medium depth (2.6m) > bottom depth (5.2m) > surface depth (0.2m). Higher metal concentrations were observed in October. All observed trace metals except Cu, Zn and Fe exhibit high contamination ratio. The trace metals distributions in the study area are majorly controlled by; precipitations, dilution, anthropogenic activities and Sea/fresh water incursion from the Atlantic Ocean and the adjourning creeks. A strong regulation in the indiscriminate waste dump and a check in the socioeconomic activities in the Lagos Harbour are very essential.

Keywords: Lagos Harbour, Marine Debris, Atomic Absorption Spectrophotometer, Trace Metals, Contamination Ratio, Anthropogenic Activities

1. Introduction

Trace metals such as chromium, cobalt, copper, iron, nickel, manganese, lead, copper and zinc, are elements that normally occur at very low levels (<1000ppm) in the environment (Dara and Rashmi, 2009). Living things are in need of a very small amount of these trace metals, as high concentrations can become toxic. They are present in seawater in trace concentrations, whereas excessive concentration can affect marine biota, pose risk to consumers of sea food and change the physical and chemical characteristics of the water [1]

Pollution cases such as sewage sludge, wood burning / log transportation, feacal disposals, oil tankers transportations, landfill site, vehicular emissions, sand dredging, and agricultural effluents have been reported in the Lagos lagoon [2,3,4]. The continuous growth in human population and materials consumed in Lagos have led to the generation of substantial wastes of unprecedented quantities which are no more mere nuisance, but toxic and hazardous to the environment [5].

Although Lagos harbor (an extension of the Lagos lagoon axis) figure 1, serves as a seaport and centre for recreational sailing, it has also been regarded as a sink for domestic and industrial wastes [6]. However, there is paucity of information on the extent of pollution resulting from coastal activities such as marine debris deposition, oil contamination, ships/commercial boating spills, naval vessel discharges, dredging, sitting and construction of socio-economical activities within the Lagos harbor[7-10]; hence it becomes imperative to continually monitor the pollution status of the Lagos Lagoon axis. This paper therefore aims at assessing the vertical variations in the concentration of some trace metals: Pb, Mn, Zn, Cr, Fe, Ni and Cu in the study area and also seeks to examine their sources and environmental consequences.

2. Description of the Study Area

Lagos is situated in south-western Nigeria on the West Coast of Africa and is undoubtedly the commercial nerve-center of Nigeria. Geologically; it falls within the eastern part of the Dahomey Basin, bounded to the north by then Precambrian Basement complex of southwestern Nigeria. It is bounded by the Gulf of Guinea to the south and eastward by the Okitipupa ridge [11]. The harbour (figure 1) situated

in Lagos is Nigeria's most important seaport and the first inlet from the Atlantic Ocean, beyond the Republic of Benin[12]. It is one of the three main segments of Lagos Lagoon Complex; other segments are: Metropolitan and the Epe Division Segments. It is 2 km wide and receives inland waters from the Lagos Lagoon in the north, and from Badagry Creek in the west. Oil depots are located along the shore of western parts coupled with the proliferation of urban and industrial establishments on the shore of eastern part. NIOMR jetty station (Latitude 6° 25′ 14, 88° N, Longitude 3° 24′ 24, 42° E) is located in the commodore channel of Lagos harbour with Jetty facilities awaiting rehabilitation. Subsistence fishing takes place at this part on the water body by artisanal fishermen [13].

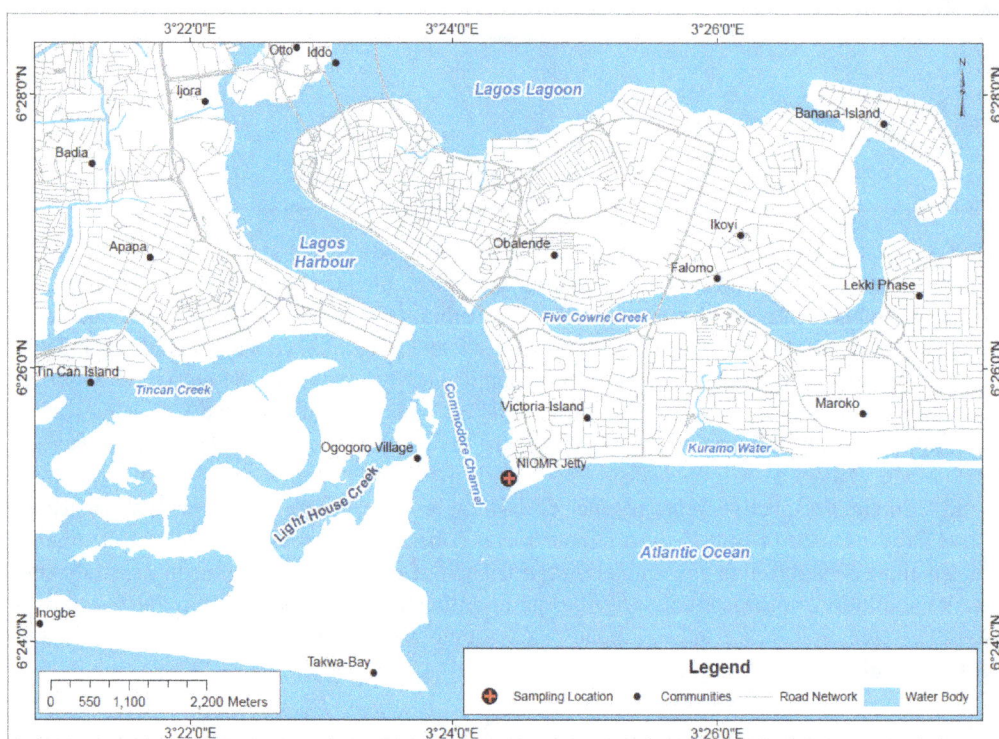

Fig. 1. Map of Lagos Harbor showing the NIOMR Jetty platform.

3. Material and Methods

Surface water samples from the Nigerian Institute for Oceanography and Marine Research Jetty, henceforth termed NIOMR Jetty were collected with the aid of 500ml container at about 0.2m depth. Water samples at 2.6 meters and 5.2meters depth were collected with the aids of 1dm^3, 0.1meters diameter water sampler attached to a graduated rope of 10meters long. Samples were collected from August to October 2014.The samples were acidified with 2 to 3ml of nitric acid (HNO$_3$) in pre-cleaned 1 litre plastic containers at the sampling point so as to ensure that the respective ions remain in solution pending analysis. Trace metal contents (Pb, Cr, Cu, Ni, Fe, Mn and Zn) in the respective depth profiled were analysed with argillent 200 A model, Atomic Absorption Spectrophotometer (AAS), after the samples had

been digested with Nitric/Hydrochloric acid (1:3), aqua regia using standard digestion procedure [14]. The results obtained from the analysis were subjected to various contamination ratios and standards to ascertain the metal's natural or anthropogenic enrichment in the NIOMR Jetty water body.

The sampling point was inaccessible at a certain time (late August and mid September) during the sampling period due to high precipitation and tidal fluctuations.

4. Statistical Analysis

All descriptive statistics and graphs were executed using Statistical 7 software and Microsoft office Excel 2010. Data were further subjected to Correlation test to find significant relationship between the measured variables at 0.05 levels of significance

5. Results

The mean concentrations of some trace metals in water samples are shown in figure 2. Chromium has the highest value (7.06 ± 0.81mg/l) at depth 5.2meters in October and least value (3.21 ± 0.69mg/l) in August at 2.6meters depth. The peak concentration for Lead (6.98 ± 0.59mg/l) is at 2.6meters depth and the least value (0.04 ± 0.01mg/l) was at 0.2meters in October. Iron exhibits the highest values (14.88 ± 5.6mg/l) in October at depth 5.2meters and lowest concentration (3.09 ± 0.58mg/l) at 2.6 meters in August. The highest concentration for Mn (2.59 ± 0.37mg/l) and Ni (1.57 ± 0.03mg/l) were recorded in October and September at 0.2meters and 5.2meters while, the lowest values 0.09 ± 0.02mg/l and 0.47 ± 0.08mg/l were recorded in August and October at 0.2meters respectively. Copper and Zinc have lowest concentration in all the water samples however, it has peak values of 0.43 ± 0.02mg/l at 2.6meters and 1.04 ± 0.07mg/l in August at 2.6meters and their respective lowest values of 0.03 ± 0.02mg/l and 0.12 ± 0.06mg/l in October and August at 5.2meters.

6. Contamination Ratios

Other contamination ratios such as-: contamination factor, contamination degree and pollution load index were applied to evaluate the contamination status of the Lagos Harbour at the NIOMR Jetty sampling points (see table 3&4).The contamination factor was used in this study to evaluate the degree of trace metal contamination in water samples [15, 16]. The permissible levels, the element co ncentration in the water considered safe for marine organisms [17, 18.] were used as background level for the contamination factor, contamination degree and the pollution index ratio. Contamination factor, Cf = C / Cn, where, Cf = contamination factor; C = mean concentration of each metal in water sample; Cn = permissible level of trace metal

concentration in water (table 5&6).Contamination degree (DC) is the sum of the contamination factors of all the elements examined, degrees of contamination were calculated using the formulae-:mDC = ΣCf[17], Where Σ is the summation of the contamination factors examined.

The Pollution load index (PLI) is a result of the contribution of several trace metals and it is defined as the root of the ratio of concentration factor to the number of heavy metal. PI = Root Σ (Heavy metal concentration in water/ (Tolerable level)/Number of Heavy metals [18, 19].Values of PLI = 1 indicate heavy metal loads close to background, and values above 1 indicate progressive pollution [20].

7. Correlation Matrix

Correlation Matrix was employed for the data set to discover similarities in geochemical behaviour and basic relationship among the trace metals [21]. Correlation coefficient measures the strength of a linear relationship between two variables on a scale of -1 (perfect inverse relationship) through 0 (no relation) to +1 (perfect sympathetic relation). The correlation coefficients developed for trace metals in the study area was based on 21 samples (7 samples each of 0.2meters, 2.6meters and 5.2meters depth) and it's significant from 0.5 at 0.05 confidence level see table 4.

Various ranges of 'r' were observed between the trace metals. The correlation that exists among the trace metals contents of the study area can be summarised as follows; All trace metals are non significantly correlated(P>0.05). Cr exhibits a negative correlation with other trace metals; Mn is positively correlated with Zn, and negative correlated with Cr and Fe. Pb is strongly correlated with Cu and Ni while, Cu correlated strongly with Fe.

Scattered plots were plotted for those metals showing some reasonable level of correlations and (see fig 3-5).

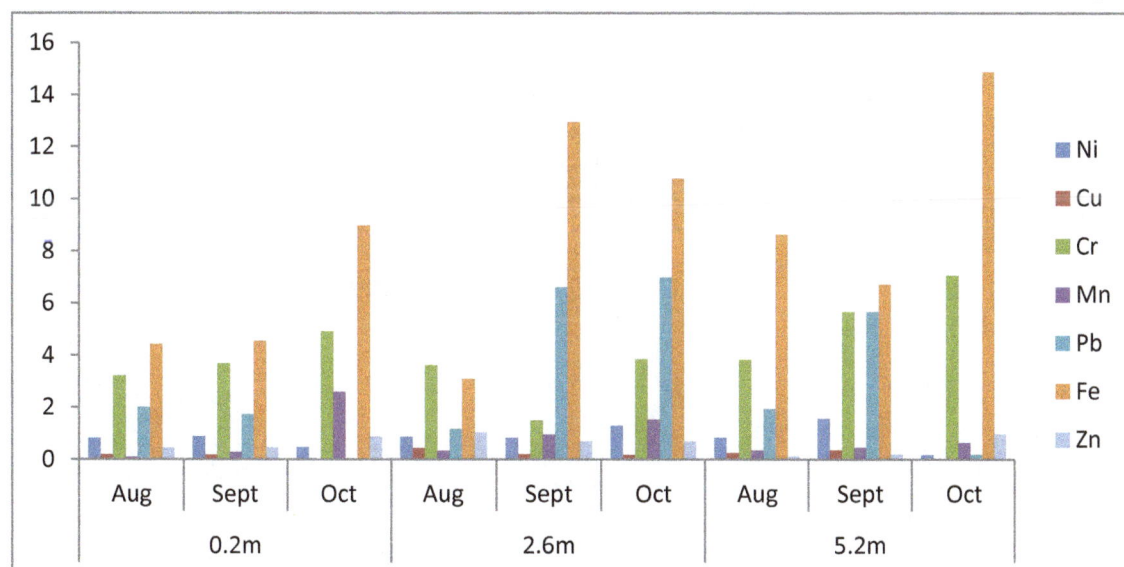

Fig. 2. Showing the monthly concentration of the trace metals (mg/l) sample.

Table 1. *Mean values ± SD of Major/trace metal concentration in the study area.*

Months	Ni(mg/l)	Cu(mg/l)	Cr(mg/l)	Mn(mg/l)	Pb(mg/l)	Fe(mg/l)	Zn(mg/l)
			0.2meters				
Aug,2014	0.82±0.09	0.19±0.10	3.21±0.69	0.09±0.07	2.00±0.35	4.42±1.07	0.45±0.36
Sept,2014	0.88±0.29	0.17±0.01	3.68±0.37	0.29±0.131	1.73±0.39	4.54±0.90	0.46±0.45
Oct,2014	0.47±0.28	ND	4.92±0.42	2.59±0.37	0.04±0.01	8.98±2.59	0.88±0.16
			2.6meters				
Aug,2014	0.86±0.12	0.43±0.02	3.62±0.05	0.33±0.08	1.17±0.62	3.09±0.58	1.04±0.37
Sept,2014	0.83±0.2	0.20±0.1	1.49±0.4	0.95±0.3	6.60±2.2	12.94±3.52	0.69±0.15
Oct,2014	1.29±0.83	0.17±0.1	3.86±1.50	1.53±0.05	6.98±0.59	10.78±1.94	0.69±0.39
			5.2meters				
Aug,2014	0.84±0.06	0.25±0.04	3.83±0.41	0.35±0.009	1.94±0.39	8.63±3.16	0.12±0.06
Sept,2014	1.57±0.88	0.36±0.20	5.67±1.51	0.46±0.25	5.67±1.51	6.71±2.55	0.20±0.1
Oct,2014	0.18±0.09	0.03±0.02	7.06±0.81	0.64±0.22	0.19±0.27	14.88±5.62	0.97±1.05

ND: Not detected

Table 2. *Contamination factor/degree and pollution index of trace metals with depth.*

Trace metal	CF(0.2Meters)	CF(2.6Meters)	CF(5.2Meters)	Background limit
Ni	18.00	24.75	21.50	0.04
Cu	0.06	0.14	0.11	2.00
Cr	39.4	29.9	55.20	0.10
Mn	24.75	23.50	12.00	0.04
Pb	12.60	49.60	26.00	0.10
Fe	0.10	0.06	0.05	0.60
Zn	0.12	0.16	0.09	5.00
DC	94.93	128.05	114.90	
PLI	4.88	5.64	5.35	

Cf-contamination factor, DC-degree of contamination, PLI-pollution load index, Background limit (mg/l).

Table 3. *Standard ratio of contamination factor /degree, Hakanson (1980).*

Contamination factor		Contamination degree(C deg)	
Class	Indication	Class	Indication
Cf<1	Low contamination	DC<8	low degree of contamination
1<Cf<3	Moderate contamination factor	8≤DC<16	Moderate degree of contamination
3<Cf<6	Considerable contamination factor	16≤DC<32	Considerable degree of contamination
6<Cf	Very high contamination	32≤DCdeg	Very high degree of contamination

Cf-contamination factor, DC-degree of contamination

Table 4. *Correllation coefficient matrix of analysed trace metals.*

	Cr	Mn	Pb	Cu	Ni	Zn	Fe
Cr	1.00					-	
Mn	-0.89	1.00					
Pb	-0.51	0.06	1.00				
Cu	-0.22	-0.25	0.95	1.00			
Ni	-0.35	-0.11	0.98	0.99	1.00		
Zn	-0.98	0.77	0.69	0.42	0.55	1.00	
Fe	0.40	-0.77	0.58	0.81	0.71	-0.19	1.00

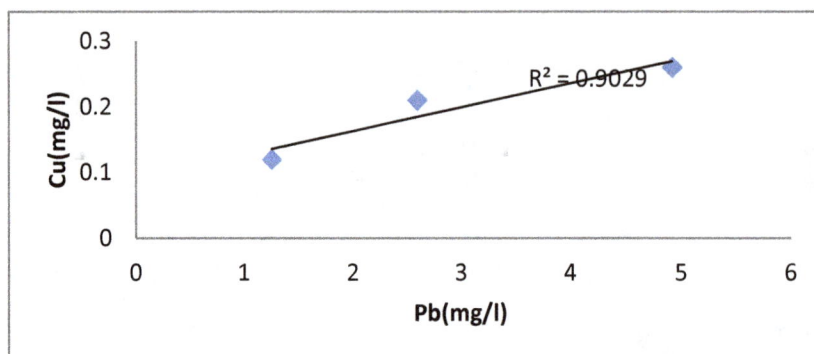

Fig. 3. *Plot of Cu against Pb concentration.*

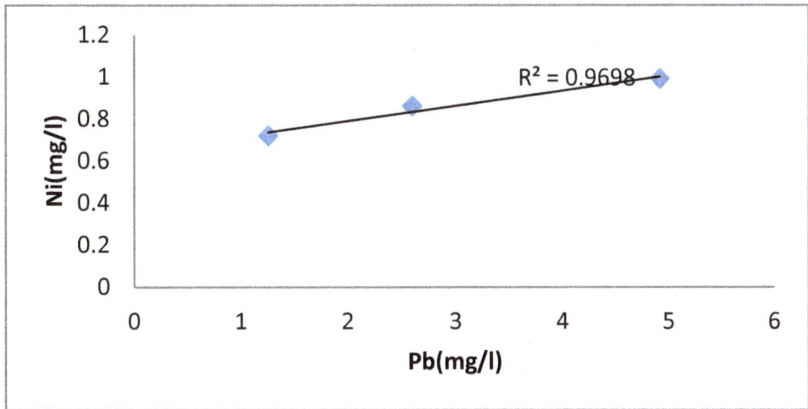

Fig. 4. *Plot of Ni against Pb concentration.*

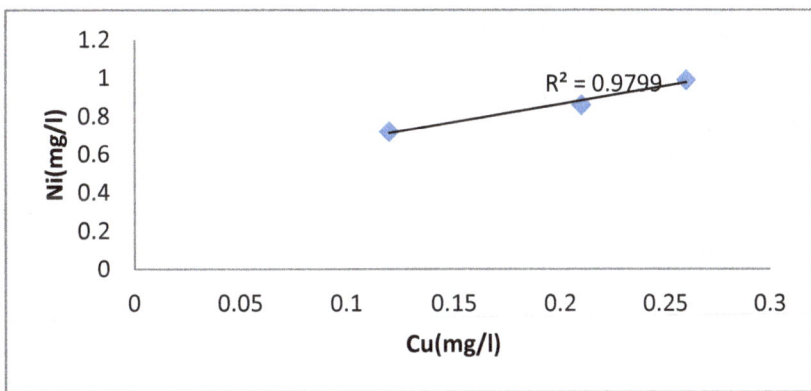

Fig. 5. *Plot of Ni against Cu concentration.*

8. Discussions

The contamination factor (CF) for Copper (Cu) and zinc (Zn) for the surface, medium and bottom depth profile (0.2meters, 2.6meters and 5.2meters) at the NIOMR Jetty are: 0.06, 0.14 0.11 and the latter 0.12, 0.16 and 0.09 respectively. These values are less than 1 (CF < 1), hence exhibit low degree of contamination (see table2&3).However, Lead (12.0, 49.6, 26.0), Chromium (39.4, 29.9, 55.2), Manganese (24.75, 23.50, 12.00) and Nickel (Ni, 18.00, 24.75, and 21.50) at 0.2meters, 2.6meters and 5.6 meters respectively all fall above a highly contaminated index ratio, Cf> 6. This is not unconnected to various metals and metallic compounds released from anthropogenic activities that add up to their natural background values. Chromium shows a non significant negative correlation with most of the trace metals at p>0.5 this further confirm its high contamination ratio (table 4) and distinct anthropogenic source from other trace metals. The correlation coefficient matrix relationship of other trace metals; Mn, Ni, Pb affirmed a common anthropogenic source.

Cr contamination is majorly associated with effluents from leather and textiles productions [22]while, Ni and Pb have been reported to be associated with effluents from electroplating, batteries storage, land disposals, sewage sludge, paint and dyes, effluents from crude oil transportations, fertilizers and vehicular emissions [22,4]. These effluents are traceable to solid wastes, allochtonous

deposits and marine debris such as high and low density polythene, empty cans of food/pesticide sprays, glass bottles, used car tyres, worn clothes and a host of others that moved alongside sea hyacinth in the NIOMR Jetty point. These features were highly conspicuous in early October (plate1). Marine debris has contributed to the high level of pollutants found virtually around the harbour and this is in agreement with past researchers [7, 23]. However, most of these wastes are non-biodegradable and continuously leach heavy metals into the water body [9].

Judging from the monthly variation in the trace metal concentration, there is an observed higher concentration in Cr, Mn and Pb in October. The high trace metal concentration in the month is attributed to reduced rain events (towards the end of wet season) and increased fresh water incursions from adjourning creeks and Badagry Lagoon; this is in agreement with the assertion by past researchers [24, 25]. The dilution from floodwater and the introduction of allochtonous material prevalent in October and high rate of tidal fluctuations(sea/fresh water incursion) which has been reported to have a tremendous influence on the coastal waters of the Lagos harbour (experienced in late September-early October) may not be unconnected to the high concentrations of Cr, Pb and Mn. This is also in agreement with the work of past researchers [26; 27-29] that affirmed that the dilution and enriching effects of floodwater governs the biota distribution of the Lagos harbour.

The overall degree of contamination and the pollution

index ratio trends are: medium depth profile (2.6meters) > bottom depth profile (5.2meters) > surface depth profile (0.2meters) fig 7 and 8. The highest contamination factor at the 2.6 meters depth is in agreement with the earlier proposed dilution factor which led to the introduction of allochtonous materials into the water body.

The low abundance of Cu and Zn may be as a result of uptake, and their low pollution index tends to relate them to background values (fig 8).However, Zinc shows a non-significant positive correlation with Mn(very high CF values), this factor can be reliably linked to a dilution in the background concentration of zinc metal. This same factor can be adduced for Copper (low CF) which exhibit a strong correlation with Pb(very high CF).

The percentage contributions of the toxic metals (Cr, Mn, Pb and Ni) to pollution index are shown in fig 9, 10 and 11.The peak percentage contributions of Cr (42%) at the 0.2meters profile, Pb (39%) at 2.6meters and Cr (48%) at 5.2meters is in agreement with the calculated pollution index ratios (PI), degree of contamination (DC) and the descending order of the trace metal concentration in the NIOMR Jetty (see table 1&2).

Iron, copper and Zinc were not used in the contamination degree and pollution index calculations based on relatively high natural concentrations of Fe metal in marine water and sediments; it is not expected to be substantially enriched from anthropogenic sources [30] and the low CF values of the latter.

Plate 1. *Showing water sampler and marine debris flowing alongside water hyacinth at the NIOMR Jetty.*

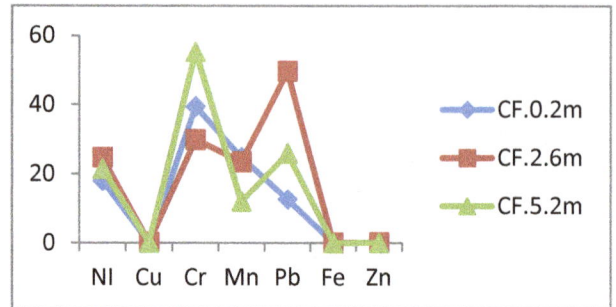

Fig. 6. *The plots of contamination factor of trace metals with depth Cf-contamination factor, m-meter.*

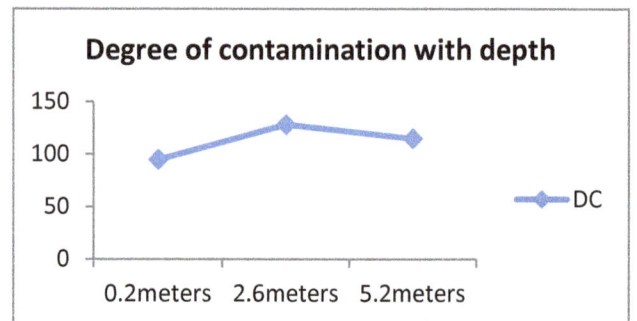

Fig. 7. *Showing the degree of contamination of the NIOMR Jetty point with depth DC-Degree of contamination.*

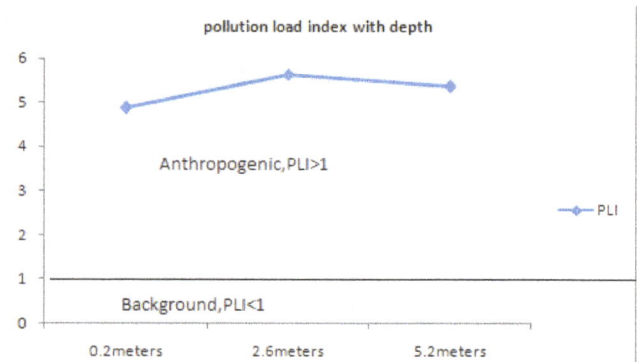

Fig. 8. *Showing the pollution index of the NIOMR Jetty point with depth PLI-Pollution load index.*

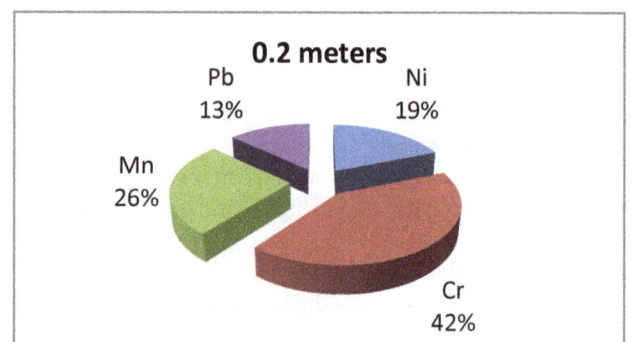

Fig. 9. *Percentage of toxic metal contributions to the pollution index in the surface depth (0.2meters) at the NIOMR Jetty.*

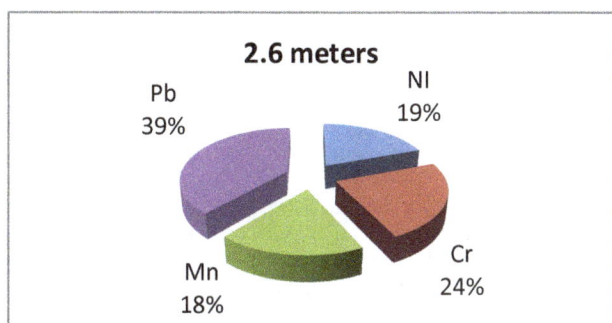

Fig. 10. *Percentage of toxic metal contributions to the pollution index at the depth (2.6meters) at the NIOMR Jetty.*

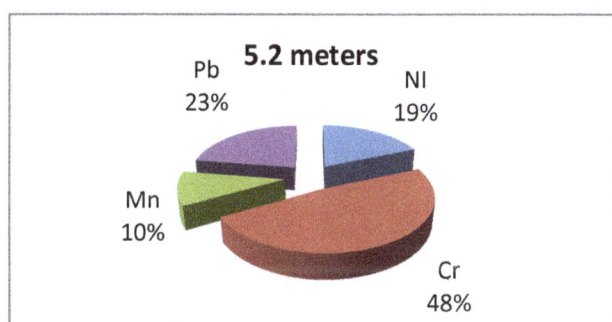

Fig. 11. *Percentage of toxic metal contributions to the pollution index at (5.2meters) in the NIOMR Jetty.*

9. Conclusion

The depth profiling sampling exercise in the NIOMR Jetty point has partially exposed the contamination pathways of trace metals around the Lagos Harbour. Copper and zinc in the NIOMR Jetty point fall below the anthropogenic enriched metals, and tend to be naturally derived, while Pb, Cr, Mn and Ni metals fall within the high contamination index ratio, hence, shows the dominance of anthropogenic enrichment. The contamination degree and the pollution index which signifies the overall contamination and toxicity rate of these trace metals also indicate progressive contamination in the NIOMR Jetty point. Low trace metal concentrations are not necessarily "natural" just because the levels are indeed low (Low trace metal concentrations may represent a mixture of small quantity of pollutants diluted by a large amount of natural water with a low trace metal content) and elevated concentrations of metals do not necessarily pose a threat as they may not be available for excessive aquatic organism's uptake. Therefore, contamination index determination is a very useful tool in the trace metal evaluation rather than the sole dependence on its high/low concentration above the permissible level.

Fresh/seawater incursions, anthropogenic effluents, precipitations, allochtonous deposit from marine debris, tidal fluctuations and mixing (ocean surge) are among the factors that control the concentrations and cycling of trace metals around the Habour. The highest contamination of toxic metals at the medium depth profile (2.6 meters) asserted the dominance of mixing and dilution in the NIOMR Jetty.

A strong regulation in the anthropogenic activities

(indiscriminate dumping of sewage and domestic waste) in the Lagos Lagoon axis is highly essential. Though dredging is inevitable to maintain sufficient water depth in shipping channels and harbors, which are continually filled in by deposition.

This project is in progress and will be beneficial if it's extended to other Lagoons, adjourning creeks and Nigerian continental waters so as to further expose the vertical distributions and natural/anthropogenic pathways of trace elements in these axes.

Acknowledgement

The author acknowledge the Head of department, Physical and Chemical oceanographic section of the Nigerian Institute for oceanography and marine research, Victoria Island, Lagos, Nigeria, Dr Unyimadu John Paul, for his innovations, quality guidance and assistance in the course of doing this project.

References

[1] Turner, A. (1996), Trace metals portioning in estuaries importance of salinity and particulate concentration. Marine Chemistry, 54, pp27 – 39.

[2] Okoye, C.O,. Onwuka,S.U. and Obiakor, M.O.(2011). Pollution studies in Lagos and its environmental consequences. A review. Tropical built environment journal. 1(1) pp42-57.

[3] Amaeze, N.H. (2009). Comparative account of the Microfauna and Planktonic Communities of two polluted creeks associated with the Lagos Lagoon System. MSc. Thesis, University of Lagos, 2009.

[4] Popoola, S.O,.Hauwa, M,.Oyeleke P.O.and Odeyemi,O.E.(2014).The effect of Ambient particulate metal concentrates in Lagos metropolis and the adjoining water bodies, Northwest f Lagos Lagoon. Scientific Research Journal (SCIRJ) vol(11), pp29-57.

[5] Olatunji, A. S. and Abimbola, A.F.(2010).Geochemical Evaluation of the Lagos Lagoon Sediments.World Applied Sciences Journal 9 (2): 178-193.

[6] Balogun, K. J., Ladigbolu, I.A. and Olajire, E.D. (2011).Hydrochemistry of a Tropical harbor: Influence of Industrial and Municipal inputs. Journal of Appl. Science and Environmental management, Vol. 15 (4) pp 575 – 581.

[7] Ajao, E. A.(1996) .Review of the state of pollution of the Lagos Lagoon. NIOMR Tech. Paper No. 106.

[8] Onyeama, I. C. and Popoola, R.C.(2013). The physico-chemical characteristics, chlorophyll a levels and phytoplankton dynamics of the east mole area of the Lagos harbor Lagos. Journal of Asian Scientific Research, 3(10) pp 995-1010.http://aessweb.com/journal-detail.php?Id=5003.

[9] Nubi,O.A., Ajao,E.A.and NubiA.T. (2008). Pollution assessment of the impact of coastal activities on Lagos Lagoon, Nigeria Science World Journal, vol. 3(2), pp. 83-88.

[10] Nubi,O.A., Oyediran, L.T. and Nubi, A.T. (2011).Inter-annual trends of heavy metals in marine resources. African Journal of Environmental Science and Technology Vol. 5(2), pp. 104-110

[11] Omatshola, M.E. and Adegoke, O.S.(1981). The tectonic evolution and Cretaceous a stratigraphy of the Dahomey Basin. J. Mining and Geol., 54: 65-87.

[12] Balogun,K.J. and Ladigbolu, I. A. (2011). Nutrients and Phytoplankton Production Dynamics of a Tropical Harbor in Relation to Water Quality Indices ; journal of America science 6(9) 2010 pp. 261 – 275.

[13] Balogun, K.J., Adedeji,A.k. andLadigbolu,I.A.(2014). Primary production estimation in the euphotic zone of a tropical harbor ecosystem, Nigeria. International Journal of Scientific and Research Publications, Volume 4, Issue 8, August 2014 1 ISSN 2250-3153.

[14] APHA. (1998). Standard method for the examination of water and waste water. 20th Edn., New York: American Public and Health Association.

[15] Nishida,H.,   Miyai,M,.Tada,F.   and   Suzuki S.(1982) .Computation of the Index of Pollution Caused by Heavy Metals in River sediment, Environmental Pollution Series B, Chemical and Physical, Elsevier, Vol. 4, No. 4, 1982, pp.   241-   248.   http://dx.doi.org/10.1016/0143-148X (82)90010-6.

[16] Hakanson, L.(1980). An ecological risk index for aquatic pollution control a sediment logical approach. Water Res.14, (2), pp 975.

[17] Lee, J.S,. Chon, H.T,. Kim, J.S,. Kim, K.W. and Moon, H.S.(1998) .Enrichment of Potentially Toxic Elements in Areas Underlain by Black Shales and Slates in Korea," Environmental Geochemistry and Health, Vol. 20, No. 30, pp. 135-147.

[18] World Health Organisation, (1993). Revision of WHO Guidelines for Water Quality. WHO Geneva.

[19] Taylor, S.R. and McLennan, S.M. (1995). The geochemical evolution of the continental crust. Reviews of Geophysics, 33, pp241-265.

[20] Tomlinson, D.L,. Wilson, J.G., Harris, C.R. and Jeffney,D.W. (1980). Problems in the assessment of heavy metals levels in estuaries and the formation of pollution index. Helgol. Wiss. Meeresunters, pp33, 566.

[21] Davis, J.C. (1986).Statistics and Analysis in Geology. John Wiley and Sons, New York, 64pp.

[22] Dara, S.S. and Srivastava R. (2009). Energy environment ethics and society. Revised edition, S.Chand& company limited, Ramnagar, New Delhi, 426p.

[23] Onyema, I.C.(2013). The Physico-Chemical Characteristics and Phytoplankton of the Onijedi Lagoon, Lagos. Nature and Science, 11(1) pp 127-135.

[24] Onyema, I.C. (2010). Phytoplankton diversity and succession in the Iyagbe lagoon, Lagos. European Journal of Scientific Research, 43(1), pp61-74.

[25] Onyema, I.C. (2011). The Water Chemistry and Periphytic Algae at a Cage culture Site in a Tropical Open Lagoon in Lagos. actaSATECH, 4(1): 53-63.

[26] Brown, A.C. and. Oyenekan,J.A(1998). Temporal variability in the structure of benthic macro-fauna communities of the Lagos lagoon and harbour, Nigeria. Pol Arch Hydrobiol, 45: 45-54.

[27] Chukwu, L. O. and Nwankwo, D. I. (2004). The impact of land based pollution on the hydro-chemistry and macro benthic community of a tropical West African Creek from the Nigerian territorial waters. The Ekologia.(2): pp19. African Journal of Environmental Science and Technology Vol. 5(2), pp. 104-110.

[28] Onyema, I.C. (2008). Phytoplankton biomass and diversity at the Iyagbe lagoon Lagos, Nigeria. University of Lagos, Akoka. Department of Marine Sciences.

[29] Onyema, I.C,.Nwankwo, D.I. and Owolabi, K.O.(2008). Temporal and spatial changes in the phytoplankton dynamics at the tarkwa-bay jetty in relation to environmental characteristics. Journal of Ecology Environment and Conservation, 14(4): pp1-9.

[30] Niencheski, L.F., Windom, H.L. and Smith, R. (1994). Distributions of particulate trace metal in Patos Lagoon Estuary (Brazil). Marine Pollution Bulletin, 28, pp96-102.

25

Combined Heat and Mass Transfer Steady Flow of Viscous Fluid over a Vertical Plate with Large Suction

S. M. Arifuzzaman, Md. Manjiul Islam, Md. Mohidul Haque

Mathematics Discipline, Khulna University, Khulna, Bangladesh

Email address:

arifsm42@gmail.com (S. M. Arifuzzaman), manjiul.math@gmail.com (Md. M. Islam), mmhaque@math.ku.ac.bd (Md. M. Haque)

Abstract: Combined heat and mass transfer of a viscous fluid along a semi-infinite vertical plate with large suction is studied analytically. Perturbation technique is used as main tool for the analytical approach. Viscous fluid behavior of heat and mass transfer over a vertical plate with large suction has been considered and its similarity equations have been obtained. Similarity equations of the corresponding momentum, temperature and concentration equations are derived by employing the usual similarity technique. The dimensionless similarity equations for momentum, temperature and concentration equation solved analytically by perturbation technique. The obtained numerical values of fluid velocity, temperature and concentration are drawn in figures. The results are discussed in detailed with the help of graphs to observe the effects of various parameters on the flow variables. Lastly, the important findings are concluded here.

Keywords: Mass Transfer, Heat Transfer, Eckert Number, Schmidt Number, Large Suction

1. Introduction

The heat transfer flows play a decisive role in many engineering applications as distillation, condensation, evaporation, rectification and absorption of a fluid as well as in fluids condensing or boiling at a solid surface. The heat transfer processes are of great interest in power engineering, metallurgy, astrophysics and geophysics. A natural convective heat transfer flow of fluid was first studied by *Finston* (1956) [1]. *Sparrow* and *Gregg* (1958) computed a similar solution for laminar free convection from a non-isothermal vertical plate [2]. A Finite difference solution of transient free convective flow over an isothermal plate was obtained by *Soundalgekar* and *Ganesan* (1981) [3]. A numerical study on the natural convective cooling problem of a vertical plate was completed by *Camargo et al.* (1996) [4].

The processes of mass transfer are of great interest in the production of materials in order to obtain the desired properties of a substance. Separation processes in chemical engineering such as the drying of solid materials, distillation, extraction and absorption are all affected by the process of mass transfer. Chemical reactions including combustion processes are often decisively determined by the mass transfer. *Callahan* and *Marner* (1976) studied a free convective

unsteady flow with mass transfer past a semi-infinite plate [5]. An investigation on free convective unsteady flow with mass transfer past an infinite vertical porous plate with constant suction was completed by *Soundalgekar* and *Wavre* (1977) [6]. Transient free convection flow on a semi-infinite vertical plate with mass transfer was observed by *Soundalgekar* and *Ganesan* (1980) [7].

The combined heat and mass transfer flows play a special role in power engineering, metallurgy, condensation, evaporation and rectification of a fluid. In the separation processes as drying of solid materials, distillation, extraction and absorption; the combined heat and mass transfer occur due to buoyancy forces caused by temperature difference and concentration difference. *Pera* and *Gebhart* (1971) was the first author to study the natural convective heat and mass transfer problem [8]. *Singh et al.* (2003 investigated the combined heat and mass transfer hydromagnetic flow of a viscous incompressible fluid along an infinite vertical porous plate [9]. *Chamkha* and *Khaled* (2001) investigated the problem of coupled heat and mass transfer from an inclined plate in the presence of absorption [10]. Hence the main aim of this research project was to study a combined heat and mass transfer steady flow of a viscous fluid along a semi-infinite vertical plate with large suction.

2. Mathematical Model of Flow

A heat and mass transfer steady flow of a viscous fluid along a semi-infinite vertical plate was considered. The flow was assumed to be in the $x-$ direction, which was chosen along the plate in upward direction and $y-$ axis was normal to it. Initially, it was considered that the plate as well as the fluid particle was at rest at the same temperature $T(=T_\infty)$ and the same species concentration level $C(=C_\infty)$ at all points, where C_∞ and T_∞ are fluid concentration and temperature species of uniform flow respectively. It was also assumed that the fluid particles outside the boundary layer moved with a uniform velocity U_0. The suitable physical configuration with co-ordinate systems are shown in Fig 2.1.

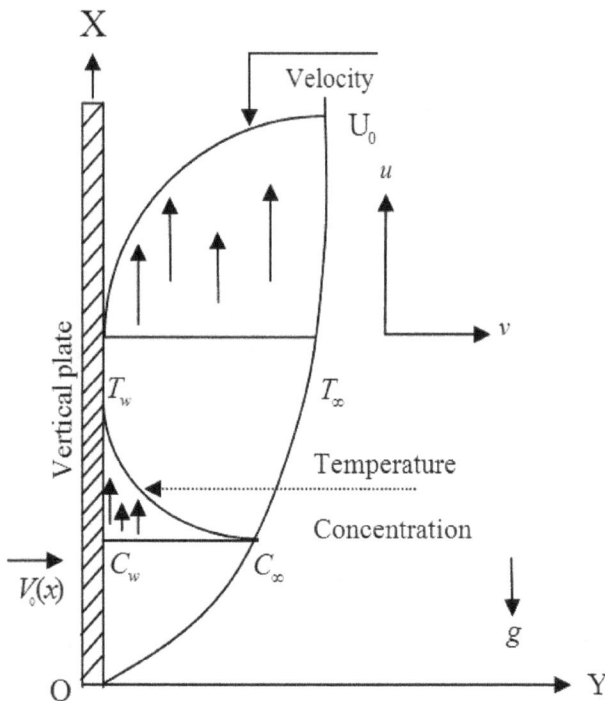

Fig. 2.1. *Physical Configuration of Flow.*

The following is the system of coupled governing non-linear partial differential equations.
The continuity equation

$$\frac{\partial u}{\partial x} + \frac{\partial v}{\partial y} = 0 \qquad (1)$$

The momentum equation

$$u\frac{\partial u}{\partial x} + v\frac{\partial u}{\partial y} = g\beta(T - T_\infty)$$
$$+ g\beta^*(C - C_\infty) + \upsilon\left(\frac{\partial^2 u}{\partial y^2}\right) \qquad (2)$$

The Energy Equation

$$u\frac{\partial T}{\partial x} + v\frac{\partial T}{\partial y} = \frac{k}{\rho\, c_p}\frac{\partial^2 T}{\partial y^2}$$
$$+ \frac{\upsilon}{\rho c_p}\left(\frac{\partial u}{\partial y}\right)^2 \qquad (3)$$

The Concentration Equation

$$u\frac{\partial C}{\partial x} + v\frac{\partial C}{\partial y} = D_m\left(\frac{\partial^2 C}{\partial y^2}\right) \qquad (4)$$

and the boundary conditions are,

$$u = 0\,,\ v = V_0(x),\ T = T_w \quad C = C_w \ \text{at}\ y = 0$$
$$u = U_0,\ v = 0,\qquad T \to T_\infty \quad C \to C_\infty \ \text{at}\ y \to \infty$$

where, x, y be the Cartesian coordinate; u, v are velocity components, g is the local acceleration due to gravity, β is the is the thermal expansion coefficient, β^* is the is the concentric expansion coefficient, T_w denotes the wall temperature, C_w is the species concentration at the wall, υ is the kinematic viscosity, ρ is density, k is thermal conductivity, C_p is specific heat at constant pressure, D_m is coefficient of mass diffusivity, m is mass per unit area.

3. Mathematical Formulation

Since the researchers goal was to attain similarly solutions of the problem the researchers introduced the following variables,

$$\left.\begin{array}{l} \eta = y\sqrt{\dfrac{U_0}{2\,\upsilon\,x}} \\[3mm] \psi = \sqrt{2\,\upsilon\,U_0\,x}\ f(\eta) \\[3mm] u = \dfrac{\partial\psi}{\partial y} \quad \text{and} \quad v = -\dfrac{\partial\psi}{\partial x} \end{array}\right\} \qquad (5)$$

Where,

$$u = U_0\, f'(\eta) \qquad (6)$$

$$u\frac{\partial u}{\partial x} = -\frac{U_0^{\,2}}{2x}\eta\, f'(\eta)\, f''(\eta) \qquad (7)$$

$$\frac{\partial^2 u}{\partial y^2} = \frac{U_0^{\,2}}{2\,\upsilon\,x}\, f'''(\eta) \qquad (8)$$

$$\therefore\ v = \sqrt{\frac{\upsilon\,U_0}{2x}}\left[\eta f'(\eta) - f(\eta)\right] \qquad (9)$$

Above values we get these;

$$f''' + ff'' + G_r\theta + G_m\varnothing = 0 \qquad (10)$$

$$\theta'' + P_r f \theta' + P_r E_c f''^2 = 0 \qquad (11)$$

$$\varnothing'' + S_C f(\eta) \varnothing'(\eta) = 0 \qquad (12)$$

where, $G_r = \dfrac{2x}{U_0^2} g\beta(T_w - T_\infty)$ is the Grashof Number and

$G_m = \dfrac{2x}{U_0^2} g\beta(C_w - C_\infty)$ is the Modified Grashof

Number. $P_r = \dfrac{\mu c_p}{k} = \dfrac{\rho \upsilon c_p}{k}$ is the Prandtl Number and

$E_C = \dfrac{U_0^2}{c_p(T_w - T_\infty)}$ is the Eckert Number.

With the boundary conditions

$$f = f_w, \ f' = 0, \ \theta = 1, \ \varnothing = 1 \quad \text{at} \quad \eta = 0$$

$$f' = 1, \quad \theta = 0, \quad \varnothing = 0 \qquad \text{at} \quad \eta \to \infty$$

Where, $f_w = -V(x)\sqrt{\dfrac{2x}{U_0 \nu}}$ the transpiration parameter

and prime is denotes derivatives with respect to η. Here $f_w > 0$ indicates the suction and $f_w < 0$ the injection.

4. Mathematical Analysis

$$\left.\begin{array}{l} \xi = \eta f_w, \\ f(\eta) = f_w F(\xi), \\ \theta(\eta) = f_w^2 H(\xi) \\ \varnothing(\eta) = f_w^2 G(\xi) \end{array}\right\} \qquad (13)$$

By using the conversion in the equation (10) and (11) we have,

$$\begin{array}{l} f_w^4 F''' + f_w F f_w^3 F'' + G_r f_w^2 H \\ + G_m f_w^2 G = 0 \end{array} \qquad (14)$$

$$\begin{array}{l} f_w^4 H'' + P_r f_w F . f_w^3 H' \\ + P_r E_C f_w^6 F''^2 = 0 \end{array} \qquad (15)$$

$$f_w^4 G'' + S_C f_w F f_w^3 G' = 0 \qquad (16)$$

Therefore the equation (13) with boundary conditions as given below

$$\left.\begin{array}{l} F''' + F F'' + \varepsilon G_r H + \varepsilon G_m G = 0 \\ H'' + P_r F H' + P_r E_C \dfrac{1}{\varepsilon} F''^2 = 0 \\ G'' + S_C F G' = 0 \end{array}\right\} \qquad (17)$$

$$F = 1, \ F' = 0, \ H = \varepsilon, G = 0 \quad \text{at} \quad \xi = 0$$

$$F' = \varepsilon, \quad H = 0, G = 0 \qquad \text{at} \quad \xi \to \infty$$

where, $\varepsilon = \dfrac{1}{f_w^2}$

Now for the large solution $f_w > 1$. If f_w is very large then ε will be very small. Therefore following Bestman (1990) and Singh & Dikshit (1988), F, H and G can be expended in terms of the small perturbation quantity ε as,

$$F(\xi) = 1 + \varepsilon F_1(\xi) + \varepsilon^2 F_2(\xi) + \qquad (18)$$

$$H(\xi) = \varepsilon H_1(\xi) + \varepsilon^2 H_2(\xi) + \qquad (19)$$

$$G(\xi) = \varepsilon G_1(\xi) + \varepsilon^2 G_2(\xi) + \qquad (20)$$

Taking the order $O(\varepsilon)$ the system (17) becomes,

$$\left.\begin{array}{l} F_1''' + F_1'' = 0 \\ H_1'' + P_r H_1' + P_r E_C F_1''^2 = 0 \\ G_1'' + S_C G_1' = 0 \end{array}\right\} \qquad (21)$$

With boundary condition

$$\left.\begin{array}{l} F_1 = 0, \ F_1' = 0, \ H_1 = 1, G_1 = 1 \quad \text{at} \quad \xi = 0 \\ F_1' = 1, \ H_1 = 0, G_1 = 0 \qquad \text{at} \quad \xi \to \infty \end{array}\right\}$$

Again for order $O(\varepsilon^2)$ the system (17) becomes,

$$\left.\begin{array}{l} F_2''' + F_2'' + F_1 F_1'' + G_r H_1 + G_m G_1 = 0 \\ H_2'' + P_r(F_1 H_1' + H_2' + 2E_C F_1'' F_2'') = 0 \\ G_2'' + S_C(F_1 G_1' + G_2') = 0 \end{array}\right\} \qquad (22)$$

with boundary condition

$$\left.\begin{array}{l} F_2 = 0, \ F_2' = 0, \ H_2 = 0, G_2 = 0 \quad \text{at} \quad \xi = 0 \\ F_2' = 0, \ H_2 = 0, G_2 = 0 \qquad \text{at} \quad \xi \to \infty \end{array}\right\}$$

5. Solution

The first order solution becomes;

$$F_1 = -1 + \xi + e^{-\xi}, \ H_1 = e^{-P_r \xi} + A_{11} e^{-2\xi}$$

$$\text{and} \ G_1 = e^{-S_C \xi}.$$

and the second order solution becomes

$$F_2 = -3\xi e^{-\xi} - \xi e^{-\xi} - \frac{1}{2}\xi^2 e^{-\xi}$$

$$+ A_{21}e^{-P_r\,\xi} + A_{22}e^{-2\xi} + A_{23}e^{-S_c\,\xi}$$

$$H_2 = A_{27}e^{-S_c\,\xi} + A_{28}e^{-3\xi} + \left(B_{12} + B_{13}\xi + B_{14}\xi^2\right)\,e^{-2\xi}$$

$$+ \left(A_{30}\xi - B_{11}\xi^2\right)e^{-P_r\,\xi} + A_{31}\,e^{-\xi\,(P_r+1)}$$

$$G_2 = (C_{11} + C_{12}\xi + C_{13}\xi^2)e^{-S_c\,\xi} + C_{11}e^{-\xi}$$

6. Results and Discussions

The analytical solutions were obtained by using the perturbation technique. In order to analyze the physical situation of the model, the researchers computed the numerical values of the flow variables for different values of suction parameter $\left(f_w\right)$, Grashof number $\left(G_r\right)$, Modified Grashof Number $\left(G_m\right)$, Schmidt number $\left(S_C\right)$, Eckert Number $\left(E_c\right)$ and the Prandtl Number $\left(P_r\right)$. The fluid velocity, temperature and concentration versus the non-dimensional coordinate variable η are displayed in Figs. 6.1 - 6.15. It is noted in Fig. 6.1 that the velocity remains unchanged with the increasing of Eckert Number (E_c). Fig.6.2 represents the velocity increase with the increase of the value of the suction parameter $\left(f_w\right)$. From Fig. 6.3 it can be seen that the velocity decreases with the increase of the value of the Grashof number $\left(G_r\right)$. Fig. 6.4 shows that the velocity remains unchanged with the increase of the value of the Prandtl Number $\left(P_r\right)$. In Fig. 6.5, the velocity decreases with the increase of the value of the Schmidt number $\left(S_C\right)$. In Fig. 6.6, the fluid temperature decreases with the increase of the value of the Eckert Number (E_c). Fig. 6.7 shows that the temperature decreases with the increase of the value of the suction parameter $\left(f_w\right)$. It is found in Fig. 6.8 that the temperature remains unchanged with the increase of the value of the Grashof number $\left(G_r\right)$. Fig. 6.9 shows that the temperature decreases with the increase of the value of the Prandtl Number $\left(P_r\right)$. Fig. 6.10 shows that the temperature increases with the increase of the value of the Schmidt number $\left(S_C\right)$. Fig. 6.11 shows that the concentration remains unchanged with the increase of the value of the Eckert Number (E_c). It is shown in Fig. 6.12 that the concentration decreases with the increase of the value of the suction parameter $\left(f_w\right)$. Fig. 6.13 shows that the concentration remains unchanged with the increase of the value of the Grashof number $\left(G_r\right)$. It is observed from Fig. 6.14 that the concentration remains unchanged with the increase of the value of the Prandtl Number $\left(P_r\right)$. Fig. 6.15 shows that the concentration decreases with the increase of the

Schmidt number $\left(S_C\right)$.

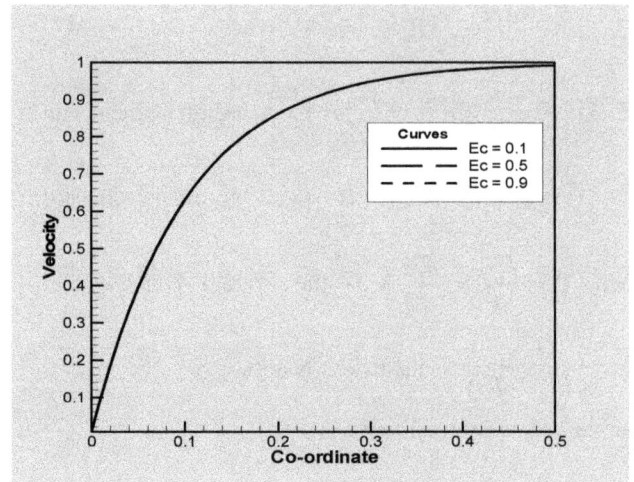

Fig. 6.1. *Velocity profiles for different values of* E_C *with* $G_m = 0.2$, $G_r = 0.2$, $P_r = 0.5$, $f_w = 10.0$ *and* $S_c = 0.3$.

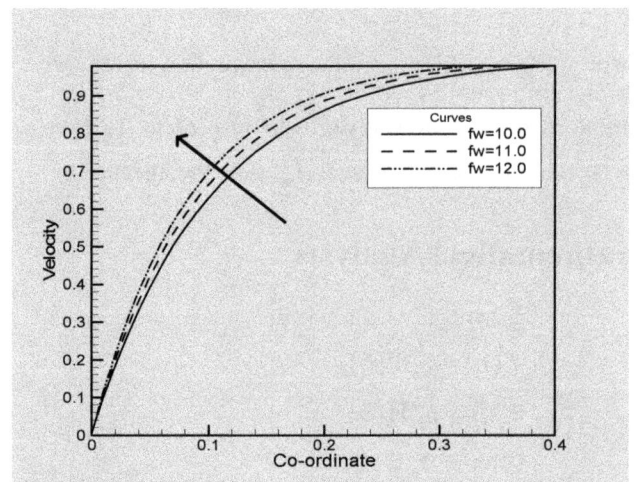

Fig. 6.2. *Velocity profiles for different values of* f_w *with* $G_m = 0.2$, $G_r = 0.2$, $P_r = 0.5$, $E_c = 0.1$ *and* $S_c = 0.3$.

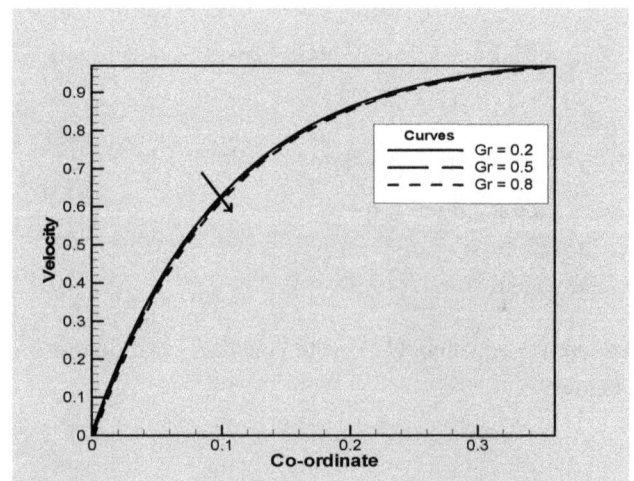

Fig. 6.3. *Velocity profiles for different values of* G_r *with* $G_m = 0.2$, $E_c = 0.1$, $P_r = 0.5$, $f_w = 10.0$ *and* $S_c = 0.3$.

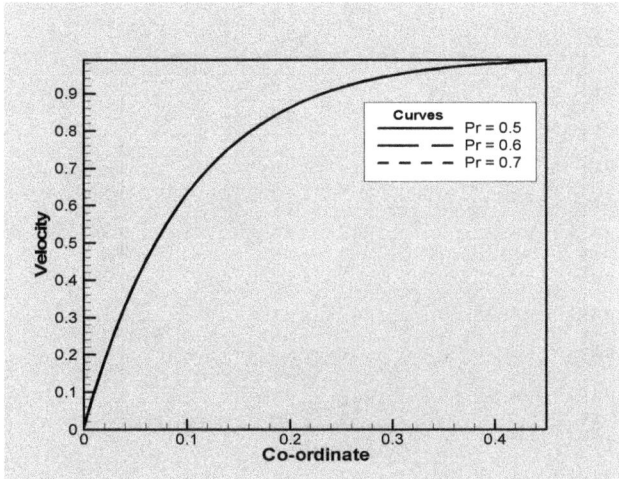

Fig. 6.4. *Velocity profiles for different values of* P_r *with* $G_m = 0.2$, $E_c = 0.1$, $G_r = 0.2$, $f_w = 10.0$ *and* $S_c = 0.3$.

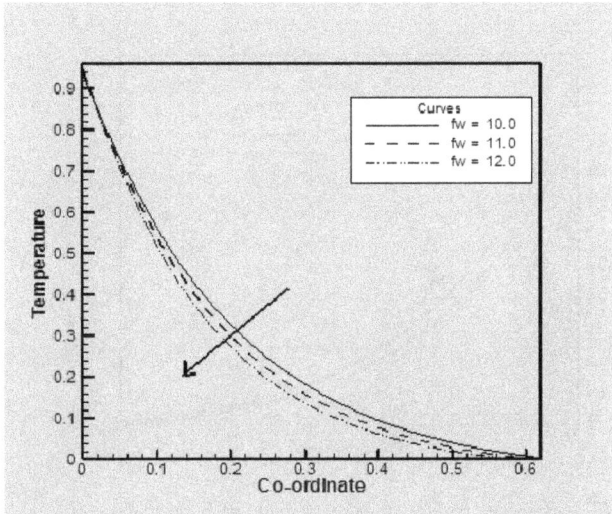

Fig. 6.5. *Velocity profiles for different values of* S_c *with* $G_m = 0.2$, $E_c = 0.1$, $G_r = 0.2$, $f_w = 10.0$ *and* $P_r = 0.5$.

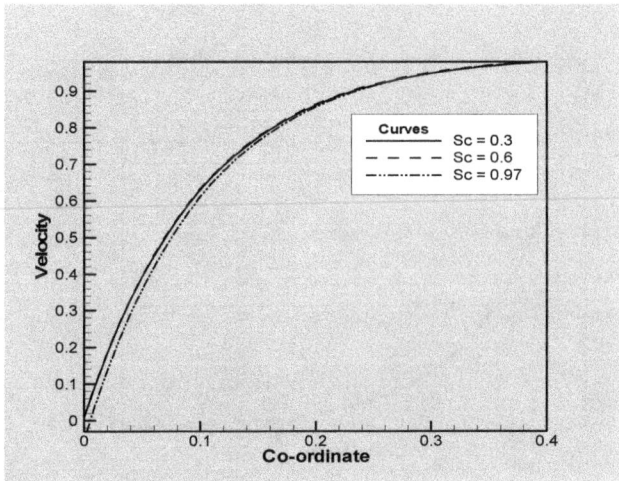

Fig. 6.6. *Temperature profiles for different values of* E_c *with* $G_m = 0.2$, $S_c = 0.3$, $G_r = 0.2$, $f_w = 10.0$ *and* $P_r = 0.5$.

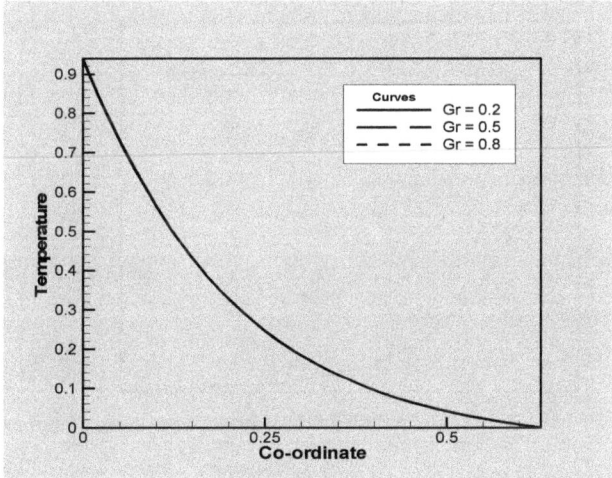

Fig. 6.7. *Temperature profiles for different values of* f_w *with* $G_m = 0.2$, $S_c = 0.3$, $G_r = 0.2$, $E_c = 0.1$ *and* $P_r = 0.5$.

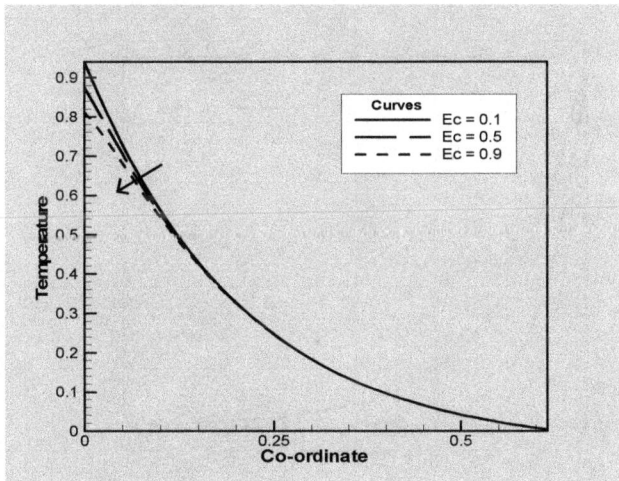

Fig. 6.8. *Temperature profiles for different values of* G_r *with* $G_m = 0.2$, $S_c = 0.3$, $E_c = 0.1$, $f_w = 10.0$ *and* $P_r = 0.5$.

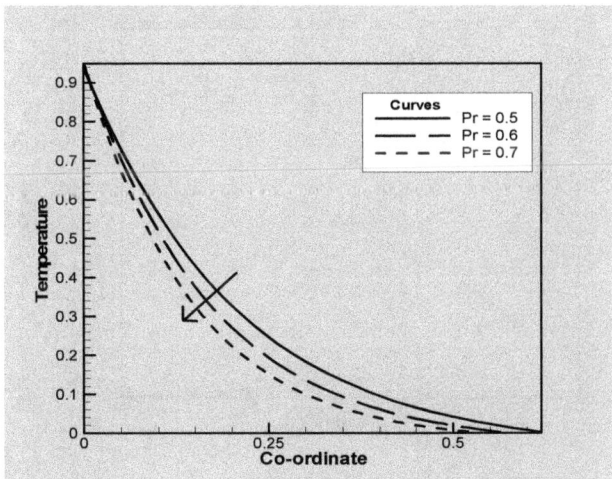

Fig. 6.9. *Temperature profiles for different values of* P_r *with* $G_m = 0.2$, $S_c = 0.3$, $G_r = 0.2$, $f_w = 10.0$ *and* $E_c = 0.1$.

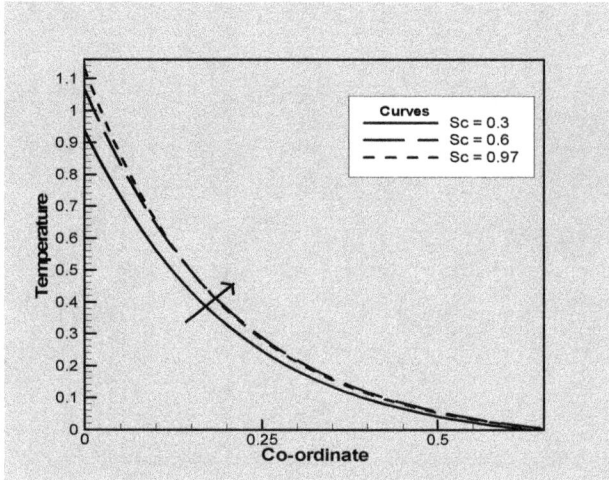

Fig. 6.10. Temperature profiles for different values of S_C *with* $G_m = 0.2$, $E_C = 0.1$, $G_r = 0.2$, $f_w = 10.0$ *and* $P_r = 0.5$.

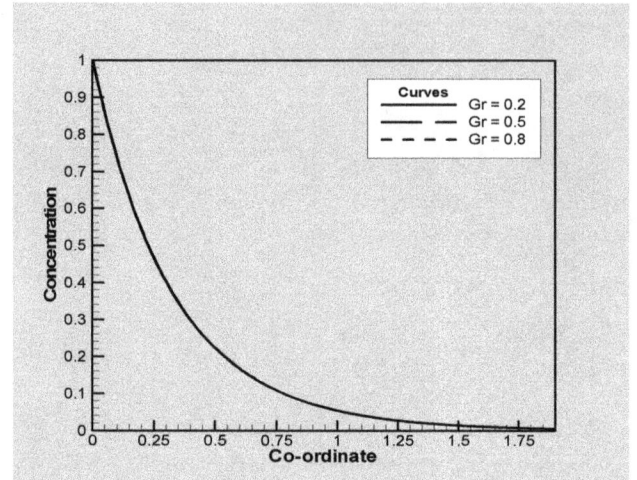

Fig. 6.13. Concentration profiles for different values of G_r *with* $G_m = 0.2$, $S_C = 0.3$, $f_w = 10.0$, $E_C = 0.1$ *and* $P_r = 0.5$.

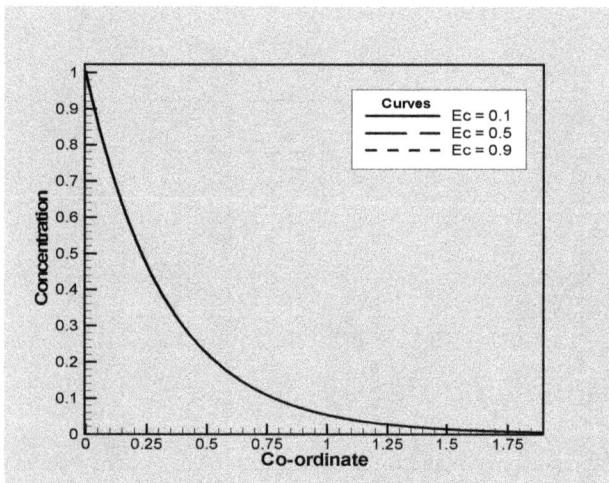

Fig. 6.11. Concentration profiles for different values of E_C *with* $G_m = 0.2$, $S_C = 0.3$, $G_r = 0.2$, $f_w = 10.0$ *and* $P_r = 0.5$.

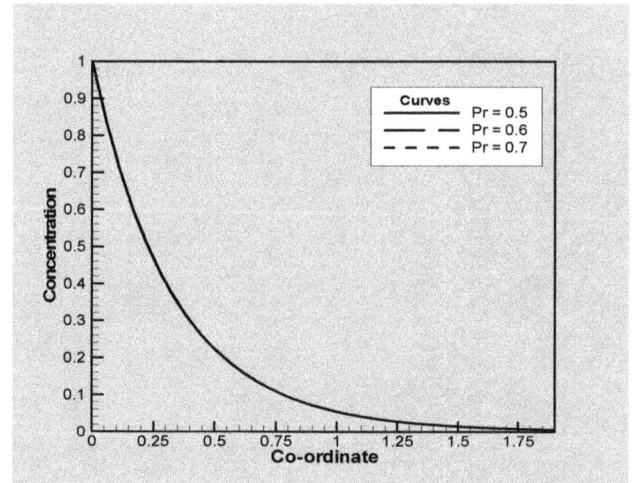

Fig. 6.14. Concentration profiles for different values of P_r *with* $G_m = 0.2$, $S_C = 0.3$, $G_r = 0.2$, $E_C = 0.1$ *and* $f_w = 10.0$.

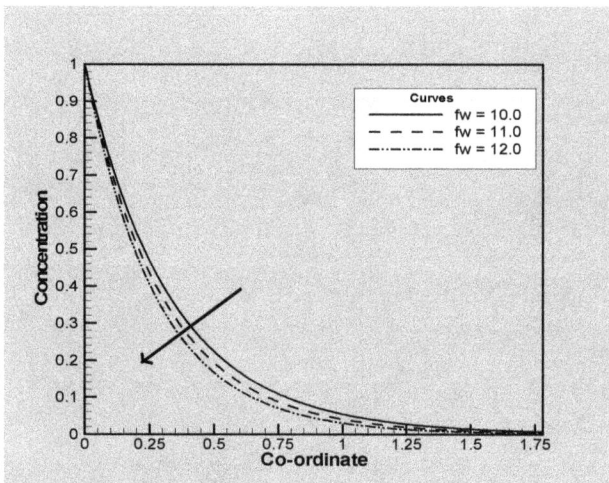

Fig. 6.12. Concentration profiles for different values of f_w *with* $G_m = 0.2$, $S_C = 0.3$, $G_r = 0.2$, $E_C = 0.1$ *and* $P_r = 0.5$.

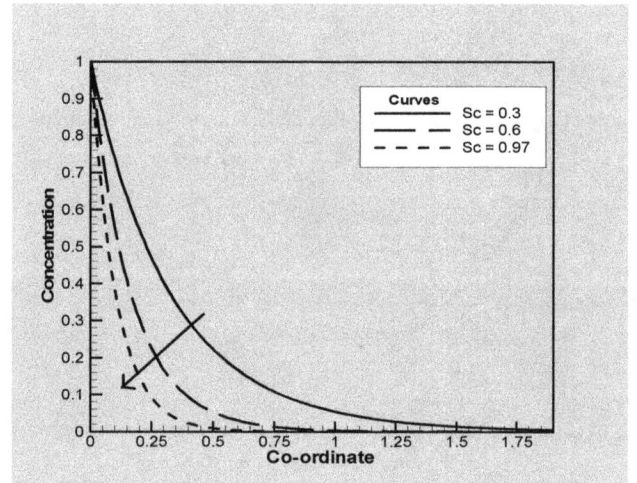

Fig. 6.15. Concentration profiles for different values of S_C *with* $G_m = 0.2$, $f_w = 10.0$ $G_r = 0.2$, $E_C = 0.1$ *and* $P_r = 0.5$.

7. Conclusions

Viscous fluid behaviors of heat and mass transfer over a vertical plate with large suction have been observed in this work. The resulting governing dimensionless coupled non-linear partial differential equations have been solved analytically by perturbation technique. The results are shown for different values of important parameters as the Eckert number, suction parameter, Prandtl number, Grashof number and Schimdt number. Some of the important findings obtained from the graphical representation of the results are listed below:

1. The velocity profiles increases with the increase of suction parameter f_w while the velocity profile decreases with the increase of grashof number G_r.

2. The temperature profiles increases with the increase of suction parameter f_w while the temperature profiles decrease with the increase of Prandtl number P_r.

3. The concentration profiles decreases with the increase of Grashof Number (G_r) and Schimdt number S_C.

Acknowledgement

The research project has supported Mathematics discipline, Khulna University. Therefore we have expressed gratefulness to Mathematics discipline, Khulna University for providing lab facilities.

Nomenclature

S_c	Schmidt number
G_r	Grashof number
G_m	Modified Grashof number
C	Concentration
D_c	Mass Diffusivity
u	Velocity component in the x-direction
v	Velocity component in the y-direction
w	Velocity component in the z-direction
∇	Differential operator
P_r	Prandtl number
k	Thermal Conductivity
f_w	Suction parameter
β	Co-efficient of thermal expansion
β^*	Concentric expansion coefficient
δ	Boundary layer thickness
μ	Co-efficient of viscosity
υ	Co-efficient of kinematic viscosity
ρ	Density of the fluid

References

[1] M. Finston, (1956), Free convection past a vertical plate, *J. Appl. Math. Phy.*, 7:527-529.

[2] E. M. Sparrow and J. L. Gregg, (1958), Similar solutions for free convection from a non isothermal vertical plate, *ASME J. Heat Trans.*, 80:379-386.

[3] Soundalgekar and Ganesan P (1981), Finite-differential analysis of transient free convection with mass transfer on an isothermal vertical plate, *International journal of engineering and science*, 19:757-770.

[4] R. Camargo, E. Luna, C. Trevino, (1996), Numerical study of the natural convective cooling of a vertical plate, *International journal of Heat and Mass Transfer*, 19(1-2):89-95.

[5] G.D. Callahan, W.J. Marner, (1976), Transient Free Convection with Mass Transfer on an Isothermal Vertical Flat Plate, *International Journal of Heat and Mass Transfer*, 19(2):165-174.

[6] V.M. Soundalgekar, P.D. Wavre, (1977), Unsteady free convection flow past an infinite vertical plate with constant suction and mass transfer, *International Journal of Heat Mass Transfer*, 20(12):1363-1373.

[7] V.M. Soundalgekar, P. Ganesan, (1980), Transient free convective flow past a semi-infinite vertical plate with mass transfer, *Journal of Energy Heat and Mass Transfer*, 2(1):83.

[8] B. Gebhart and L. Pera, (1971), The nature of vertical natural convection flows resulting from the combined buoyancy effect of thermal and mass diffusion, *International Journal of Heat and Mass Transfer*, 14:2025-2050.

[9] Singh, A.K., A.K. Singh and N.P. Singh, (2003), Heat and mass transfer in MHD flow of a viscous fluid past a vertical plate under oscillatory suction velocity, *Indian J. Pure Appl. Math*, 34:429.

[10] Chamka and Khaled, (2001), Simultaneously heat and mass transfer in free convection, *Industrial Engineering Chemical*, 49:961-968.

26

Design for Fabrication of Effective Seed Cane Hot Water Treatment Plant for Ethiopian Sugar Estates/Projects

Endale Wondu[1, *], Tadesse Negi[2], Wendimu Weldegiorgis[3], Alemayehu Dengiya[2]

[1]Mechanical and Vehicle Engineering (Manufacturing Engineering), Design and fabrication, Wonji/Shoa, Ethiopia
[2]Agronomy, Agronomy and plant protection, Wonji/Shoa, Ethiopia
[3]Electrical Engineering, Design and fabrication, Wonji/Shoa, Ethiopia

Email address:

endalew2010@gmail.com (E. Wondu)

Abstract: Sugarcane is prone to infection by a large number of diseases that can impact significantly on productivity. Diseases caused by systemic pathogens (those that occur within the plant tissues) can introduce into a new crop by the planting of infected seed cane. So as to avoid such disease causing systemic organisms treatment of the seedcane is mandatory. The initial treatment of the seedcane often includes the routine application of thermotherapy to eliminate fungal and bacterial pathogens as well as pests. Especially for the control of smut, RSD and Albino (Mycoplasma) hot water treatment is preferable; because most of infections are killed at a temperature slightly lower than that lethal for sugarcane. Therefore, Hot water Treatment is an effective method for the elimination of pathogenic infections that are seed piece transmissible. Uniformity of temperature, proper circulation of water, Keeping 1: 4 and avoiding heavy packing of seed materials are the main requisites for successful treatment. Keeping the above facts in to consideration a rotary circular pattern treatment tanker and an automatically controlled heating tanker has been designed at Wonji/showa. The plant is designed to use both steam and electric power independently or simultaneously as a source of heat by making simple adjustment on the heating tanker. The treatment plant incorporates three successive tankers namely; 1. Water heating tanker on which the heating, limit switch and water circulating mechanisms are installed. 2. Seed cane treatment tanker on which water circulating, temperature control and avoiding heavy dumping mechanisms are installed. 3. Fungicide dipping tanker. The three tankers have been designed to plant with chronological order of the work flow and a bulk of planting material will be successfully treated by this system.

Keywords: Treatment, Effective, Pathogenic, Transmissible, Fungicide, Designed

1. Introduction

Sugarcane is prone to infection by a large number of diseases that can impact significantly on productivity, as in [1]. This is partly due to the nature of the crop and the manner in which it is propagated, cultivated and managed. In commercial practice sugarcane is propagated vegetatively, by planting pieces of stalk, termed setts or seed cane. The tillers that emerge from the axial buds on the planted stalk pieces develop into the new crop of stalks. Diseases caused by systemic pathogens (those that occur within the plant tissues) can therefore be readily introduced into a new crop by the planting of infected seed cane.

For this reason the control of diseases is a critically important aspect of the management of sugarcane and must

largely be achieved through a policy of prevention. Because there are many hazardous diseases caused by systemic pathogens that are readily spread by the planting of infected seedcane, the procurement of healthy seedcane is a prerequisite for the successful production of sugarcane.

The initial treatment of the seedcane often includes the routine application of thermotherapy, such as hot water treatment (HWT), to eliminate fungal and bacterial pathogens as well as pests. Especially for the control of smut, Ratoon Stunting Disease (RSD) and Albino (Mycoplasma) disease of vegetatively propagated crop, sugarcane; because most of infections are killed at a temperature slightly lower than that lethal for sugarcane.

The treatment time and temperature is a compromise between the need to eliminate pests and pathogens without severely impacting on the subsequent germination of the treated seedcane. The standard treatment in most sugarcane industries is 2 h at 50 °C, which has been shown to achieve a high degree of elimination of the RSD pathogen L. xyli subsp xyli, as in [2]. HWT is usually followed by a fungicide dip to prevent infection of the treated seedcane by soil-borne diseases after planting, as in [3].

An important aspect of Hot water treatment plant design includes having a relatively high volume of water to seedcane (approximately 4:1) so that the water temperature does not decline substantially when each batch of seedcane is added. Also important is the need for accurate control of the water temperature, which is usually controlled by thermostats. The water in the tank must be efficiently circulated to avoid 'hot or cold spots' and the water must be replaced regularly; otherwise it soon becomes acidic and contaminated, with adverse effects on germination, as in [4].

In subjection of germination, an experiment has been conducted to know the status of germination of seed cane, at South Africa sugar Association experiment station, on three samples viz hot water treated cane, hot air treated cane and untreated whole stalks. The result indicates that hot water treated cane germinated first in most varieties [5].

The project is designed to use both steam and electric independently or simultaneously as a source of heat. Therefore, it can be implemented at any of the sugar factories, where steam heat is available, or any of sugar projects, where electric power is available, by making simple adjustments on the heating device. "Temperature control is critical, with temperatures over 50°C adversely affecting germination, and temperatures below50°C ($< 49.8^0$C) reducing the effectiveness of disease control", as in [6].

The economic analysis was done by considering the impact due to reduction of sugar cane yield loss by using effective Hot water treatment. As indicated with different researchers the cane yield loss due to infection of different pathogens, especially Smut and RSD are shown bellow: Smut yield losses from 12% –75% have been reported [7, 8]. In Ethiopia it causes 19.3 to 43% in sugar yield and 30 to 43% in cane yield [9]. Estimated yield loss of the commercially available cane varieties due to RSD with artificial inoculation of the pathogen at different concentration in the three Ethiopian sugar estates shown that 17.21 to 27.76% in cane yield and 19.14 to 27.83% in sugar yield [10].

Therefore, when this design realize and lays to the ground can eliminate the disease causing organisms and save the above mentioned sugar and cane losses.

The main objective of this paper is to design of effective seed cane hot water treatment plant for fabrication.

2. Methods and Materials

2.1. Description of the Study Area

Various designs and sizes of HWT plant are used in different industries. Ethiopian sugar estates have also their own HWT plants in different styles at Wonji, Metehara and Finccha sugar Factories. Even though, the treatment plants are

constructed as per the standard criterion their output is not as such effective, because the treated seed canes are not cured of RSD and smut. The main reasons of this problem are lack of proper circulation of hot water between setts, lack of uniform temperature at the upper and bottom of the treatment tank and heavy packing of setts in the cages. As confirmed practically at the existing plants, the temperature of hot water at the upper and bottom in the treatment tankers are: At Wonji/shoa treatment plant 50°c and 40°c, at Metehara treatment plant 50°c and 45°c, and at Finccha treatment plant 50°c and 47°c.

In response to the above problems, this design suggests: to design for fabrication an effective hot water treatment plant in the nearby workshops and distribute to the existing and other newly established sugar estates. The designed project will enable to solve the problems (lack of proper circulation of hot water, heavy dumping of setts, bulk compactness of setts in the cage, improper temperature control and lack of temperature uniformity in the treating tank) seen in the existing seed cane hot water treatment plants which are the main causes for the failure of curing the seed cane to be treated.

2.2. Methodology

A basic understanding of the plant and the operations will be achieved through first hand observations on the site. Information will be gathered from primary data sources such as, interviews and direct observation on the site, and from secondary data sources like, books, journals, websites and etc.

In the meantime, both quantitative and qualitative data types will be collected from the existing hot water treatment plants and other written documents. The quantitative data includes size of the treatment tanker, size and number of cages, size of the heating tanker, size of disinfectant and cane cooling tank, etc where as the qualitative data includes type of temperature controller (sensor), tolerances for dimensional accuracy of each parts, type of materials used to fabricate each part, assembly methods such as welding, riveting, bolt and nut, etc. Having this information on hand, the drawbacks and best technologies of each existing hot water treatment plant was identified. Finally, a solution was set for the troubles seen in the existing hot water treatment plants and best technologies were adopted and incorporated in to the new design.

2.3. The Proposed Design of the Plant

The treatment plant incorporates three successive tankers namely water heating tanker, seed cane treatment tanker and fungicide dipping tanker. The three tankers were designed to plant with chronological order of the work flow.

2.3.1. Water Heating Tanker

The heating tanker is designed to fabricate from 2mm thick of mild steel sheet in a rectangular pattern of 3m x 3m x1.5m = $13.5m^3$. Either resistors or 2.996m long perforated ½ inch pipes are installed for the path of heat distribution when the heat sources are electric power and steam respectively. The numbers of pipes are 10 and each has perforation throughout the entire surface of their length for the exit of the steam. The heating tanker should be cover with asbestos and laminated sheet externally to prevent heat dissipation to the surrounding. The tanker is put on a frame of equal height with the treatment tanker and a 50^0c hot water will circulate continually from the heating tanker to the treatment tanker and back to the heating tanker with the help of gravity and centrifugal pipe respectively. Tap water drainage pipe is linked with the tanker, which is weld carefully being leakage free. Limit switches are also fitted to this tanker so as to keep 50^0c temperature constantly by switching on and off the power sources. A full assembly drawing of the heating tanker is shown on fig.1 bellow.

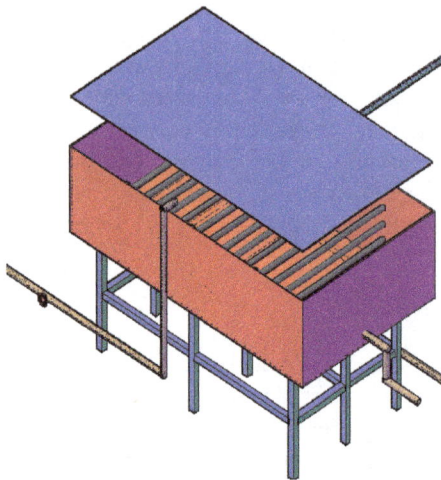

Fig. 1. *Assembly drawing of the new heating tank.*

2.3.2. Seed Cane Treatment Tanker

The main treatment tanker is designed to fabricate in circular profile with diameter 3 meters, height 1.5 meter ($10.6m^3$) and 4mm thick mild steel. The unit consists of a centrally fitted long (1.6 meter) shaft of 80mm diameter with double ball and trust bearings at the bottom and one ball bearing at the upper end. The tank accommodates six equal conical cages, supported by the sit frame attached to the shaft through the bearing house at both ends. An electrical motor is coupled with the shaft at the top of the tanker to agitate the water by rotating the cages and sustain the distribution of uniform hot water to each cane setts in each cages. Timer control switches are used to on and off the motor with an interval of 15 minutes.

The tanker has an inlet for the hot water at the top and outlet at the bottom. The inlet pipe is connected with a ½ inch circular pipe rolled around the main tanker externally possessing six nozzles facing downward at an angle of 20^0 into the tanker to spray hot water in the direction of each cage.

Two Thermostats (sensors) are also fitted to the body of the tanker, one at the upper side and the other at the lower side, to read the temperature inside the treatment tanker and transfer off and on order to the limit switch of the heating tanker when the temperature inside is above and below 50^0c respectively. Besides to this a temperature gauge is installed on the out let pipe before the pump and a manual check up with thermometer will be also carried out during treatment for

additional confirmation. A full assembly drawing of the treatment tanker is shown on fig.2 bellow.

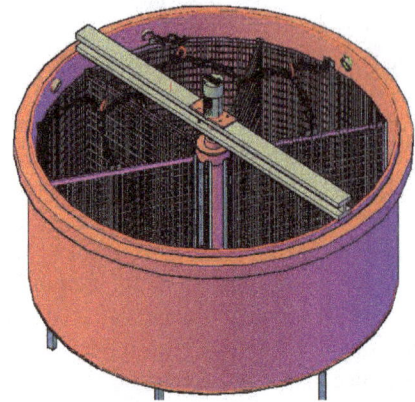

Fig. 2. *Assembly drawing of the new seed cane treating tank.*

2.3.3. Fungicide Dipping Tanker

In some instances a fungicide dip treatment is required after hot-water treatment to prevent re-infection with sugarcane smut during transport and germination. Hence, a rectangular shape of 2m x 2m x 1m = $4m^3$ fungicide dipping tanker is designed to manufacture from 2mm thick mild steel to dip the cane for 5 minutes in the fungicide. Fungicides applied through planting machines are not effective at controlling smut because a lower rate is used and the cane is not in contact with the fungicide for sufficient time for effective control of smut.

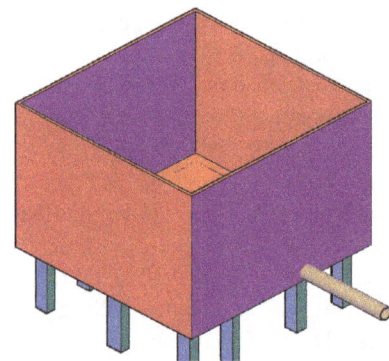

Fig. 3. *Assembly drawing of the new seed cane fungicide tanker.*

2.3.4. Seed Cane Cages

Six truncated triangular cage frames were made from L-structural steel (angle iron) and flat iron (at the top) with a curved shape from one side. On each frame round iron ø6mm were welded with an interval of 20mm in vertical pattern and 50mm apart in horizontal pattern, this helps to facilitate the removal of the water during unloading. The volume of each cage is approximately about $1m^3$ as shown on fig. 4.

The shape and size of the designed cage will avoid heavy dumping and compactness or allows loose dumping of setts during treatment, accelerates hot water circulation between each stalk, conducive to agitate the water and create uniform temperature throughout the treatment tanker. A cage placed in a cage sit frame is shown on fig.5.

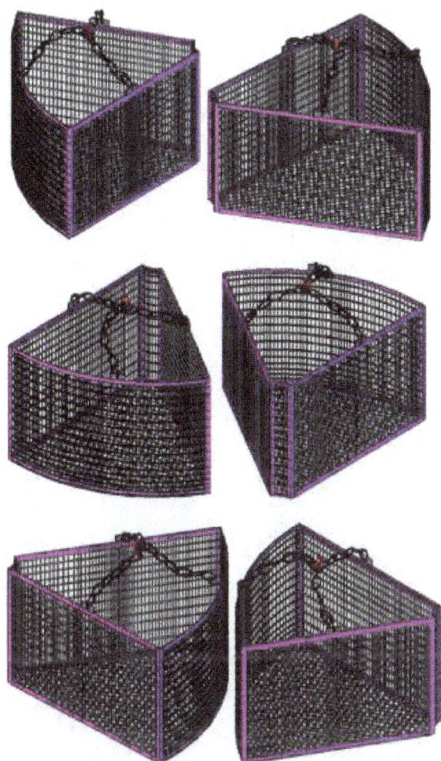

Fig. 4. Cages at different views.

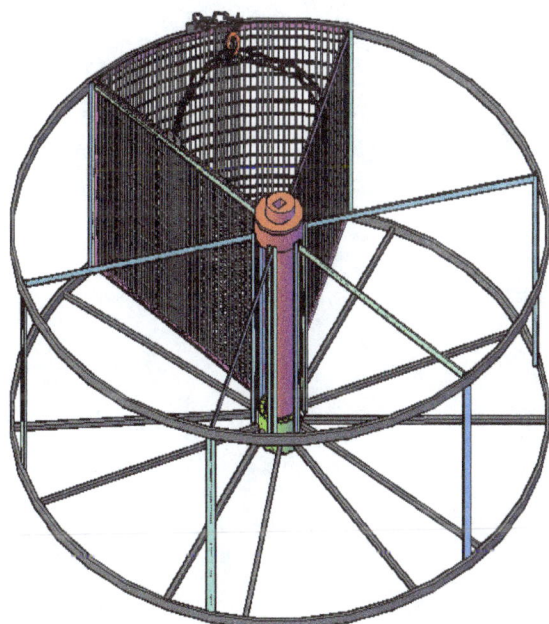

Fig. 5. A cage placed in a cage sit frame.

2.3.5. Loading and Un Loading Mechanism

An overhead gantry electrical driven crane is used to load and unload as well as to move right and left the cages containing seed cane. The gantry frame is made up of 160mm x300mm I- section structural steel and an electric driven hoist is fitted on it. This mechanism enables to accomplish the given task continuously with tireless for a long time. A full assembly drawing of the plant is shown on fig.6 bellow.

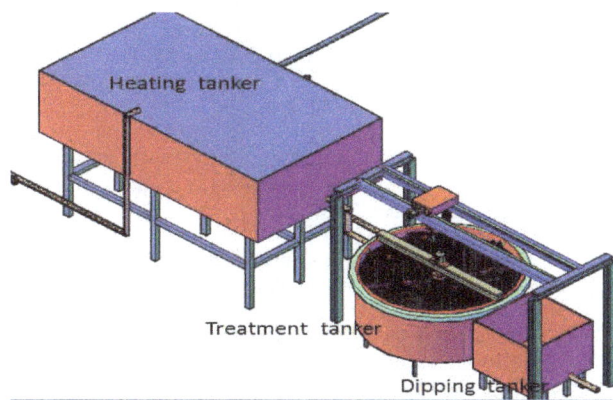

Fig. 6. Assembly drawing of the new seed cane hot water treatment plant.

3. Conclusion

This design is the result of comprehensive analysis of the existing seed cane treatment plants and detail Analysis of the critical parameters for effective hot water treatment. Even though the design is modified and somehow new, it is practical and realistic. Therefore, when this design lays on the ground and begin to serve, it will accomplish the following facts:

1. Temperature uniformity will keep by thermostats throughout the standard treatment time for 2 h at 50 ^{0}C.

2. Water to seedcane ratio (approximately 4:1) will maintain by filling water in both treatment and heating tankers.

3. The water in the tankers will circulate from heating tanker to treatment tanker and back to heating tanker by the installed circulating pipes and pumps.

4. Heavy dumping and high compactness of seedcane will resolve by the small size and shape irregularities of the cages.

5. Loading and unloading mechanism is fast and easy by using an overhead crane.

Acknowledgment

First off all I would like to express my heart full gratitude to sugar corporation Training and Research division staff members for providing the necessary information and feed backs. Then my pleasure goes to science publishing group and to those who have willing to support this paper in publication and realization of the design in fabrication.

References

[1] Bailey RA, Comstock, JC, Croft BJ, Saumtally AS, Rott P (2000). Procedures for the safe movement of sugarcane germplasm. In: Rott P, Bailey RA, Comstock JC, Croft BJ and Saumtally AS (eds) A guide to sugarcane diseases. CIRAD/ISSCT, CIRAD Publications, Montpellier, pp. 311-315.

[2] Davis MJ and Bailey RA (2000). Ratoon stunting. In: Rott P, Bailey RA, Comstock JC, Croft BJ and Saumtally AS (eds). A guide to sugarcane diseases. CIRAD/ISSCT, CIRAD Publications, Montpellier, pp. 4 9-54.

[3]   Girard JC and Rott P (2000) Pineapple disease. In: Rott P, Bailey RA, Comstock JC, Croft BJ and Saumtally AS (eds) A guide to sugarcane diseases. CIRAD/ISSCT, CIRAD Publications, Montpellier, pp. 131 -135.

[4]   Good management practices manual for the cane sugar industry (final), produced for the international finance corporation (IFC), principal author and editor: Jan Meyer

[5]   http://www.sasta.co.za/wpcontent/uploads/proceedings/1960s/1967_thomson_the%20effects%20of%20hot.pdf

[6]   http://www.sugarresearch.com.au/icms_docs/163506_Introduction_to_sugarcane_quarantine_and_disease_control.pdf

[7]   Sandhu S. A., D. S. Bhatti, and B.K. Rattan, 1969. "Extent of losses caused by red (Physalosporatucumane NSis Speg.) and smut (Ustilago scitaminea Syd.)," Journal of Research (Punjab Agricultural University), vol. 6, pp. 341–344,

[8]   Whittle A. M., 1982. "Yield loss in sugar-cane due to culmicolous smut infection," Tropical Agriculture, vol. 59, no. 3, pp. 239–242,

[9]   Abera T. and H, Mengistu. 1992. Effect of smut on yield of sugarcane in Ethiopia. Proceedings of the joint conference Ethiopia Phytopathological Comitee and Comitee of Ethiopian

[10] Yohannes Zekarias, Firehun Yirefu, Leul Mengistu, Teklu Baissa and Abera Tafesse Yield Loss Assessment due to Ratoon Stunting Disease (*Leifsonia xyli* subsp. *xyli*) of Sugarcane in the Ethiopian Sugar Estates

27

Embedded ECG Based Real Time Monitoring and Control of Driver Drowsiness Condition

M. Sangeetha[1], S. Kalpanadevi[2], M. Rajendiran[2], G. Malathi[2]

[1]Embedded System Technologies, Knowledge Institute of Technology, Salem, India
[2]Electrical and Electronics Engineering, Knowledge Institute of Technology, Salem, India

Email address:
msangeethaece82@gmail.com (M. Sangeetha), skeee@kiot.com (S. Kalpanadevi)

Abstract: In recent years preventing accidents under drowsiness state has become a major focus for active safety driving. To reduce the accidents rate, it is needed to provide an efficient safety measure. Literature says that, the drowsiness condition of driver is best monitored by using an eye blink sensor (or) fabric electrode (or) ECG Sensor. But by monitoring the drowsiness condition alone, the accidents cannot be avoided, unless vehicle speed is controlled. To overcome this problem, the proposed system is implemented with a handheld hardware model designed with ARM and an embedded electrocardiogram electrode which measures the heart rate of the driver. The electrode is fixed in the right thumb and left index finger of the driver to monitor the drowsiness condition. If the system identifies that the driver is in drowsy state, it follows the sequence of operation like alters the driver by a buzzer, sends a warning signal to the control room, monitor and control the speed of the vehicle by cutting the fuel supply to the engine. In addition to that, the LED fixed at the backside of vehicle alerts the rear vehicles to reduce its speed. The proposed idea verified using Lab VIEW software is also discussed.

Keywords: ARM, ECG Electrode, DC Motor, GSM

1. Introduction

Now-a-days, the increasing number of transportation accidents has become a serious problem in society. Literature says that, the driver drowsiness state is one of the important factors for causing accidents. Hence drowsiness detection system finds an potential application in intelligent vehicle systems. Previous approach to drowsiness detection primarily makes pre-assumptions about the relevant behavior, focusing on blink rate, eye closure, and yawning from reference [1] and [2]. This paper, proposes a real time ECG based system for drowsiness detection.

All the above said works are recommended for monitoring the drowsiness state and alerting the driver by buzzer. Simply by monitoring the drowsiness condition of the driver alone is off no use, unless some control strategies is used for controlling the vehicle speed, due to which accidents can be avoided. To execute this, a GSM based embedded control system is proposed to monitor and control the parameters selected using its built-in input and output peripherals. The use of GSM is to receive the information transmitted and processing it further as required to perform several operations.

Depending on the nature of the information send the sequence of operations are to be performed like, the transmitted information is stored and polled from the receiver control station and then the required control signal is generated. This signal is send to the intermediate hardware that is designed using ARM and the signal is processed accordingly to perform the required task.

2. Literature Survey

2.1. Sang-Joong Jung et al

This paper describes about the driver health condition monitoring system and it alerts the driver under drowsiness condition. They proposed a new embedded electrocardiogram (ECG) sensor with electrically conductive fabric electrodes which can be fixed on the steering wheel of a car to monitor the driver's health condition.

The ECG signal were measured at a sampling rate of 100 Hz from the driver's palm as they stay on the pair of conductive electrodes located on the steering wheel. Practical tests were conducted using an embedded ECG sensor with a wireless sensor node, and the performances were assessed under non-stop two hours driving test. The driver's health condition such as normal, fatigued and drowsy states were analyzed by evaluating the heart rate variability in the time and frequency domains. But the vehicle speed is not controlled to avoid accidents caused by drowsiness state. [1]

2.2. D. Jayanthi et al

This system proposes an integrated method to detect driver's low alertness by considering both the performance of vehicle steering and the inattention. They proposed a new image processing algorithm designed to efficiently estimate driver's gaze direction, which is the facial inattention indicator adopted in this work.

Driver performance is measured according to the manipulation behavior of steering wheel. These two different aspects are generally complementary and thus can be integrated to obtain accurate detection. It can be implemented as a road device to the use of non-intrusive techniques. However, it is difficult to estimate driver's distraction or drowsiness accurately from the observation of those aspects.[2]

2.3. Swapnil et al

This paper proposed driver's fatigue approach for real-time detection of driver towards the driver's fatigue .The system consists of a sensor directly pointed towards the driver's face. The input to the stream is a continuous stream of signals from the sensors. It monitors the driver eyes to detect micro-sleeps, and monitors the driver jaw movement and it also detects the driver pulse from finger using LED and LDR assembling for detecting the driver illness. Driver fatigue levels based on the response signals were obtained which alerts the driver. It is expensive to be commercialized and needs complex noise processing. [7]

2.4. Daniel Haupt et al

This paper describes about reliable system for driver's drowsiness recognition. Unfortunately, majority of research work are carried out with data acquired in laboratory under ideal or simulated conditions. Therefore it is difficult to implement their results to real car and prove its reliability and accuracy. In this paper data taken for analysis is acquired from real traffic and therefore it overcomes all disadvantages partially modeled in laboratory. Here data acquisition has been chosen as an in-direct measurement from car CAN bus. All data are preprocessed according to assumptions about driver's behavior and transformed to frequency domain by means of orthogonal transform (STFT, CWT and DWT). Subsequently, data is analyzed by data mining methods including features extraction and filter feature selection. The performance of the feature is measured by the area under the receiver operating characteristics. Especially in Noninvasive systems the accuracy of driver's drowsiness detection is not sufficient. [4]

3. Materials and Methods

The overall block diagram of the proposed system for transmitter section and receiver section is shown in fig 1 and 2 respectively. This system comprises of ECG electrode, GSM, DC motor, alarm and control circuitry. It has transmitter and receiver section. In the transmitter section, an embedded electrocardiogram electrode is fixed on the right thumb and left index fingers of the driver to detect the ECG signal. An electrocardiogram electrode is a recording of the electrical activity of the heart over time produced by an electrocardiograph. Since the measured ECG signal is off low level, the amplifier amplifies the signal and sends to ARM 8 processor.

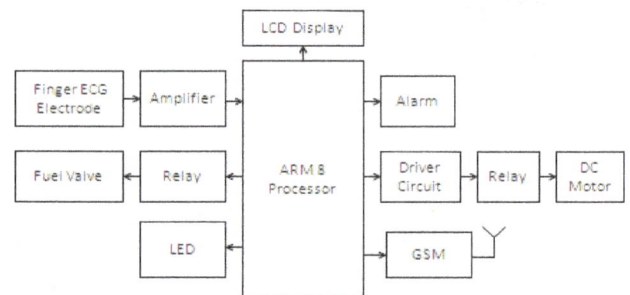

Fig. 1. *Transmitter Section.*

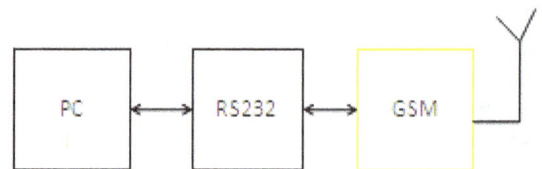

Fig. 2. *Receiver Section.*

3.1. Block Diagram Description

The ARM 8 processor supports a wide range of performance and it has the ability to scale in speed from 600MHz to greater than 1GHz. It receives the amplified signal and process it accordingly. Here global system for mobile communication is used which has high performance, low consumption and long distance communication. By using GSM, these signals are transmitted to control room continuously. Based on the comparison between the received signal and the stored database in the control station, the signal will be send back to the vehicle through GSM to carry out further action. If the received signal indicates the driver is in drowsiness state, then the processor will activate the driver circuit and it alerts the driver by producing an alarm.

The drowsiness prediction system uses DC motor instead of vehicle engine for to control the speed of the vehicle. The DC motor is connected to the processor using the relay control. When any abnormalities in the sensed values are detected by the processor, the relay is open and the motor stops running.

The LCD monitors the condition of the driver and LED fixed at the back side of the vehicle alert the following vehicles to reduce it speed. At the same time the relay will pass the control signal to the fuel valve to cut-off the fuel given to the engine to stop it operation temporarily.

3.2. Algorithm

Step 1: Start the program.

Step 2: Initialization the system. Check whether the driver is in normal state or in abnormal state.

Step 3: If the driver is in drowsiness state the alarm will be intimate otherwise again go to step 2.

Step 4: Check the heart rate, if it in drowsiness state it displays condition of the driver.

Step 5: Stop the Program.

3.3. Flow Chart

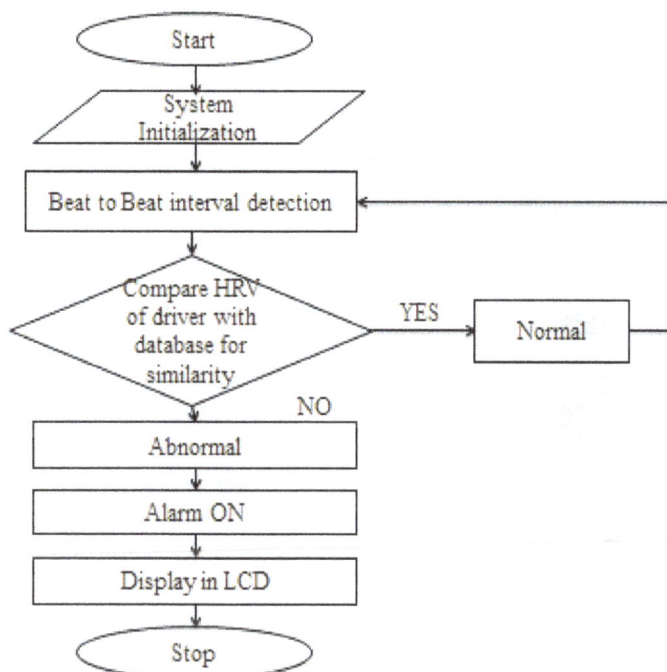

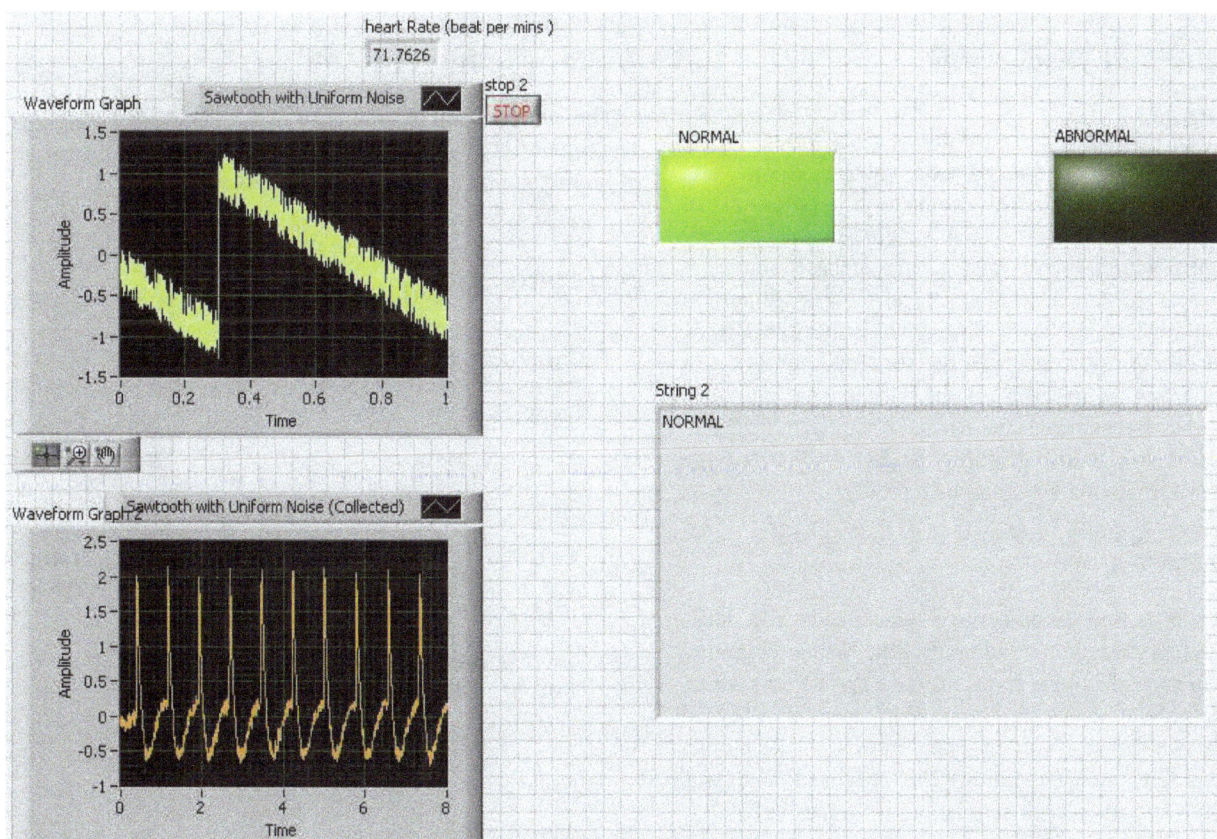

Fig. 3. Normal State.

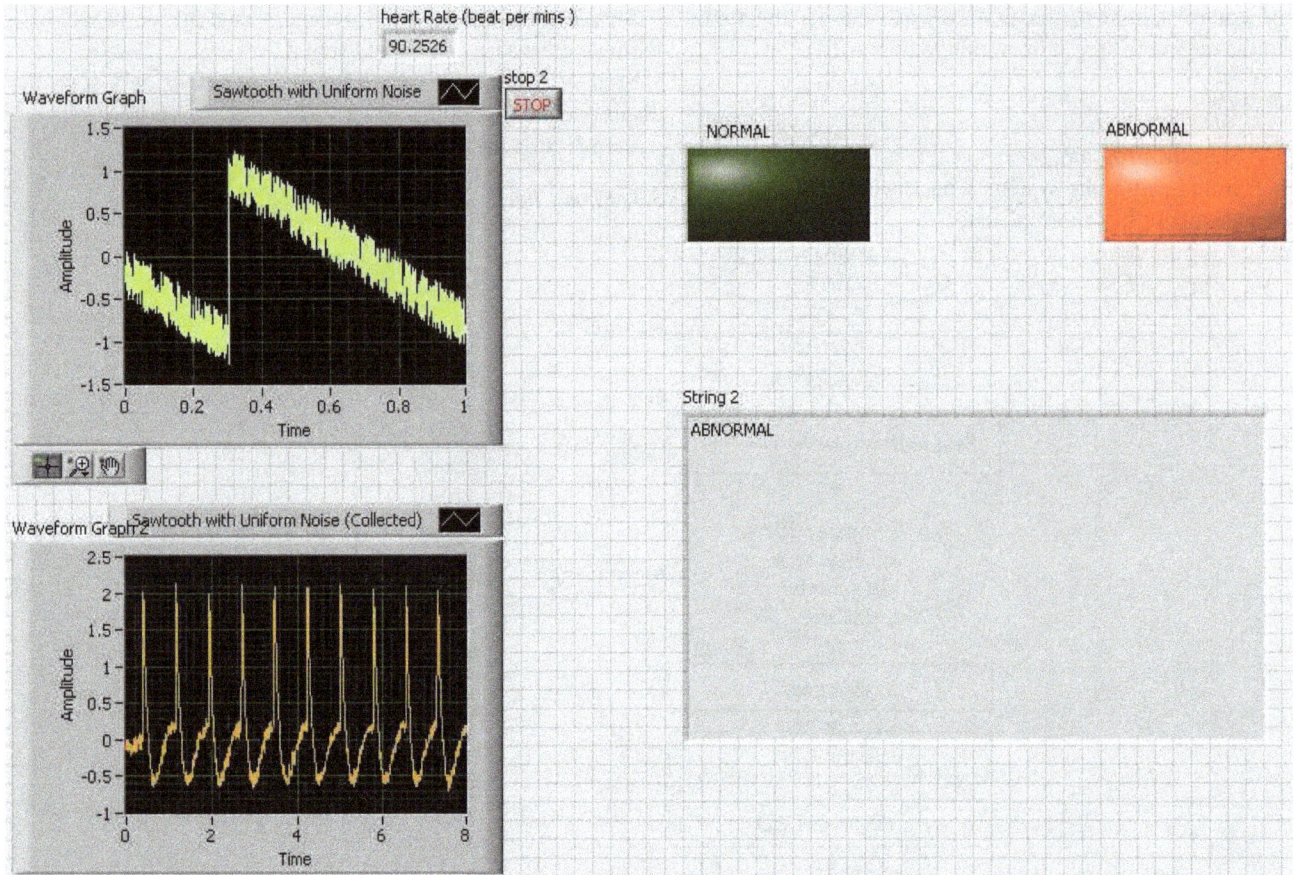

Fig. 4. *Abnormal State.*

4. Results and Discussion

Software Results

The proposed idea is simulated using Lab VIEW software. The ECG signal used in this simulation is the normal heartbeat signal of the driver which is stored in database. By analyzing the actual HRV signal with database signal. It indicates whether the driver is in normal or abnormal condition and it is graphically represented using Lab VIEW. If the comparison result indicates the abnormal condition, it is displayed using the graphical method. If the heart rate is 72 beats per second, it indicates the normal condition. Fig.3 and 4 which is shown below displays the ECG signal of the driver under normal and abnormal condition.

5. Conclusion

The idea proposed gives a novel embedded drowsiness detection system to monitor and control the human cognitive state and to provide biofeedback to stop the vehicle when drowsiness state is detected. This system is suitable for all automobile applications with an intention to save the human life by avoiding accidents rates. The proposed idea was successfully verified through simulation using Lab VIEW.

Acknowledgment

The authors wish to thank the management of Knowledge Institute of Technology and Head of the department for providing the facilities to carry out this work.

References

[1]   Sang-Joong, Heung-Sub Shin, Wan-Young Chung "Driver fatigue and drowsiness monitoring system with embedded electrocardiogram sensor on steering wheel" IET intell. Transp. Syst, 2014, Vol.8.

[2]   D. Jayanthi, M. Bommy "Vision based real time driver fatigue detection system for efficient vehicle control" (IJEAT, Vol-2, 2012).

[3]   DeRosario. H, Solaz. J. S., odríguez: "Controlled inducement and measurement of drowsiness in adriving simulator" (IETIntell. Transp) 2010, 4, pp. 280–288).

[4]   Daniel Haupt, petrHonzix, Peter Raso"Steering wheel motion analysis for detection of the driver's drowsiness".

[5]   Jia-Xiu Liu, Ming-KuanKo"Detection of driver's low vigilance by analyzing driving behaviour from multi-aspects" (National Conference, Vol-14, 2012).

[6]   Chin-Teng Lin, wen-Hung Chao, Yu-Jie Chen"EEG-based drowsiness estimation for safety driving using independent component analysis" (IEEE Transactions, Vol-52, 2005).

[7] Swapnil, Deepak, Kapil "Driver fatigue detection using sensornetwork" (IJEST, 2011).

[8] Federico Baronti, Roberto Roncella, Roberto Saletti" Distribetded sensor for steering wheel grip force measurement in driver fatigue detection" (EDAA, 2009).

[9] Betsy Thomas, Ashutosh Gupta "Wireless sensor embedded steering wheel for real time monitoring of driver fatigue detection"(International Conference, 2012).

28

Assessment of the Effect of Urban Road Surface Drainage: A Case Study at Ginjo Guduru Kebele of Jimma Town

Getachew Kebede Warati[1], Tamene Adugna Demissie[2]

[1]Jimma University Institute of Technology, Civil engineering department, Jimma, Ethiopia
[2]Jimma University Institute of Technology, Hydraulic and water resource engineering department, Jimma, Ethiopia

Email address:

getchitaly@yahoo.com (G. K. Warati), tamene_adu2002@yahoo.com (T. A. Demissie)

Abstract: Drainage is one of the most important factors to be considered in the road design, construction and maintenance projects. It is generally accepted that road structures work well and last longer to give the desired service. When a road fails, whether it is concrete, asphalt or gravel, inadequate drainage is often a major factor to be considered. Researchers have shown that poor drainage is often the main cause of road damages and problems with long term road serviceability. Though provision of proper road surface drainage systems have such a great importance for the urban road to give the intended use and thereby contribute to the overall development of a nation, in particular in road sector, the practice of the construction of proper integrated drainage structures did not get due attention in our country in general and Jimma town in particular for many years. Therefore the problems and achievements on the design, construction and maintenance of surface road drainage systems need to be assessed to provide remedial measures for the better performance of the road infrastructure. The objective of this study was to assess road surface water drainage problems and its net-work integration systems in Ginjo Guduru Kebele of Jimma town. A cross-sectional study was conducted in Ginjo Guduru Kebele of Jimma town from January to August 2014.The data collected was then be analyzed quantitatively and qualitatively, and the result of the study thus presented in tables and in themes. From the study made, generally it was observed that the road surface drainage found to be inadequate due insufficient road profile, insufficient drainage structures provision, improper maintenance and lack of proper interconnections between the road and drainage infrastructures thereby resulting damages to road surface material and flooding in the area.

Keywords: Road Drainage System, Urban Road, Maintenance, Integration

1. Introduction

Though Water is very essential for all life on earth, it can also cause devastation through erosion and flooding. Due to the development of infrastructures as a result of urbanization, the surface runoff water greatly increased in the town damaging the roads. The contributed runoff water thus need to be safely disposed to the rivers/outlet channels so that the functional utility of the road infrastructure maintained and thereby avoid the damages which otherwise occurred to the road and property.

Adequate drainage is very essential in the design of highways since it affects the highway's serviceability and usable life. If ponding on the traveled way occurs, hydroplaning becomes an important safety concern. Drainage design involves providing facilities that collect, transport and remove storm water from the highway.

Inadequate urban storm water drainage problems represent one of the most common sources of compliant from the citizens in many towns of Ethiopia (GTZ-IS,2006), and this problem is getting worse and worse with the ongoing high rate of urbanization in different parts of the country.

The pattern of urbanization and modernization in Ethiopia has meant increase densification along with urban infrastructure development. This has led to deforestation, use of corrugated roofs and paved surfaces. The combined effect of this results in higher rainfall intensity and consequently accelerated and concentrated runoff in the urban areas. Due to inadequate integration between road and urban storm water drainage infrastructure provision, many areas are exposed to flooding problems and road damages in urban roads (Belete, 2011).

There are two major road drainage systems. These are

surface drainage and subsurface drainage. This study concerned with the surface drainage since this type of drainage is most advantageous on low permeability soils, the case of the study area soil, restricting the soil layers prevent the ready infiltration of high intensity rainfall. In addition surface drainage is cost effective to implement per unit area

on such type of soil and has significant benefit with ease of maintenance. It is essential that adequate drainage systems provisions are made for road surface to ensure that a road pavement performs satisfactorily. Thus a drainage system which includes the pavement and the water handling system must be properly designed, built, and maintained.

Figure 1. Ponding of water on gravel road in Ginjo Guduru Kebele in front of Jehovah's Witness hall.

1.1. Problem Statement

Jimma town which is surrounded by steeply or hilly topography is subjected to frequent flooding in rainy season. The occurrence of road over flooding on October 7, 2013 in the town which forced vehicles or automobiles to stop for some hours to give usual service is due to the lack of proper and interconnected drainage systems in the town. The over flooding of road by storm water runoff resulting in the accumulation of silt on the road leading to overturning of cars as happened in the town two times in the year 2013.

The storm water runoff is mainly contributed from the very steep mountain surrounding the town. Ginjo Guduru is one of the 13 Jebeles of Jimma town located at a lower elevation which is directly affected by surface runoff water contributed from Jiren mountain of Guduru sub-catchment water shed area.

1.2. Objectives of the Study

1.2.1. General Objective

To assess road surface water drainage problems and its net-work integration challenges of their provision in Jimma town, in particular in Ginjo Guduru Kebele

1.2.2. Specific Objectives

1. To assess the pavement damage due to improper drainage

2. To identify areas most prone to flooding problems.

3. To assess the existing condition of road and surface drainage infrastructure.

4. To examine the impacts of road surface drainage structures integration on road Performance and related social as well as environment issues

5. To make recommendations on urban road and drainage structures integration, their provision and management.

1.3. Significance of the Research Study

1) To minimize the possible damage of pavement through proper drainage structure provisions.
2) The Jimma town/kebele can use it as reference while they are preparing their annual plans in relation to spatial and financial plans for roads and urban storm water drainage infrastructure.
3) To reduce the environmental and health safety problems.
4) Concerned body and organizations working in the area of roads and urban storm water drainage infrastructures can use it as a reference for proper design, implementation and maintenance of urban road surface drainage.

1.4. Scope of the Study

This research was geographically limited to Guduru sub catchment and the outlet of the Guduru water shed is found in Ginjo Guduru Kebele of Jimma town.

Generally, the study addresses issues related to urban road surface drainage and the integration between drainage and road infrastructures in the Kebele. The specific focus of it includes: existing condition of road and drainage structures, their net-work condition, maintenance of road and drainage infrastructures, impacts of road & drainage infrastructures integration on road performance and associated flood prone areas in the study area.

1.5. Description of the Study Area

The study area is found in Jimma town, which is the largest city in south western part of Ethiopia. It is found in Oromia region at a distance of 335 kilometers from Addis Ababa, capital city of Ethiopia having an average latitude of 7°40' N and longitude of 36°50'E.The altitude of Jimma town varies from 1718 m to 1842 meters above the mean sea level. The town is surrounded by high steeply mountain in the north and north east. There are three streams; Awetu, Kitto and Guduru which cross the town.

2. Literature Review

Different valuable materials published by various authors have been employed to reinforce this research work. The main valuable materials used were:
1. Road and urban storm drainage network integration in Addis Ababa (Belete D.A ,2011).
2. Roadway and road side drainage (David P.Orr.P.E (2003
3. Highway Drainage and storm water management, road design manual (DelDOT (2008).
4. Local road assessment and improvement drainage manual (Donald Walker (2000).
5. Effects of water on the structural support of the pavement system (AASHTO ,1993).
6. Earth and Gravel roads (Penny state University, 1997).
7. Drainage manual. Addis Ababa, Ethiopia (ERA (2000).
8. Urban storm management planning (Fabian P.Barry JA (2003).
9. Urban drainage design manual (Federal Highway administration, FHWA (1996).
10. Urban storm drainage design manual of Ethiopia (Federal urban coordinating Bureau (2008),.
11. Storm water system design.CE58, University of Colorado (Guo,J.C.Y.(1999).
12. Urban drainage manual series on infrastructure.(GTZ-IS (2006).

3. Research Methodology

3.1. Study Setting/Area

Both descriptive and exploratory types of research methods were employed. The descriptive type was used to describe the existing condition and coverage of roads and urban storm water drainage facilities. Whereas, the exploratory type was particularly used to explore the existing condition by making some required physical measurements, and compare with standards.

The study area is limited to Guduru sub-catchment which encompasses three Kebeles namely Mendera Kochi , Ginjo and Ginjo Guduru. The Guduru sub-catchment watershed comprises an area of about 1102 ha .There are three streams in this watershed namely Samiche, Jiren and Guduru streams.

3.2. Study Design/Data Types

Quantitative as well as qualitative data types were employed. Of the total data about 90% of the research data was collected from primary sources of surveying field measurement. Whereas the rest 10% was collected from secondary data sources in order to reinforce the primary data sources.

3.3. Data Collection Methods

Two data collection systems or methods were employed for the study. These are Questionnaire and field measurement of surveying operations. Questionnaire and Interview was employed to collect data related to flooding hazards. The Field survey measurement was done using surveying equipments such as Tape meter, engineering level, total station with their accessories and GPS.

3.4. Data Processing and Analysis

The data collected were checked and analyzed. Software like Auto CAD, Eagle point and EP SWMM5 were used in the analysis of the data besides the common Microsoft office software.

4. Result and Discussion

4.1. Existing Road Types and Drainage Structures

Generally there are four categories of road types observed in the area. These are arterial, sub arterial, collector streets and local road. Based on the surfacing material, four types of road were identified in the study area. These are asphalt (20%), cobble stone (16%), gravel (23%) and earth road (41%). The lengths of these roads are indicated in the table 1.

Table 1. The distribution of the road pavement for each category of road type.

Type	Principal arterial(m)	Sub arterial(m)	Collector(m)	Local(m)
Asphalt	3973.120			
Cobble stone			2295.390	719.750
Gravel		2119.920	174.790	2151.390
Earth road				7962.820

Figure 2. Road net-work in Ginjo Guduru Kebele of Jimma town.

4.1.1. Road Profile of CBS 1

The longitudinal profile of cobble stone road (CBS 1) is

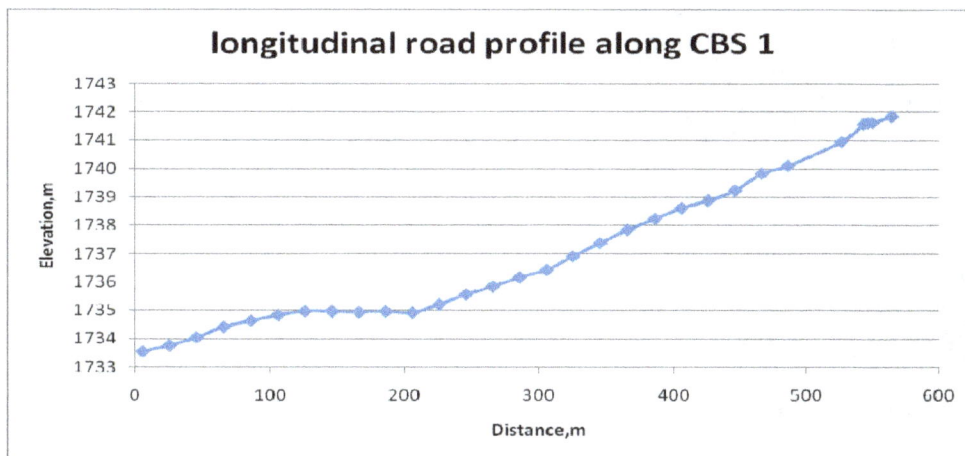

Figure 3. CBS1 road profile.

Table 2. The distribution of longitudinal and cross fall slopes of CBS 1 road.

Distribution or coverage ,%	Slope in percentage				
	< 0.5	0.5 - 1.0	1.1 - 2.5	2.6 -4.0	4.1-6.0
Longitudinal slope	14.80	3.70	62.93	18.50	-
Cross fall	7.41	11.12	44.44	25.93	11.10

As it can be seen from table 2, about 15% of the road length have less than 0.5 percent longitudinal slope and

shown in Figure 3. As per to the master plan of the town, this road is categorized as collector street.

62.97 % of cross fall slope being less than 2.5 %, which is below the required minimum slope for adequate drainage.

4.1.2. Road Profile of CBS 2

Figure 4 shows the longitudinal and cross sectional profile of cobble stone road (CBS 2), which starts from the junction in front of rural mechanization center towards Aramyik hotel having a length of about 900m.This road also categorized as Collector Street.

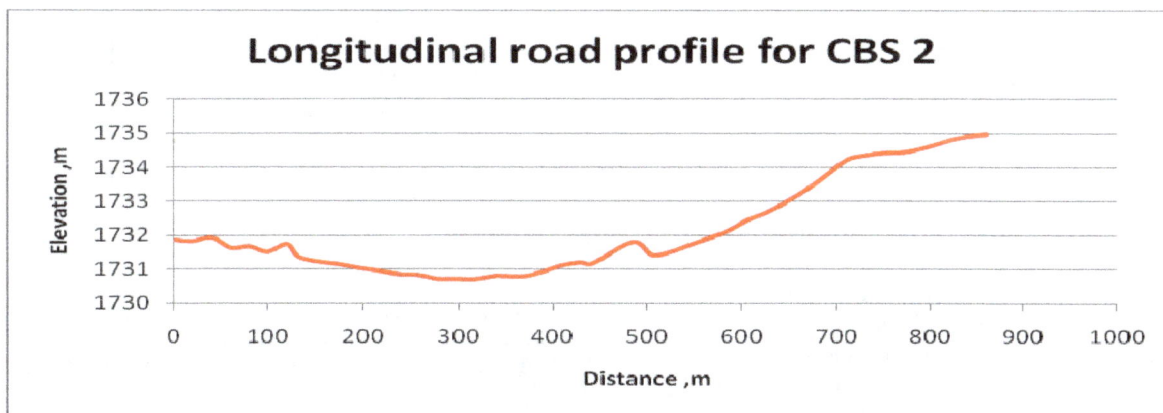

Figure 4. Profile of CBS2 road.

Table 3. The distribution of longitudinal and cross fall slopes of CBS 2 road.

Distribution or coverage ,%	Slope in percentage				
	< 0.5	0.5 - 1.0	1.1 - 2.5	2.6 -4.0	4.1-6.0
Longitudinal slope	48.93	31.92	17.02	2.13	-
Cross fall	10.64	2.13	38.30	31.91	17.02

In this part of road segment of 0.9 km, about 48.93 % of the road likely to have less than 0.5 percent longitudinal slope and 51.07 % of the road have less than 2.5 % cross slope. The distribution of the longitudinal as well as cross fall slope between stations of the road is also not uniform and

this retards the safely disposal of water from the road.

4.1.3. Road Profile of CBS 3

Table 4. The distribution of longitudinal and cross fall slopes of CBS 3 road.

Distribution or coverage ,%	Slope in percentage				
	< 0.5	0.5 - 1.0	1.1 - 2.0	2.0-4.0	4.1-6.0
Longitudinal slope	14.29	28.57	57.14	-	-
Cross fall	74	26	-	-	-

In this cobble stone road, the cross fall is totally less than 1.0 percent and there is no uniform longitudinal slope as it can be seen from the table which gives favorable condition for accumulation of water on the road.

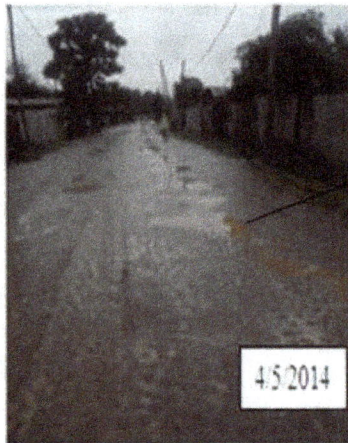

As one can see from the photo, the water film spreads almost along the longitudinal profile of the road as there is no crown. In addition, there is no side drain provided for the road.

Figure 5. Photo of CBS 3 road.

4.1.4. Profile of Asphalt Road

The profile of Asphalt road segment from Dipo to Kidene Mihert church junction was also determined from the surveying field measurement of the study. This road categorized as Arterial Street on the Jimma town master plan with relatively high traffic volume.

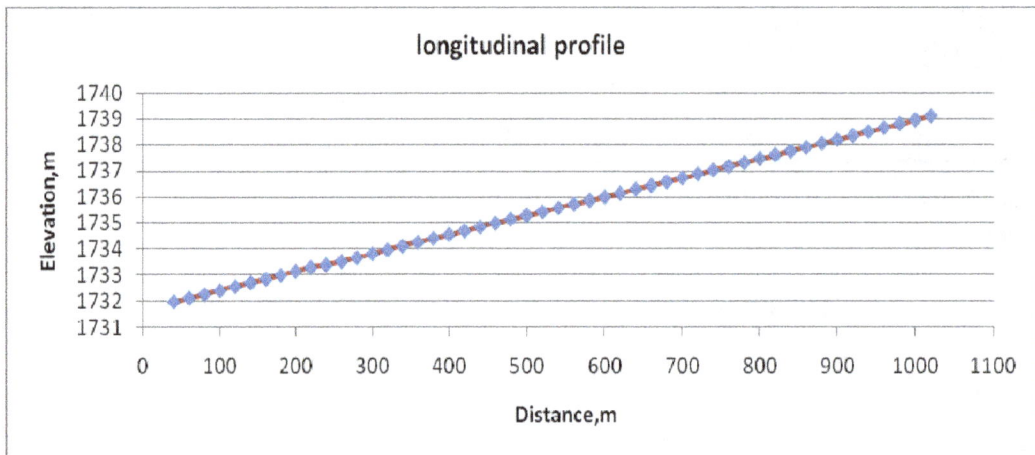

Figure 6. Longitudinal Profile of Asphalt road.

The profile of the asphalt road shown in Figure 6 indicates a uniform longitudinal slope of about 0.86 percent and a cross fall of 3.98 % respectively, which provides rapid removal of storm water from the carriage way of the road to the side drain ditch. The width of shoulder provided for this road ranges from 1.8 m to 2.4 m but the shoulder material is not sealed and subjected to erosion. The erosion of this shoulder is one of the main sources of material that fills the side ditch and results clogging. No manhole provided for the drain ditch of the road to regulate the flow of runoff water.

4.1.5. Road Surface Drainage Structures

As it was observed, there are nine small bridges and twenty two culverts for road crossing drainage structure in the Kebele .There are also eight drainage manholes for the surface drainage system along the newly constructed sub arterial gravel road (GR 2) built in this year and these manholes provide proper connection at the junctions of drainage ditches for regulating the water flow in the drainage structure

Table 5. Road side ditch drainage structure distribution.

Type of road	Side ditch (one side),m
Principal arterial	2741.570
Sub arterial	1124.290
Collector street	1403.680
Local road	1514.060
Total	6783.600

As it can be seen from table 5, about 6783.600 meters of side drain ditch were constructed on each side of the roads. From this, one can infer that only about 35 % of the road provided with defined side drain drainage structure. Masonry trapezoidal and rectangular types of side drain ditch were provided along the roads. From the constructed side ditches, only about 37 % provided with reinforced concrete cover to protect the ditches from any intrusions or garbage.

4.1.6. Road Surface Damage

Different types of damages to the roads in the Kebele were observed during the study. These are potholes, washing and deformations of the road pavements. About 5 cm to 30 cm deformation were observed on cobble stone and gravel road respectively while washing of asphalt road shoulder in the

order 5 to 7 cm also observed due to improper slope. Figure 7 shows different damages occurred to roads in the study area as it was observed in the field investigation.

Potholes and water detention

Deformations

Over flooding

pavement Erosion

Figure 7. *Photo showing Road damages of the study area.*

4.2. Runoff Water and Hydraulic Capacity

In this study, the runoff water generated from the drainage basin was determined based on urban storm water drainage design manual of our country prepared by ministry of works and urban development in 2008. The hydraulic capacities of the open channels in the study area were determined using the Manning's equation. Accordingly, the peak rate of runoff and hydraulic capacities of the channel constructed were computed by the formulae stated and the obtained result presented in table 6.

Table 6. Peak runoff rate and hydraulic capacity of channels in the study area.

Sub-catchment	Area (ha)	Peak runoff rate (m^3/s)	Hydraulic capacity, (m^3/s)
1	2.598	0.347	0.748
2	5.306	0.708	0.814
3	3.719	0.496	1.024
4	5.524	0.737	1.783
5	15.353	2.049	3.587
6	5.254	0.701	1.024
7	7.298	0.974	0.706
8	19.902	2.324	2.381

As it can be seen from table 6, all the channels except that of sub catchment 7, are sufficient to carry the runoff water contributed to them with regard to their hydraulic property.

4.3. Road and Drainage Network Integration

As it was observed during the field investigation of the study, proper connections were made along the newly constructed gravel road (GR2) in which the curbstone properly constructed and inlet spacing was provided every 2m to 3 m interval. In addition drainage manholes were constructed at required locations along this road. However on the rest of the roads proper connections were not provided. In CBS1 road, non-uniform curbstone was provided with no inlet or opening to dispose water from road to the side ditch.

In the junction of CBS1 to CBS3, improper (under- sized) pipe culvert was provided and this create an obstruction to convey the water along the ditch and thus over flooding of water occurs at road crossing junction after every rainfall event. Figure 8 shows flooding due to improper connections or integrations at this location.

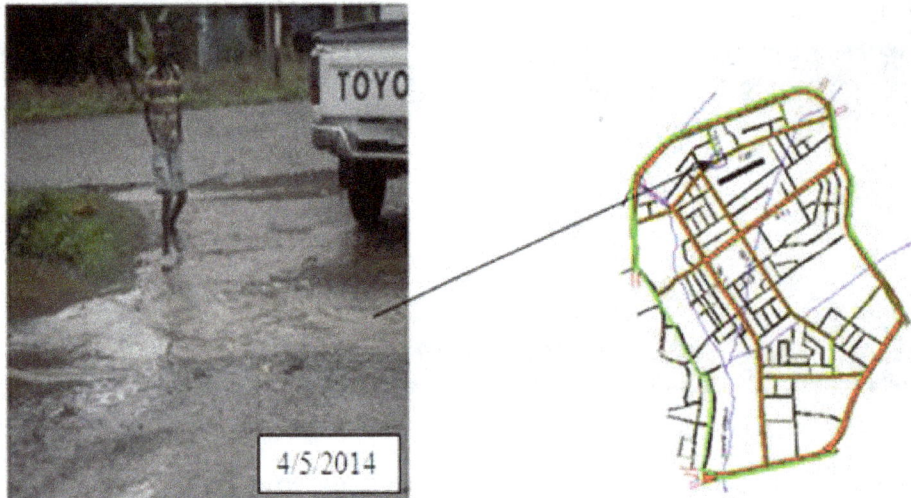

Figure 8. Over flooding of water from side drain ditch at road crossing.

According to field observation made, some of the side drain ditches were constructed for nothing as there is no inlet or opening to collect storm water from the adjacent surrounding area or road. In some cases, the inverted levels of the ditches are above the elevation of the adjacent surrounding area and thus water cannot enter to the ditch. The drain ditch along road besides Jimma Town Water Supply office compound can be cited as the case and no water flow in this channel. Rather, runoff water accumulated on the road damaging the pavement.

4.4. Maintenance of Road Surface and Drainage Structures

There is no regular maintenance of road and drainage structures as it was investigated during the study. Most of the side drain ditch is full of garbage and sediment at many places which obstruct the normal flow of water in the channel. Some drain ditches are also covered totally with grasses and shrubs and thus not giving the desired function for which it was constructed.

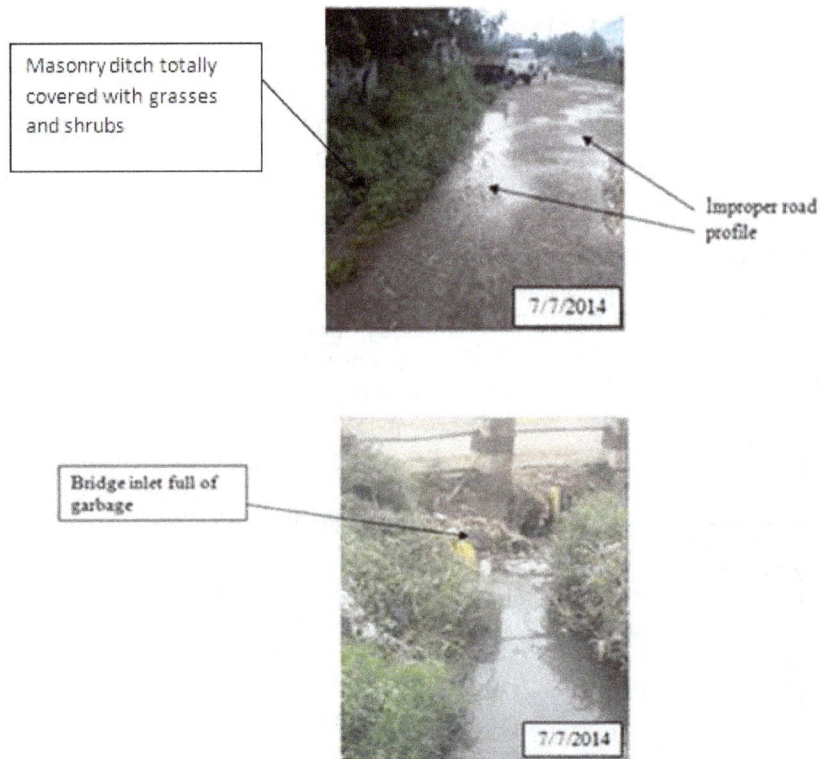

Figure 9. Poor maintenance of drainage ditch, bridge and road surface.

4.5. Rating and Evaluation of Road Drainage Condition

Table 7. *Rating and evaluating roadway drainage.*

Type of road	Side ditch (one side), m	Drainage condition
Principal arterial	2741.570	Good
Sub arterial	1124.290	Fair
Collector street	1403.680	Fair
Local road	1514.060	Poor
Total	6783.600	

From the investigations made concerning the road and drainage structure infrastructures in the study area as presented in the previous sections of this document, the surface drainage condition of the road was analyzed by adopting drainage evaluation system of *Wisconsin–Madison transportation center* as our country drainage manual does not have such rating. Accordingly about 59 % of the constructed drainage structures likely to have fair to poor drainage condition as per to the required standards and thus needs major improvement.

4.6. Flooding Problems

The result of the study shows that though the drainage problem is common in the area, the hazard of the flooding problem is dominant for about 38.6 % of the area and this flooded prone area is located at downstream reach of the Guduru sub-catchment along the streams. Two small bridges over flooded with runoff water in July 2013 due to blockage by debris and the flooding extended to the surrounding residential buildings causing damages or loss to their property. In 2014 rainy season, over flooding of runoff water also occurred to the bridge three times in a season.

Figure 10. *Over flooding of road due to bridge and pipe culvert blockage by debris.*

Generally, separate type of storm water drainage ditch constructed in the study area. But it was observed that liquid wastes released to the storm water drainage ditch & streams from some residential buildings and JU compound which affected the proper functioning of the drainage structures and creating environmental pollution.

As per to the interview made with key personnel, before 18 years the depth of Jiren and Guduru streams were about 0.7 m and 0.8 m below natural ground level respectively, which currently increased to 1.68 m and 1.84 m at the cross section stated. This shows that there is erosion of the stream bed every year in the order of 5.44 cm and 5.78 cm per year on the average respectively.

The sides of the embankment also eroded at different section along the streams by the high driving force of the surface runoff water contributed from the surrounding upstream reach of steeply area.

4.7. Storm Water Management

The study area was divided in 14 sub-catchment areas for modeling with EPA SWMM 5.0 version. The required parameters determined for each sub catchment to input for running the simulation and the obtained results are indicated in figures 11 and 12.

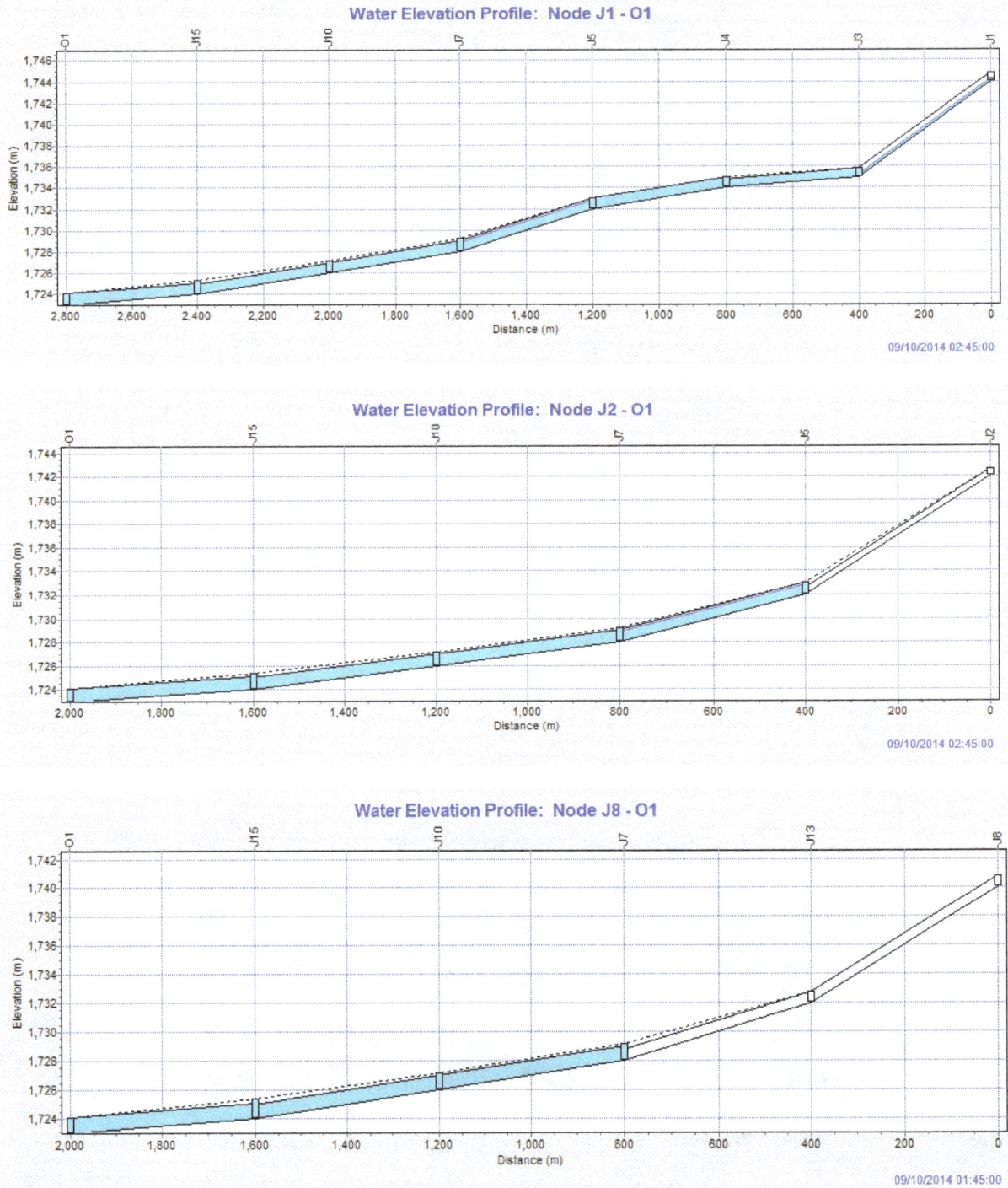

Figure 11. Water elevation profile for three different paths of the modeling.

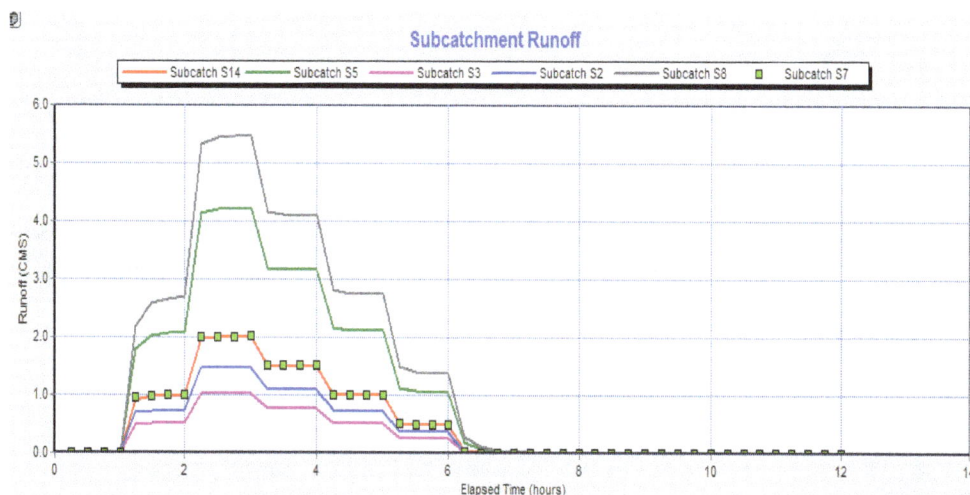

Figure 12. Runoff water for different sub-catchment for the given rainfall event.

5. Conclusion and Recommendation

5.1. Conclusion

Generally it can be concluded that road surface drainage of the study area found to be inadequate due insufficient road profile, insufficient drainage structures provision, improper maintenance and lack of proper interconnection between the road and drainage infrastructures thereby resulting damages to road surfacing material and flooding problems in the area.

5.2. Recommendations

- Proper road geometry need to be maintained to provide required crown and proper side drain drainage structures need to be provided for roads without drainage structures.
- Provision of proper connections or integrations between the road network and drainage network systems is required with regular maintenance.
- Improvement of surface drainage system layout as per to the master plan of the town and implementing the urban storm water drainage design manual of the country to improve storm drainage systems of the area.
- Developing the skill of SWMM soft ware for planning, analysis and design of storm water runoff and drainage systems in urban areas and monitoring the infrastructures.
- Use of the research study results for further study of other sub catchment of the Jimma town in order to have a standardized and harmonized urban drainage systems.

References

[1] Animashaun, B. (1976).Topography and Urban form. A case of Calabar ADRA conference, Nigeria.

[2] Belete D.A (2011).Road and urban storm drainage network integration in Addis Ababa.Jouranal of Engineering and technology research.Vol.3 (7).pp.217-225.

[3] Berbee, R., Rijs, G., de Brouwer, R., & van Velzen, L. (1999). Characterization and Treatment of Runoff from Highways in the Netherlands Paved with Impervious and Pervious Asphalt. Water Environment Research, Vol. 71, No. 2.

[4] David P.Orr.P.E (2003).Roadway and road side drainage. Newyork LATP center, Newyork.

[5] Department of Planning, (New South Wales) (1993), Better Drainage: Guidelines for the MultipleUse of Drainage Systems, Sydney

[6] DelDOT (2008).Highway Drainage and storm water management, road design manual.

[7] Donald Walker (2000).Local road assessment and improvement drainage manual. Transportation information center, University of Wisconsin, Madison.

[8] ERA (2000).Drainage manual. Addis Ababa, Ethiopia.

[9] Fabian P.Barry JA (2003).Urban storm management planning.USA, PP.57-89.

[10] Federal Highway administration, FHWA (1996).Urban drainage design manual. Hydraulic engineering circular 22, FHWA, Washington, DC.

[11] Fedral urban coordinating Bureau (2008). Urban storm drainage design manual, Addis Ababa, Ethiopia.

[12] Guo, J.C.Y. (1999).Storm water system design.CE58, University of Colorado at Denever, co.

[13] GTZ-IS (2006).Urban drainage manual series on infrastructure. Addis Ababa, Ethiopia, PP.17-54.

[14] Seppo Tuisku, Annele Matintupa (2010). ROADEX drainage analysis technique and guidelines on the Irish road network.

[15] Young, C.P. and Prudhoe, J. (1973). Transport and Road Research Laboratory Report LR 565, The Estimation of Flood Flows from Natural Catchments.

29

Development of Indigenous Engineering and Technology in Nigeria for Sustainable Development Through Promotion of Smes (Case of Design of Manually Operated Paper Recycling Plant)

Adefemi Adeodu[1, *], Ilesanmi Daniyan[1], Charles Omohimoria[2], Samuel Afolabi[1]

[1]Department of Mechanical and Mechatronics Engineering, Afe Babalola University, Ado Ekiti, Nigeria
[2]Department of Petroleum and Chemical Engineering, Afe Babalola University, Ado Ekiti, Nigeria

Email address:
adeodua@abuad.edu.ng (A. Adeodu), afolabiilesanmi@yahoo.com (I. Daniyan), charles_4real@yahoo.com (C. Omohimoria),
afowolesam@yahoo.com (S. Afolabi)

Abstracts: Existence of small and medium scale businesses are essential to the growth and sustenance of any economy. SMEs serve as propellant to the development of large existing industries. Nevertheless, the rate of failure of SMEs globally is alarming. Notable in Nigeria as a case, the economy is suffering from structural defects and remains a consumer economy as a result of failure of indigenous engineering personnel to understand and take the lead role to process and utilize abundant natural resources for industrial development of the country. This paper presents an overview of development of indigenous technology in Nigeria through promotion of engineering based SMEs as a capacity building strategy for sustainable development and poverty alleviation. It is concluded that in order to achieve meaningful economic development and sustenance in developing nations like Nigeria, application of indigenous technology through promotion of engineering based SMEs should be considered. Also increase awareness among the public, policy maker and industrialist on the pivotal role of science and technology plays in national development.

Keywords: Commercialization, Indigenous Engineering, Innovations, Sustainable Development, SMEs

1. Introduction

1.1. Indigenous Technology Knowledge as an Engine for Sustainable Development

Sustainability addresses effectively the equity deficit of environmental sustainability [1]. Sustainability is a process which tells of a development of all aspect of human life affecting sustenance. It involves resolving the conflict between the various competing goals and simultaneous pursuit of economic prosperity, environmental quality and safe technology [2]. Sustainable development seeks to ensure better quality of life now and into the future in a just and equitable manner whilst living within the limits of supporting ecosystem [3]. On the other hand, capacity building involves diagnosing several challenges that prevent people, governmental and non-governmental organizations from achieving their developmental goals while helping them acquire competent and versatile skills that allow them achieve sustainable results.

Some of the challenges bedeviling third world have been identified as: inefficient use of available resources (human, financial, material etc.), low human capacity building, absence of good and safe infrastructure, and epileptic power supply amongst others. The resultant plaques are poverty and disease, environmental degradation and pollution, mass unemployment, poor quality of life, mass crime and insecurity just to mention a few. Great uncertainties surround the survival of both present and future generations in the third world as a result of growing population, food shortages, poor resource control, severe effects of environmental changes due to environmental pollution, fast depletion of available resource, fuel scarcity, fluctuation of fuel prices etc. Hence, the need to stem the tide and increase security in critical sectors so as to sustain and increase the standard of living while averting its devastating effect on the economy, human health, quality of life etc.

The use of indigenous engineering and technologically based viable alternatives to transform key sectors for wealth creation in the third world while developing competent, resourceful and skilful man power that can add value cannot be over emphasized.

Sustainable development is an evolving process which involves the judicious use of available resources to create wealth, raise the standard of living of people, and enhance economic and social prosperity for both present and future generations in a secured environment [4]. It means the balancing of economic, social, environmental, and technological considerations, as well as the incorporation of a set of ethical values [5]. Modern indigenous engineering and technological inventions if prudently exploited will minimize waste and maximize value in critical sectors [4]. The goal of building human capacity is to inhibit challenges relating to policy making, sustainable development and method of development while considering the potentials, limits and needs of the people. The gap between what higher institutions offer and the demands of the labour market is widening by the day. Graduates are more exposed to theory than the practical aspects of their training. A careful appraisal of university education in the third world shows that there are not enough facilities for students. Engineering students are confronted with obsolete training equipment. Laboratories are either not well equipped or are unable to meet modern standard. The world is embracing a knowledge-based economy. Therefore, third world should toe the path of countries in Europe and the United States in the quest for sustainable development and also in the area of acquiring more knowledge for economic growth. Some of such technologies developed include post harvest food processing, industrial minerals beneficiation, textile and ceramic utilization, water treatment; electrical and electronic design and fabrication, plant tissue culture, application of nuclear energy and energy generation from biomass.

1.2. Challenges of the Application of Indigenous Engineering and Technological Discoveries

The following were the challenges experienced with regard to application and development of indigenous engineering and technological discoveries:

- Lack of appreciation of the role of indigenous engineering and technology in national development by policy makers and the general public as they were viewed to be inferior to foreign technologies
- Inadequate financial and material support to institutions mandated by Government to facilitate the application of scientific and technological discoveries
- Most research and development activities were conducted in isolation from industry and this resulted in research outputs not being taken up by industry.

1.3. The Way Forward

In order to solve problems militating against the application of indigenous engineering and technology, the following considerations should be made:

- increase awareness amongst the public, policy makers and industrialists on the pivotal role engineering and technology plays in national development;
- Government must strength the capacity of its institutions to promote the application of indigenous engineering and Technology in the nation; and
- Lobby Government to increase its financial obligation to the engineering and technology sector.
- A comprehensive proactive policy framework is the best way to conserve indigenous knowledge that has helped produce, use and maintain diversity in the region.
- A strong sustainability connection exists between indigenous and modern knowledge

1.4. Converting the Challenges into Opportunities for Growth Through Innovation

The above scenario can be overcome by looking forward for a fruitful collaboration with advanced countries for technology commercialization. Since technology failure is a global phenomenon. Even in the most advanced countries the success rate is 30-35%. In addition, the obsolescence of technology is a real threat to technology failure. Furthermore, the release of alternate products to the market as well as the power, capability and penetration of the market by promoters are also threats. In Nigeria, tying up the balance finances with other financial institutions is hog in the' development and commercialization of indigenous technology in Nigeria. The speed of innovation, promotion of indigeneity in new ventures cannot be stopped or slowed down just on the plea of threats of speed breakers being on the road. The technology commercialization gets acceleration due to globalization. There is an apparent dearth of engineering and technology entrepreneurship capital in Africa, a situation that has led to the near non-existent productive capacity of the continent, with very minimal potentials for value addition [6], [7]. The result of the foregone scenario is low capacity for wealth creation and increasing levels of unemployment. Entrepreneurship (especially technological entrepreneurship) and innovation (technological innovation) are the twin pillars of socioeconomic development in this modern era.

Since, knowledge and innovation are two key drivers for sustaining economic growth in the 21 century. Nigeria will have to key in to harness its strong engineering and technology foundational ecosystem to industry-relevant research with a strong focus on commercialization, and to extend the ecosystem to facilitate innovation and enterprise development. The areas expected to be covered in this paradigm include: human capital development, agriculture, industrial growth, health, environment, energy, banking and finance, information and communications technologies, women and youth empowerment, job creation, tourism, trade, science acculturation, natural resources management, building and construction, national security, nuclear science and technology, sports and recreation, diplomacy and transport management among others.

This can be achieved through three broad priorities include:

• Boosting skills in every job via a comprehensive national effort to boost productivity and make enterprise innovation pervasive, supported through both broad based and targeted sectoral programmes.

• Deepening corporate capabilities to seize opportunities in Africa, entrenching Nigeria as the essential base in Africa for both multi-national companies and global small and medium enterprises through increasing private sector research and development (R&D) expenditure, developing stronger alliance to promote technology transfer, test-bedding and commercialization, and helping SMEs develop multi-national companies and global SMEs.

Small and Medium Enterprises constitute a significant part of most economies valuable contributions to its growth through innovation and competition. SMEs form proportion of enterprises of most developed and rapidly developing countries because of their contribution to GDP, employment and socio-economic development. Given their limitations of size and resources, SMEs need special attention and assistance to survive and compete in the global market place. Technology Business Incubation therefore becomes a constructive intervention process to establish a positive environment that can nurture technology-based SMEs for sustainable development.

1.5. Concept of Small and Medium Scale Enterprise

An enterprise is a project, an undertaking, a company, or an individual that is engaged in one form of economic activity or the other, with the aim of producing some goods or services for sale to others [8]. The definition of the size of the enterprise and their classification into micro, small, or medium has been generally based on criteria such as volume of sales turnover, number of workers in employment, or value of assets and investments [9]. Other definitions of the term small and medium enterprises (SMEs) vary from country to country. And also varies between the sources reporting SME statistics. The commonly used criteria at the international level to define SMEs are the number of employees, total net assets, sales and investment level [8]. If employment is the criterion to define, then there exists variation in defining the upper and lower size limit of a SME [10]. For instance in Australia, a small business is defined by the Fair Work Act 2009 as one with fewer than 15 employees. By comparison, a medium sized business or mid-sized business has fewer than 500 employees in the US, and fewer than 200 in Australia [8].

In Nigeria, there have been different definitions of SMEs by different institutions. These institutions include the Central Bank of Nigeria (CBN), the Small and Medium Industries Equity Investment Scheme (SMIEIS), the Nigerian Institute for Social and Economic Research (NISER), Federal Ministry of Industry (FMOI), the National Association for Small and Medium Enterprises (NASME) and the Small and Medium Enterprises Development Agency of Nigeria (SMEDAN). For instance, a SME has been defined by CBN as an outfit with a total capacity outlay (excluding land) of between N2 million and N5 million, while SMIEIS recognizes an SME as any industry with a maximum asset base of N200 million, excluding land and working capital, with the number of staff employed by the enterprises not less than 10 and not more than 300 [11]

Table 1. Classification of Nigerian SMEs adopted by the National Policy on MSMEs.

S/N	Size Category	Employment	Assets (N Million) (Excluding Land And Building)
1	Micro Enterprises	Less Than 10	Less Than 5
2	Small Enterprises	10-49	5 Less Than 50
3	Medium Enterprises	50-199	50 Less Than 500

Source: [8]

1.6. Role of SMEs in Nigeria

The role of small and medium scale industries in developing countries, Nigeria in particular is progressively becoming elaborate. It is highly imperative that all available resources in any given situation in the economic well-being of a nation must be developed for industrialization and ultimately consumption through the small and medium scale businesses [8]. Hence the impact and potential contribution of small and medium scale business base as well as their accelerated effect in achieving macro-economic objectives pertaining to full employment, income distribution and the development of local technology makes the existence most inevitable [8]. Therefore, the importance of small and medium scale industries in particular to the general economic development of any nation especially a developing one like Nigeria cannot be overemphasized [12]

According to SMEDAN the following give some key roles of the small scale enterprises sector in Nigerian in actual terms.

i. Economic Contribution: The overall economic activities of small and medium enterprises in Nigeria have been estimated to amount to less than ten per cent of the country Gross Domestic Product (GDP) (SMEDAN 2006).

ii. Enterprise Creation: As in most parts of the world, micro, small and medium enterprises currently represent about 87% of enterprises in the country (Chemonics International report for PRISM for USAID)

iii. Employment Generation and Poverty Reduction: in spite of valid data, Nigeria lacks adequate census on relevant economic indices, it is estimated that Small and Medium Enterprises in Nigeria currently account for over 75% of employment in the country [13].This relatively high percentage is however a paradox as 60% of Nigerians still lives below the poverty level [14]. When the 26 percent of Nigerians that are unemployed and 60 percent living below the poverty line are taken into account the share those gainfully employed in the SME sector is more likely to be in the region of 10% as recorded by US Industry Small Business Administration (SBA)

iv. Export Earnings: The contribution of SMEs to the nations export earnings is a dismal 2%. This shows the lack of competitiveness of Nigeria's SME sector in this regard

1.7. Problems Facing the Growth and Development of Nigeria Engineering Base Small and Medium Scale Enterprises

Although SMEs are seen as veritable and viable engines of economic development, the growth and development of SMEs in Nigeria have been hindered due to challenges confronting this all-important sub- sector of the economy. The poor performance of SMEs in Nigeria relative to their counterparts elsewhere by revealing that although about 96% of Nigerian businesses are SMEs compared to 53% in US and 65% Europe, average they contribute approximately 1% of GDP compared to 40% in Asian countries and 50% in both US and Europe [15].

The major challenges that SMEs are faced with, in bracing up to operating in a challenging business environment are briefly enumerated below:

i. Poor Infrastructure: Poor infrastructure is definitely the most significant contributing factor to the critical state of the SME sector. Energy supply is epileptic, bad road networks; rail lines are non-functioning. In addition to high operating costs arising from infrastructural deficiencies, SMEs also have to contend with the high cost of imported raw materials, equipments and spare parts. Thus SMEs are uncompetitive when it comes to pricing and are faced with stiff competition from cheaper imported goods. Thus, infrastructural inadequacies contribute well above 30 percent to the cost of doing business as revealed in a recent study by MAN.

ii. Funding Constraints:

Inadequate funds: Apart from the insufficiency of funds to match the financing needs of SMEs, specialized funding windows are non-existent. Most developed and emerging countries long ago realized the intricacies of business start-ups, the peculiarities of business slips, the imperatives of credit guarantees and export promotion and the need to make them more resilient [8]. Such countries had taken measures to ensure specialized funds and windows were available to SMEs for each of the stages and aspects of SME development. Nigeria is yet to adopt such practices in SMEs development.

Poor access to funds and advisory services: Access by SMEs to the limited funds available is impaired by their short tenure and high interest rates. This is hardly surprising since these funds, which should ideally have medium to long term tenures, are not tailored to SMEs. Securing of loans from the Banks and Financial institutions takes time and in most cases only exists on paper. Many banks require the satisfaction of much conditionality before loans are granted, and the small scale industries find it difficult to secure the loans [16]. Leading banks are not committed to granting SME loans probably because most of them are cash collecting centers with insufficient enough staff, specialized in appraisal and management of SME funding, capable of providing the requisite advisory services associated with SME funding.

iii. Weak Corporate Governance, Management and Accounting Practices: Most SMEs, especially at the micro level, are privately funded and privately owned. Management is at the whims and caprices of the owner and accounting books are not professionally kept. There is little dichotomy between personal and enterprise funds and some SME proprietors deliberately divert loans obtained for project support to ostentatious expenditure. Auditing is only functional to satisfy regulatory demand, where one exists, is often manipulated to avoid taxation and is of no value in enhancing managerial competence. Workers, in most cases, are disregarded and usually work in appalling working conditions hence motivation is low. Key management positions are not subject to merit but are usually occupied by relation of the owner. All these translates to absence of business planning and the non-existence of actual planning strategies makes it difficult to remain firm in changing, dynamic and every unpredictable economic and business condition [16]. These problems retard growth and development of Small Scale Enterprises.

iv. Social Responsibility Issues: The operations of all SMEs contribute to environmental hazards in some form or the other. Unfortunately most SMEs are ill equipped to carry-out environmental impact appraisals of their activities and fail to meaningfully engage host communities in dispute and conflict resolution consequent on their activities. In many cases, social responsibility activities are constantly dislocated resulting in loss or closure of the enterprise.

v. Poor Business Partnership/ Alliance Culture: SMEs are typically linked to more than one business or sector of the economy usually as suppliers. However there is usually a lack of mutual trust amongst business partners. SMEs should be encouraged to create strategic win-win relationships to develop certain aspects of their operations or supply chain and thus grow their activities/businesses. To increase the quality of their products and services and thus their competitiveness, they should also be encouraged to develop links with R&D institutions and partner to leverage more modern technologies.

vi. Low Human Capital development: Entrepreneurial skills are poor and insufficient to drive and sustain the sector. SMEs are not equipped with technical management, marketing and ICT capabilities as a result of the low level of training of their operatives. Regular failure to upgrade technical competencies through training results in poor utilization of available technology.

vii. Low Level of Technology: Many SMEs still employ labour intensive production processes, particularly in the agricultural sector. Also in manufacturing, the use of ICT to enhance productivity is limited. Equipment and machinery are in most cases obsolete and cannot cope with modern challenges.

viii. Institutional Support:
Poor Policy framework: Although a policy on SMEs exists, it is lacking in comprehensiveness.
Lack of coordination among various programmes: Coordination amongst the various SMEs related programmes are not coordinated and there seems to be no deliberate effort to dovetail activities to meet specific SME needs such as the commercialization of Research and Development findings, upgrading of SME products and services, enhancement of productivity, provision of assistance to participate in trade missions etc.
Absence of linkage programmes: There are yet any programmes to forge inter-firm linkages amongst SMEs and between SMEs and multinationals. The local content policy needs to be strengthened, training of SMEs in new skills needs to be encouraged and quality control facilities, targeted at suppliers to large firms, need to ensure specifications are met.
Inadequate legal framework: The country's legal framework brings under its ambit, business registration, recognition of individual property rights and dispute resolution. However provisions within this framework are not regularly updated to reflect the current reality and hence are unsupportive of SMEs development. The present framework to grow the SMEs sector in ineffective; hence there are more SMEs in the informal sector than formal.

1.8. Development of Small and Medium Scale Enterprises in Nigeria

To facilitate the growth and development of small scale enterprises in a country or in a region, it is necessary to identify those creative branches that have a potential for growth, their location in the country or region, and to quantify their potential for inducing socio-economic growth [8].

The environment in which SMEs in Europe, South East Asia and America operate provides stable power and water supply, standard road and rail network, efficient water and air transport system, advanced technology, modern communication facilities, efficient and responsive financial system, and above all good governance. Unless Nigeria puts

its policies right, many SMEs may not survive this competitive drive. The following are the suggested tips to assist in creating the enabling environment for development, competitiveness and growth of SMEs in Nigeria.

- Financing Small and Medium Scale Enterprises: Creative businesses can sometimes access funds/benefits from such sources as personal investments, grants for promoting creativity, for business start-up, private R&D spending, tax deductions, loans and philanthropy, amongst others. However, as may be expected the lack of financing is impeding the growth of the sector, even in developed economies. The capital required for the development and implementation of promising ideas is frequently lacking.

- Availability of Research Findings: The system of making available the results of research institutions in new production techniques to SMEs through extension outreach for popularization, demonstration and adoption should be further strengthened. This will reduce cost of production, distribution and marketing, which will raise competitiveness, allow expansion and create more jobs. A well established and operational Business incubator system should be supported to warehouse critical data and information on these results, as well as locally available raw materials and their uses.

- Fiscal Incentives and Support through Tax Rebate: for SMEs that put effort on local sourcing of raw materials, serious in adding value to commodities for exports and other business ethics, which government may wish to foster. Similarly, government could increase funding for the development of the sub-sector through direct budgetary allocations and enhance private sector investment opportunities that will focus on specific areas of capacity enhancement.

- Infrastructural Development: Develop and upgrade rural/urban road and rail network, water and air transport system, and communication infrastructure by Government and the private sector

- Cluster Formation: encourage networking among SMEs operators and use of shared facilities such as Common Facility Centre (CFC). This also involves development of and access to information and, communication technology, and partnership among operators, which, will help reduce cost of production and improve product quality and competitiveness. In this way, SMEs would be positioning themselves to benefit from the implementation of the proposed programme of NEPAD.

- Marketing Channels: provision of effective marketing and distribution channels for SMEs products to penetrate sub-regional and global markets.

- Vertical Integration: encourage linkages between SMEs and large-scale industries to ensure patronage rather than competition among them.

- Capacity Building: a system of technical skills and entrepreneurship training should be developed by Government and the private sector for the operators of

SMEs, so that they can improve on product quality, upgrade their operations to international standard and attract investment for expansion. With globalization, it should be noted that the SMEs that cannot meet the acceptable standards would be compelled to close down.

- Efficient Financial System: efficient and responsive financial system that could serve the economy and the introduction of delivery mechanisms of financial supports to SMEs in particular [17].

2. Case Study

2.1. Design of Manually Operated Paper Recycling Plant

Paper Recycling, which is the extraction and recovery of valuable materials from scrap or other discarded materials, is employed to supplement the production of paper. The design of a used paper recycling machine is therefore a welcome development as it will ensure that the source of raw material for paper production is multiplied and also waste paper that could have constituted into wastes are recycled for various productive purposes. Design of paper recycling machine

ensures that a cheap and non-complex method of production of paper product is guaranteed.

2.2. Description of the Recycling Machine

The design of a waste paper recycling machine includes the determination of the volume of the refiner, hydropulper and head box and also the selection of a convenient material for the construction of the individual units. The majority of the parts of the plant are to be fabricated using mild steel, this is because it is the easiest to be joined among all other metals. It is a very versatile metal, necessitating its use by many industries for fabrication of process unit equipment. Apart from its versatility, it is also very cheap and readily available compared to other metals. Some basic properties of mild steel that enhance these qualities include:

- Tensile strength: 430KN/mm;
- Yield stress: 230KN/mm;
- Percentage longatum: 20%;
- Tensile modulus: 210KN/mm^3
- Hardness: 130APLS

Figure 1. Manually Operated Paper Recycling Machine.

3. Unit Design Calculations

3.1. The Disc Refiner

The unit consists of three main parts: a hopper for charging in the pulp slurry, a screw type conveyor for moving the slurry to the treating element blade and a treating element.

Volume of hopper is a frustum of a pyramid and the volume is given by V = Ah/3, where V is volume, A is area of base of pyramid, and h is height of pyramid.

- Using similar triangle theorem, height:

$$\frac{h}{8} = \frac{H}{30}, \ H = h + 29, \ \frac{h}{8} = \frac{h+29}{30}, \ 30h = 8h + 232,$$

$$h = \frac{232}{22},$$

$$h = 10.545 \text{ cm}$$

- Total volume of pyramid V_p: $V_p = (1/3)(300)^2 \cdot (10.545 + 29) = 11863.5 \text{ cm}^3$
- Volume of truncated pyramid: $V_S = (1/3) \cdot Ah =$

$(1/3) \cdot 8 \cdot 8 \cdot 10.545 = 224.96$ cm^3

• Volume of cylinder enclosing shaft: $V_C = \pi r^2 h$, r = 2.5, h = 8cm then $V_C = 157.08$cm^3

• Total volume of hopper refiner is $V = (V_p - V_s) + V_c$ and then: $V = (11863.5 - 224.96) + 157.080 = 11795.62$cm^3

3.2. The Hydropulper

This is an open cylindrical vessel incorporating one bladed rotating element that serves both to circulate the slurry and to separate the fibre from each other. It makes the paper source become disintegrated, transformed and well blended into fibre slurry. This unit is operated manually. It follows:

• Volume of Hydropulper, V_r it result from its mass and density. Using a scale up factor of 10 (for the whole plant) mass of pulp slurry leaving hydropulper, mass is $7.14675 \cdot 10$ = 71.46750 Kg = 71467.5 g; Density of pulp is 1.172 g/cm^3, then volume of pulp slurry V_C is 60979.096 cm^3. Total Volume of Hydropulper: Vr = Vc + 0.32 Vc = 1.032Vc, and replacing the numeric values: V_r = 60979.096 + 0.032 (60979.096) = 62930.427cm^3;

• Diameter of Hydropulper (a cylindrical vessel) it result from volume of cylinder $V = \pi r^2 h$, where height of 50 cm is used i.e. h = 50 cm;

• Radius of circular cylinder: $r^2 = V/\pi h$; $r = \sqrt{\dfrac{V}{\pi h}} =$

$\sqrt{\dfrac{62930.43}{\pi \cdot 50}} = \sqrt{400.628} = 20.016$ cm.

• Diameter of cylinder D = 2r = 2·20.016 = 40.032 cm.

• The total surface area of cylinder A: A = $2\pi r(h+r)$ = $2\pi \cdot 20.016(50+20.06)$
= 8805.51 cm^3

• Circumference of cylinder, C: C = $2\pi r$ = $2\pi \cdot 20.016$ = 125.764 cm.

3.3. Blade Design for Hydropulper

The blade is design in a way that it has more mixing effect than cutting. The diameter, D_a vary from 1/2.33 D_T to 1/3 D_T.

From the lower value, Blade diameter = 1/2.33 D_T, where D_T = 40.032 cm (diameter of tank), D_a= 1/2.33·40.032 (Diameter of blade) = 17.181cm.

It follows:

• Height of blade (H) from blade of cylinder: H = 0.15 D_T - 0.12D_T

• The lower value: H_i = 0.12 D_T = 0.12·40.032 = 4.804 cm.

3.4. The Head Box

This unit is made out of an 18" gauge flat sheet into a square tank. Its purpose is to ensure that a continuous flow of stock at constant velocity across the width of the machine is provided. Its principal design involves the use of a single slice to develop a free jet of pulp that is then deposited onto the moving felt conveyor. It has an inlet medium fitted with a 2" pipe socket that allows for a continuous flow of pulp slurry. It follows:

• Volume of Head box, using a scale up factor of 10 (for the whole plant) result from mass of slurry leaving the hydropulper to the head box, equal to 7.14675·10 = 71.4675 Kg.

• The density of the pulp, 1.172 g/cm^3.

• Volume of the pulp slurry = $\dfrac{mass}{density}$ = $\dfrac{71.4675}{1.172}$ =60.979096m^3

• Volume of the headbox = 60979.096cm^3.

• Free jet area is length × breadth = 2.5 × 25.50 = 63.75 cm^2.

• Free jet displacement sheet area, A_{fj}, A_{fj} = (a+b)·J = (19.2+9)·25.5 = 71.91 cm^2

• The total area of headbox slices covering top edge: A_{hs} = (40.60·4.00)·4 = 649.60 cm^2

• Entrance area from the hydropulper to the headbox is of: Internal diameter = 4.00cm = Φ; External diameter = 6.00cm = Φ_{ex}, and then:

Internal area = $\dfrac{\pi \cdot (4.00)^2}{4} = 12.57$cm^2,

External area = $\dfrac{\pi \cdot (6.00)^2}{4} = 28.77$cm^2

3.5. Felt Blanket Conveyor

The design of the felt is to serve three (3) main purposes:
1. A conveyor to assist the sheet through the manufacturing process;
2. A porous media to provide void volume and channels for effective water removal;
3. A texture cushion for passing moist sheet without crushing or significant marking.

As a tension band to maintain sheet feltness and ultimate contact with followings:

• hot dry surface length of cylinder (50cm);
• radius of cylinder (7cm);
• circumference of cylinder ($2\pi r$ = $2\pi \cdot 7$ = 43.99cm);
• AB = 43.99/2 = CD = 21.99cm;
• Total length of felt: AB + CD +BC + DA = 21.99 + 21.99 +140 +140 = 323.98 cm.

3.6. The Dryers

This unit consists of two hollow cylinders design in the form of a roller, an external mild steel metal cylinder and an internal ceramic cylinder.

The internal cylinder 7" in diameter is made of ceramic material. It is hollow in form and serves as the heating plate. The choice of a ceramic material for the heating plate is hinged on the fact that ceramic does not conduct electricity and is resistance to heat. Each heating plate consists of two heating elements, connected to electric mains outside. The external cylinder encloses the internal cylinder as a casing.

The external cylinder has:

• Diameter of roller = 14cm, therefore radius = 14/2 = 7

cm;
- Length of cylinder = 50 cm;
- Circumference of external cylinder = $2\pi r$ = $2\pi 7$ = 43.98 cm.

Therefore, circumference of cylinder is 43.98cm.
Internal cylinder has:
- Diameter of pipe = 8 cm;
- Length of ceramic pipe = 30 cm;
- Radius = 8/2 = 4 cm;
- Circumference of internal ceramic = $2\pi r$ = $2\pi \cdot 4$ = 25.132 cm.

Therefore, circumference of internal ceramic is 25.132 cm.
Since the other rolls have same dimension as the external cylinder of the dryer, therefore the circumference and diameter of all the six cylinders have equal values. The external cylinder of the dryer i.e. 43.98 cm = $4.398 \cdot 10^{-2}$ m.

4. Conclusion

SMEs easily thrive where there is an enterprise culture or a business oriented society, that is, a society where the way of life is focused on the importance of individual creating their own wealth through their business. It is clear that engineering personnel are imbued with entrepreneurial potentials yet to be tapped. The solution to countries dismal performance and deficient economic structure is to promote workable enterprise culture. The engineering personnel need to play leading role by establishing technology based businesses, formation of consortium and adopt strategies for mergers and linkages.

References

1] Agyeman J., (2004). Sustainable Communities and the Challenges of Environmental Justice. New York, USA. New York University Press.

2] Hasna A. M., (2007). Dimensions of Sustainability. Journal of Engineering for Sustainable Development. Energy, Environment and Health. 2(1) 45 – 57.

3] Agyeman J., Bullard R. D., and Evan B., (2003). Just Sustainabilities: Development in an unequal World. Cambridge M A, USA, MIT Press.

4] Daniyan I. A., Daniyan O. L., Adeodu A. O., and Aribidara A. A., (2014). Towards Sustainable Development in the third World: Design of a Large Scale Biodiesel Plant. Journal of Bioprocessing and Chemical Engineering 1(1), Pp 1 – 9.

5] Council of Academics of Engineering and Technological Sciences (1995). The Role of Technology in Environmentally Sustainable Development. Kiruna, Sweden.

[6] Gordon M. Bubou and Festa N. Okrigwe, (2011). Fostering Technology Entrepreneurship for Socioeconomic Development: A case for Technology Incubation in Bayelsa State, Nigeria. Journal of Sustainable Development, 4(6): 5539/jsd.v4n6p138.

[7] Adelowo Caleb M., Olaopa R. O., and Siyanbola W. O., (2012). Technology Business Incubation as Strategy for SME Development: How far and how well in Nigeria? Science and Technology, 2(6): 172-181.

[8] Mohammed Sani Haruna, (2013). Development of Small Scale Enterprises: The Role of Engineering Personnel. Invited paper presented at the 22nd Engineering Assembly of Council for the Regulation of Engineering in Nigeria (COREN). International Conference Centre, Abuja. August 20-21.

[9] Joseph E. A. and Michael D. O., (2013). Promoting Small and Medium Enterprises in the Nigerian Oil and Gas Industry. European Scientific Journal. 9(1), ISSN: 1857-7881.

[10] Shambhu Ghatak (2013). Micro, Small and Medium Enterprises (MSMEs) in India: An Appraisal. www.legalpundits.com/Content-folder/SME Arti 1506010.pdf.

[11] Lawal W. A. and Ijaiya M. A., (2007). Small and Medium Scale Enterprises Access to Commercial Banks? Credits and their Contributions to GDP in Nigeria. Asian Economic Review, Journal of the Indian Institute of Economics. 49(3), 360-368.

[12] Jimah M. S., (2011). Establishing Small and Medium Scale Enterprises: Problems and Prospect. Being a paper presented at the Institute of Chartered Accountants of Nigeria (ICAN) Zonal Conference held in Jalingo, Taraba State.

[13] National Policy on Micro, Small and Medium Enterprises – SMEDAN (2006).

[14] AFDB, OECD, UNDP, UNECA (2012) 'Nigeria 2012" African EconomicOutlook.http://www.africaneconomicoutlook.org/fil eadmin/uploads/aeo/PDF/Nigeria%20Full%20PDF%20Countr y%20Note.pdf

[15] Oyelaran oyeyinka B., (2012). SMEs: Issues, Challenges and Prospects. International Conferences on Financial Systems Strategy 2020. United Nation.

[16] Ayozie Daniel Ogechukwu, (2011). The Role of Small Scale Industry in National Development of Nigeria. Universal Journal of Management and Social Sciences. 1(1) (cprenet.com/uploads/archive/UJMSS-12-1021.pdf).

[17] Olorunsola J. A. (2003). Problem and Prospects of Small and Medium Scale Industry in Nigeria. Seminar on Small and Medium Industries Equity Investments Scheme (SMIES). Publication of CBN Training Centre, Lagos, CBN No. 4.

30

Design, development and performance evaluation of an on-farm evaporative cooler

Rajendra Kenghe[1], Nilesh Fule[1], Kalyani Kenghe[2]

[1]Department of Agricultural Process Engineering, Mahatma Phule Krishi Vidyapeeth, Rahuri, India
[2]Department of Mechanical Engineering,Sanjivani College of Engineering, Kopargaon, India

Email address:

rnkenghe@yahoo.co.in (R. Kenghe), fulenilesh@gmail.com (N. Fule), kalyanikenghe@yahoo.com (K. Kenghe)

Abstract: The portable evaporative cooler was designed for storage of 50 kg fresh fruits having overall diamentions of 1220 x 860 x 787 to study the performance of an on-farm evaporative cooler with the effect of different filling materials viz., coconut coir, saw dust + gunny bag, ECC cool pad, *wala* sheet and gunny bag. Selection of filling material was based on the cost, water holding capacity, rate of evaporation of water from filled material and easy availability. Sapota (Cv. *Kalipatti*) fruits were stored in an on-farm evaporative cooler for 16 days. The effect of filling materials on Physiological weight loss (PLW), inside temperature, inside relative humidity and cooling efficiency of five filling materials were studied. The weight loss varied between 2.50 to 13.29, 14.96, 15.37, 15.36 and 15.20% in coconut coir, saw dust + gunny bag, ECC cool pad, *wala*sheet and gunny bag, respectively. The mechanical test such as cooling efficiency for each filling material was determined and it was recorded as 90%, 54%, 38%, 70% and 79% for coconut coir, saw dust + gunny bag, ECC cool pad, *wala*sheet and gunny bag respectively. Minimum inside temperature of 16.5 to 17.2^0C and maximum inside relative humidity of 97 to 90% was recorded in coconut coir of an on-farm evaporative cooler when ambient temperature was 27.66^0C and ambient relative humidity was 51%.The maximum storage life of 16 days was found in coconut coir. Coconut coir was performed better when compared with other filling material. It was most economical filing material over the other filling materials.

Keywords: Evaporative Cooler, Filling Material, Cooling Efficiency

1. Introduction

Much of the post-harvest losses of fruits and vegetables in developing countries is due to the lack of proper storage facilities. While refrigerated cool stores are the best method of preserving fruits and vegetables they are expensive to buy and run. Consequently, in developing countries there is an interest in simple low-cost alternatives, many of which depend on evaporative cooling which is simple and does not require any external power supply.

The basic principle relies on cooling by evaporation. When water evaporates it draws energy from its surroundings which produce a considerable cooling effect. Evaporative cooling occurs when air, that is not too humid, passes over a wet surface; the faster the rate of evaporation the greater the cooling. The efficiency of an evaporative cooler depends on the humidity of the surrounding air. Very dry air can absorb a lot of moisture so greater cooling occurs. In the extreme case of air that is totally saturated with water, no evaporation can take place and no cooling occurs. Generally, an evaporative cooler is made of a porous material that is fed with water. When ambient air drawn over the material the water evaporates into the air raising its humidity and at the same time reducing the temperature of the air. There are many different designs of an evaporative cooler. The design will depend on the materials available and the user's requirements.

In principle, fresh commodities need proper postharvest management to reduce loss and maintain quality. However, at present there is no improvement over traditional postharvest handling methods of fruits and vegetables. The peasants have no storage facilities at their disposal and the fruits and vegetables they harvest are usually exposed to high temperatures and low relative humidity until wholesaler or retailers collect them. The trade also does not operate any intermediate storage for carrying oversupply to obtain better prices. No cooling facilities or packaging houses at any stage of product line from farm to consumer or exports market are currently available. As a result, nutritional loss and postharvest decay are found to be the serious issues. Reduced temperature decreases physiological, biochemical, and microbiological activities, which are the causes of quality deterioration (flavour, texture, colour, and nutritive value).

Evaporative cooling systems are commonly used in countries where the climate is hot and dry. Several studies have been devoted to the application of an evaporative cooling principles in the field of fruit and vegetables preservation mostly in India and the USA. The potential energy savings envisaged by replacing conventional refrigerated systems by evaporative systems. Evaporative cooling is an adiabatic cooling process whereby the air takes in moisture which is cooled while passing through a wet pad or across a wet surface showed that evaporative cooled storage is more energy efficient than a mechanical refrigeration system.

In developed countries, methods employed for extending shelf life and minimizing post-harvest losses of perishable produce include mechanical refrigeration, controlled atmospheres, hypobaric storage, and other sophisticated techniques. These techniques are highly capital intensive and for most developing countries, the required manpower is either lacking or inadequate. These cooling methods, except adiabatic cooling, are expensive for small scale peasant farmers, retailers and wholesalers, as they require electric power. Moreover, in the existing mechanical refrigerating systems, proper storage conditions are not often put into consideration as stored items (vegetables) were normally

subjected to excessive chilling or freezing.

Low temperature and high relative humidity can be achieved by using less expensive methods of evaporative cooling [1]. Evaporative cooling has been reported for achieving a favorable environment in green houses [2] animations and the storage structure for fruit and vegetables [3] and [4]. The present study was therefore planned to design and develop a low cost, portable evaporative cooling system that could be utilized to store fruits and vegetables at their minimal storage temperature.

2. Materials and Methods

2.1. Design of Components

The portable prototype of an evaporative cooler was designed for storage of 50 kg fresh fruits having overall dimensions of 1220 x 860 x 787 mm. (Fig.1 and Fig. 2). The actual design realization was worked out based up on the literature reviewed and for simulation of results of filled material the designed prototype was used and was compared with the earlier studies.. The clearance between two layers of net was 50 mm. The container and inner frame clearance was 100 mm. The distance between two stakes of container was maintained 100 mm throughout the experiment .The clearance between two layers of net was filled with five filling materials viz., coconut coir, saw dust + gunny bag , ECC cool pad, *wala* sheet and gunny bag.

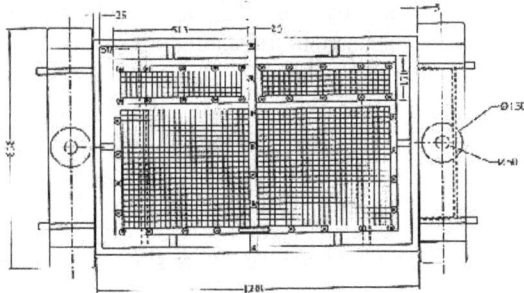

Fig. 1 (a) TOP VIEW

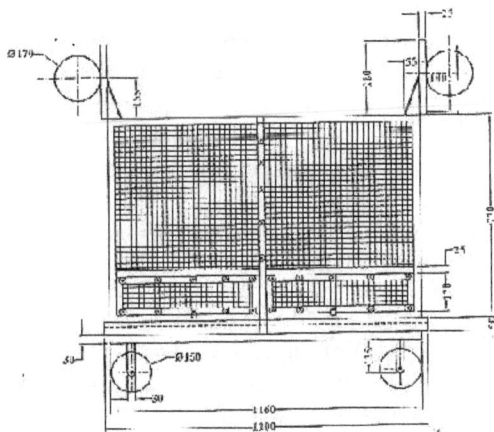

Fig. 1(b) FRONT VIEW

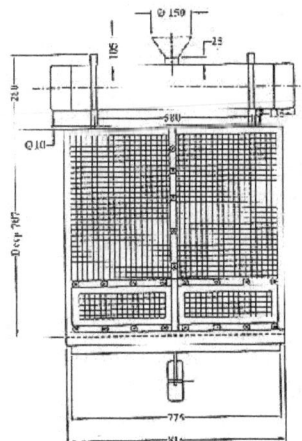

Fig. 1(c) SIDE VIEW

Figure 1. *Different views of an evaporative cooler.*

The water distribution was achieved though laterals of 12 mm diameter for trickling the water on the filled material sandwiched between two layers of net. The two water tanks made up of PVC pipe of 170 mm diameter and 910 mm in length of 20 liters capacity were used as reservoir to fulfill the requirement of water for storage. The water tanks were elevated to the height of 155 mm from top of evaporative cooler for natural circulation of water. The laterals were provided with drippers placed 150 mm apart from each other so as to trickle the water uniformly on filled material. For easy filling of pads inside the net gap a door of 170 x 513 mm with suitable handle was designed. The experiment was conducted in completely randomized design (CRD).

Figure 2. *Photographic view of evaporative cooler.*

2.2. Selection of Filling Material

As part of the general requirements, the efficiency of an active evaporative cooler depends on the rate and amount of evaporation of water from the filling material. This is dependent upon the air velocity, filling material thickness and the degree of saturation of the filling material which is a function of the water flow rate wetting the filling material[5]and [6]. Similar filling materials have been used by [7] and [8].

Table 1. *The on-farm evaporative cooler (OFEC) consists of following main parts and specifications.*

Particulars	Material	Size
Base	Galvanized iron sheet	1220 x 770 x 787(mm)
Outer frame	Mild steel	1180 x 820 x 800(mm)
Inner frame	Mild steel	1080 x 720 x 800(mm)
Door	Mild steel	980 x 650 x 50(mm)
Wheels (3 Nos.)	Caster wheel	6 inch. Dia.
Net	Galvanized iron	20 x 20(mm)
Lateral	PVC	12 mm dia.
Drippers	PVC	1 No./150 mm length of lateral
Water tank	PVC	2 nos. 170 mm dia.

The ambient and cabinet temperature was measured using digital thermometer and relative humidity by digital humidity –temperature meter. Products weight (preserved and unpreserved) was determined by digital weight balance 10.00, 14.00and18:00 hrs. The evaporative cooling system was tested over a period of 16days using 50 kg of sapota fruit. The chamber was tested for its suitability to reduce the temperature while maintaining the increased relative humidity. The experiment was carried out using the developed evaporating cooling system at no load condition for 7 days. The system was also used at loaded condition to preserve sapota fruit for the other 16daysfor storage of sapota fruits.. During the testing period, the thermometer was suspended in the chamber through a small hole in the cabinet to ascertain the variation of temperature in the chamber, while a control sample of 50 kg of sapota fruit spread on a tray were expose to the open air.

2.3. Cooling Efficiency

Analysis of the moist air properties is important to look at the suitability of a given modified air condition for fruit and vegetables storage in hot climate. Cooling efficiency is an index used to assess the performance of a direct evaporative cooler. Cooling efficiency in percentage can be defined as suggested by[8]and [9].

$$\eta = \frac{Td - Tc}{Td - Tw} x100 \tag{1}$$

where: Td and Tw are the dry and wet bulb temperatures of the ambient air and Tc is the dry bulb temperature of the cooled air in ^{0}C.

2.4. Per Cent Loss in Weight (PLW)

Per cent loss in weight (PLW) was determined by weighing the sapota fruits after 4 days interval during storage with the equation used by[8]and[10].

$$(PLW) = \frac{W_1 - W_2}{W_1} x \ 100 \tag{2}$$

where,

W_1: Weight of sample before storage

W_2: Weight of sample after storage

2.5. Statistical Analysis

Data analysis was performed using GLM procedure of SAS. Effects were considered significant in all statistical calculations (≤ 5). Graphs were plotted in MS Excel.

3. Results and Discussion

3.1. Effect of Filling Materials on Physiological Loss in Weight (PLW)

The effect of filling materials on physiological loss in weight (PLW) of sapota was found statistically significant in all filling materials used for evaporative cooler. The weight loss showed variation from 2.50 to 15.45% during storage. The weight loss found to be minimum (13.29%) in case of fruits stored in coconut coir on 16th day of storage whereas it was maximum (15.45%) in case of fruits stored at room temperature on 4th day of storage. The effect of filling materials on physiological loss in weight was plotted in Fig. 3. It was observed from Fig. 3 that the physiological loss in weight increased with increase in storage period for all filling materials. Similar trends were reported by [11] for potatoes stored in desert cooler and [12] for storage of eggplant under passive evaporative cooler.

It was observed from the data that the increase in physiological loss in weight was at faster rate in fruits stored at room temperature followed by ECC cool pad and saw dust + gunny bag respectively. Among all filling materials, coconut coir showed minimum weight loss. The increase in weight loss with storage period may be due to reduction in moisture content on respiration. The rate of respiration might have decreased due to low temperature. Similar results were reported by [13] for sapota and [14] for storage of mango in evaporative cooler.

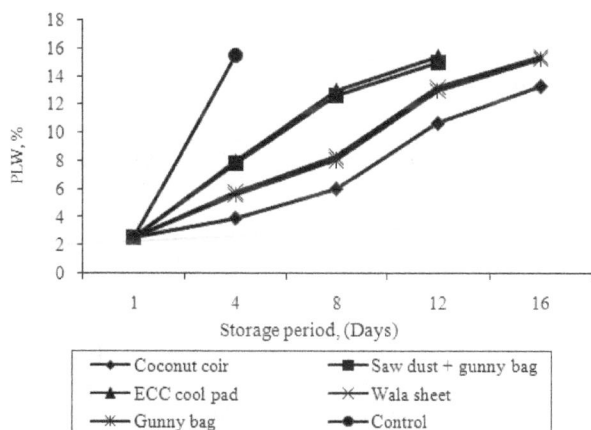

Fig 3. *Effect of filling materials on PLW.*

3.2. Effect of Filling Materials on Inside Temperature

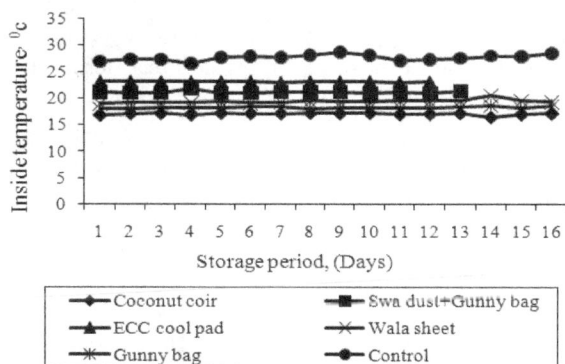

Fig 4. *Effect of filling materials on inside temperature.*

The effect of filling materials on inside temperature of on-farm evaporative cooler was found statistically significant in all filling materials. The data on effect of filling materials on inside temperature was plotted in Fig. 4. The average temperature obtained during storage period ranges from 16.5 to 23.3^{0}C throughout the storage period for all filling materials. Minimum temperature of 16.5 to 17.2^{0}C was obtained in coconut coir when ambient temperature was 26 to

29^{0}C. The higher (9.5 to 11.8^{0}C) temperature drop was obtained in case of coconut coir. Lower temperature drop was obtained in ECC cool pad followed by saw dust + gunny bag, *wala* sheet and gunny bag respectively.

Similar trend of inside temperature of the evaporative cooler have been reported by [8] for sapota fruit storage. Temperature drop of 20^{0}C was reported by [15] in acute summer for stored potatoes in cool chamber. Reference [16] reported temperature drop of 8 to 14^{0}C for fruits stored under low cost household evaporative cooler.

3.3. Effect of Filling Materials on Inside Relative Humidity

The effect of filling materials on relative humidity was found statistically significant in all filling materials. Relative humidity observed at atmospheric condition was 42 to 59% during storage. The data on effect of filling materials on inside relative humidity of evaporative cooler was plotted in Fig. 5. Inside relative humidity recorded for all of the filling material was ranging from 74 to 97%. Minimum relative humidity was recorded in ECC cool pad of evaporative cooler as compared to other. Whereas, maximum relative humidity (90 %) was recorded in coconut coir followed by gunny bag, *wala* sheet and saw dust + gunny bag of evaporative cooler. Same trend of inside relative humidity in the evaporative cooler have been reported by [8] for sapota fruit storage.

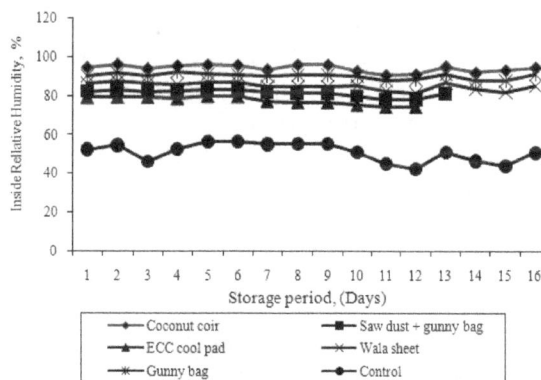

Fig 5. *Effect of filling materials on relative humidity.*

3.4. Effect of Filling Materials on Cooling Efficiency

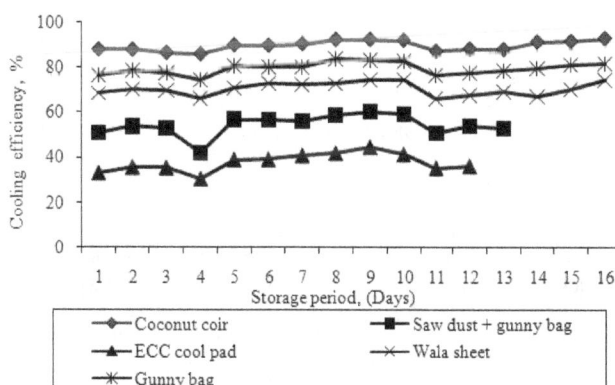

Fig 6. *Effect of filling materials on cooling efficiency.*

The effect of filling materials on cooling efficiency was found statistically significant in all filling materials. The data on effect of filling materials on cooling efficiency of evaporative cooler was plotted in Fig. 6.Maximum cooling efficiency was recorded in coconut coir followed by gunny bag and *wala* sheet. The highest cooling efficiency (90 %) was recorded in coconut coir. Minimum cooling efficiency was observed in ECC cool pad followed by saw dust + gunny bag. Cooling efficiency of ECC cool pad was ranging from 30.50 to 40.41%.

Similar results were reported by [17] for perishable products stored in evaporative cooling system and [18] for absorbent material stored in evaporative cooling.

4. Summary and Conclusion

Based on the results following conclusions could be drawn:
1. The percent weight loss (PLW) varied between 2.50 to 15.45%. Maximum weight loss was found in ECC cool pad whereas minimum (13.29%) was found in coconut coir. The PLW was found increased with increase in storage period for all the filling materials.
2. The maximum total storage life of 16 days was found in case of fruits stored in coconut coir withinthe acceptable range. Whereas, minimum storage life of 4 days found in control sample followed by 12 days in ECC cool pad.
3. Minimum inside temperature of 16.5 ^{0}C was recorded in coconut coir among all filling materials when average ambient temperature was 29^{0}C. The average temperature drop of 9.5 to 11.8 ^{0}C was found in case of coconut coir pad throughout the storage period.
4. Maximum inside relative humidity of 90% was recorded in coconut coir among all filling materials when average ambient relative humidity was 59%.
5. Highest cooling efficiency of 90% was recorded in coconut coir of an on-farm evaporative cooler followed by gunny bag 83%.
6. Coconut coir proved to be the best for increasing the shelf life (upto12days) of sapota fruits amongst all other filling material.
7. Coconut coir was found was found to be the most economical filling material amongst all other.

References

[1] Seyoum T.W. and Woldetsadik K, 2004. Forced ventilation evaporative cooling of fruits: A case study on Banana, Papaya, Orange. Lemon and Mandarin. Trop. Agric. J., 81(3): 179-185.

[2] Jain D, and Tiwari, G.N, 2002. Modeling and optimal design of evaporative cooling system in controlled environment greenhouse, Energy Conversion Manage., 43(16): 2235-2250.

[3] HelsenA, and Willmot J.J, 1991. Wet air cooling of fruits, vegetables and flowers. Current practice in Europe. Technical innovation in freezing and refrigeration of fruits and vegetables, International Institute of Refrigeration, Paris, France, 169-77.

[4] Umbarker S.P, Bonde R.S, and Kalase M.N, 1991. Evaporative cooled storage stature for oranges (citrus), Indian J. Agric. Eng., 1(1): 26-32.

[5] Wiersma F. 1983. Evaporative cooling in ventilation of agricultural structures An ASA E. Monograph 6th Series, Michigan, USA.

[6] Thakur B.C, Dhingra D.P, 1983. Parameters influencing the saturation efficiency of an evaporative rusten cooler" University of Glasgow College of Agric. Bulletin. No. 115.

[7] Igbeka J. C, and Olurin T.O, 2009. Performance Evaluation of Absorbent Materials in Evaporative Cooling System for the Storage of Fruits and Vegetables. Int. J. Food Eng., 5(3): 2.

[8] Mule, S. C. 2009. Studies on development and performance evaluation of on-farm evaporative cooler for storage of sapota fruits. Unpublished.

[9] Tilahun S.W. 2010. Feasibility and economic evaluation of low-cost evaporative cooling system in fruit and vegetables storage, Afgan J. food agri. Nutrition.

[10] Taye S. Mogaji 1. andOlorunisola P. Fapetu, 2011. Development of an evaporative cooling system for the preservation of fresh vegetables, African Journal of Food Science. 5(4). 255 – 266.

[11] Mainy, S. B., J. C. Anand, S. S. Chandran and Rajeshkumar, 1984. Evaporative cooling system for storage of potato. Ind. J. Agric. Sci. 59(3): 193-195.

[12] Islam, M. D. P. T. Morimoto, and K. Hatou, 2014. Effect of passive evaporative cooler on physico-chemical properties of hot water treated eggplant fruits. Agric. Eng. Int: CIGR , 16(2): 181-186.

[13] Banik, D., R. S. Dhua, S. K. Ghosh and S. K. Sen, 1988. Studies on extension of storage life of sapota. Indian J. Hort. 45(3-4) : 241-248.

[14] Roy, S. K. and R. K. Pal, 1991. A low cost zero energy cool chamber for short term storage of mango. Acta Hort. 291, : 519-524.

[15] Jha, S. N, 2008. Development of a pilot scale evaporative cooled storage structure for fruits and vegetables for hot and dry region. J. Food Sci. Tech. 45(2) : 148-151.

[16] Gite, R. P. and P. N. Jadhav, 1999. Development and testing of existing low cost evaporatory cooler for storage of fruits and vegetables. Unpublished.

[17] Thakral, R. V. Sangwan and D. N. Sharma, 2000. Performance evaluation of evaporative cooling systems for storage of perishable products in rural kitchens. Agric. Engg. Today. 24(4) : 40-43.

[18] William, A. O William, A. O., J. C. Igbeka, and O. O. Taiwo, 2009. studied the performance evaluation of absorbent materials in evaporative cooling system for the storage of fruits and vegetables. Energy Conversion and Management. 5 (3): 60-62.

Permissions

All chapters in this book were first published in IJSTS, by Science Publishing Group; hereby published with permission under the Creative Commons Attribution License or equivalent. Every chapter published in this book has been scrutinized by our experts. Their significance has been extensively debated. The topics covered herein carry significant findings which will fuel the growth of the discipline. They may even be implemented as practical applications or may be referred to as a beginning point for another development.

The contributors of this book come from diverse backgrounds, making this book a truly international effort. This book will bring forth new frontiers with its revolutionizing research information and detailed analysis of the nascent developments around the world.

We would like to thank all the contributing authors for lending their expertise to make the book truly unique. They have played a crucial role in the development of this book. Without their invaluable contributions this book wouldn't have been possible. They have made vital efforts to compile up to date information on the varied aspects of this subject to make this book a valuable addition to the collection of many professionals and students.

This book was conceptualized with the vision of imparting up-to-date information and advanced data in this field. To ensure the same, a matchless editorial board was set up. Every individual on the board went through rigorous rounds of assessment to prove their worth. After which they invested a large part of their time researching and compiling the most relevant data for our readers.

The editorial board has been involved in producing this book since its inception. They have spent rigorous hours researching and exploring the diverse topics which have resulted in the successful publishing of this book. They have passed on their knowledge of decades through this book. To expedite this challenging task, the publisher supported the team at every step. A small team of assistant editors was also appointed to further simplify the editing procedure and attain best results for the readers.

Apart from the editorial board, the designing team has also invested a significant amount of their time in understanding the subject and creating the most relevant covers. They scrutinized every image to scout for the most suitable representation of the subject and create an appropriate cover for the book.

The publishing team has been an ardent support to the editorial, designing and production team. Their endless efforts to recruit the best for this project, has resulted in the accomplishment of this book. They are a veteran in the field of academics and their pool of knowledge is as vast as their experience in printing. Their expertise and guidance has proved useful at every step. Their uncompromising quality standards have made this book an exceptional effort. Their encouragement from time to time has been an inspiration for everyone.

The publisher and the editorial board hope that this book will prove to be a valuable piece of knowledge for researchers, students, practitioners and scholars across the globe.

List of Contributors

Jiokap Nono Yvette and Aseaku Jude Nkengbeza
University Institute of Technology (IUT) of the University of Ngaoundere, Department of Chemical Engineering and Environment, Ngaoundere, Cameroon

Desmorieux Helene
Process Engineering and Automatic Laboratory of University Claude Bernard – Lyon, Villeurbanne Cedex, France

Degraeve Pascal
Food Processing Research Laboratory of Claude Bernard – Lyon 1 University, University Institute of Technology A Lyon 1 – Biological Process Department, Technopole Alimentec – Rue Henri de Boissieu, Bourgen Bresse Cedex, France

Kamga Richard
National Advanced School of Agro-Industrial Sciences (ENSAI) of the University of Ngaoundere, Department of Applied Chemistry, Ngaoundere, Cameroon

M. Surekha
Student, Embedded System Technologies, Knowledge Institute of Technology, Salem, India

N. Suthanthira Vanitha
Head of the Department, Department of Electrical & Electronics Engineering, Knowledge Institute of Technology, Salem, India

K. Yadhari
Assistant Professor, Department of Electrical & Electronics Engineering, Knowledge Institute of Technology, Salem, India

Md. Mazharul Islam and Md. Tanjim Hossain
Department of Textile Engineering, Northern University of Bangladesh, Dhaka, Bangladesh

Mohammad Abdul Jalil
Department of Textile Engineering, Mawlana Bhashani Science and Technology University, Tangail, Bangladesh

Elias Khalil
Department of Textile Engineering, World University of Bangladesh, Dhaka, Bangladesh

Omoniyi Omotayo Adewale.
Lecturer, Department of Petroleum Engineering Abubakar Tafawa Balewa, University Bauchi, Bauchi State, Nigeria

Iji Sunday
Department of Petroleum Engineering Abubakar Tafawa Balewa, University Bauchi, Bauchi State, Nigeria

Kutsanedzie F.
Research and Innovation Department, Accra Polytechnic, Accra, Ghana

Ofori V.
Agricultural Engineering Department, Kwame Nkrumah University of Science and Technology, Kumasi, Ghana

Diaba K. S.
Agricultural Engineering Department, Anglican University College of Technology, Sunyani, Ghana

Goshayeshi Hamid Reza and Hashemi Bahman
Department of Mechanical Engineering, Mashhad Branch, Islamic Azad University, Mashhad, Iran

Pradyumna Saripalli and K. Sankaranarayana
Mechanical Engineering, Gitam Institute of Technology, Gitam University, Visakhapatnam (A.P), India

Akpakpavi Michael
Mechanical Engineering Department, Accra Polytechnic, Accra, Ghana

Haruna Musa
Department of Electrical Engineering, Bayero University, Kano, Nigeria

N. A. Mansour and N. M. Eldebawi
Faculty of Science, Physics Department, Zagazig University, Zagazig, Egypt

Ladan Asadi and Hamid Majidi
Department of Architecture, Art and Architecture Faculty, Islamic Azad University, Mashhad, Iran

Rana Kutlu
Department of Interior Architecture and Environmental Design, Istanbul Kultur University, Istanbul, Turkey

Fikadu Getachew, Gizachew Legesse and Girma Mamo
Ethiopian Institute of Agricultural Research (EIAR), Climate and Geospatial Research Directorate (CGRD), Addis Ababa, Ethiopia

Fekadu Alemu
Department of Biology, College of Natural and Computational Sciences, Dilla University, Dilla, Ethiopia

Omoniyi Omotayo Adewale and Ubale Mustapha
Department of Petroleum Engineering, Abubakar Tafawa Balewa University, Bauchi State, Nigeria

Ngatia Christopher Mugo, Mbugua John Nderi and Mutambuki Kimondo
Kenya Agricultural Research Institute (KARI), Nairobi, Kenya

Yusuf Inusa and Ogundele Olusegun John
Department of Minerals & Petroleum Resources Engineering Technology, Auchi Polytechnic, Auchi, Edo State, Nigeria

Odejobi Yemi
Leading Edge Geoservices Ltd, Suite 55, EDPA Shopping Complex, Ugbowo, Benin City, Nigeria

Auwal Ishaq Haruna
National Engineering and Technical Company Ltd, Corporate H/Q, Plot 1460 Ligali Ayorinde Street, Victoria Island, Lagos

Nguyen Van Hop
Faculty of Chemistry, Hue University of Sciences, Hue city, Vietnam

Vu Thi Kim Loan
Department of Chemistry, Hai Phong University of Medicine and Pharmacy, Hai Phong city, Vietnam

Thuy Chau To
Faculty of Resources and Environment, Thu Dau Mot University, Thu Dau Mot city, Vietnam

Zohair Malki
Department of Information and Learning Resources, The collage of Computer Science and Engineering, Taibah University, Yanbu Al Bahar, Al Madinah Province, Saudi Arabia

Md. Akhtar Hossain, Akib Adnan and Md. Maskurul Alam
Department of Civil Engineering, Rajshahi University of Engineering & Technology, Rajshahi, Bangladesh

Nasrin Ferdous and Reashad Bin Kabir
Department of Apparel Manufacturing Management & Technology, Shanto-Mariam University of Creative Technology, Dhaka, Bangladesh

Haruna Musa
Department of Electrical Engineering, Bayero University, Kano, Nigeria

Popoola Samuel Olatunde, Nubi Olubunmi Ayoola, Oyatola Opeyemi Otolorin, Adekunbi Falilu Olaiwola, Fabunmi Gaffar Idera and Nwoko Chidinma Jecinta
Department of Physical and Chemical Oceanography, Nigerian Institute for Oceanography and Marine Research, Lagos, Nigeria

S. M. Arifuzzaman, Md. Manjiul Islam and Md. Mohidul Haque
Mathematics Discipline, Khulna University, Khulna, Bangladesh

Endale Wondu
Mechanical and Vehicle Engineering (Manufacturing Engineering), Design and fabrication, Wonji/Shoa, Ethiopia

Tadesse Negi and Alemayehu Dengiya
Agronomy, Agronomy and plant protection, Wonji/Shoa, Ethiopia

Wendimu Weldegiorgis
Electrical Engineering, Design and fabrication, Wonji/Shoa, Ethiopia

M. Sangeetha
Embedded System Technologies, Knowledge Institute of Technology, Salem, India

S. Kalpanadevi, M. Rajendiran and G. Malathi
Electrical and Electronics Engineering, Knowledge Institute of Technology, Salem, India

Getachew Kebede Warati
Jimma University Institute of Technology, Civil engineering department, Jimma, Ethiopia

Tamene Adugna Demissie
Jimma University Institute of Technology, Hydraulic and water resource engineering department, Jimma, Ethiopia

Adefemi Adeodu, Ilesanmi Daniyan and Samuel Afolabi
Department of Mechanical and Mechatronics Engineering, Afe Babalola University, Ado Ekiti, Nigeria

Charles Omohimoria
Department of Petroleum and Chemical Engineering, Afe Babalola University, Ado Ekiti, Nigeria

Rajendra Kenghe and Nilesh Fule
Department of Agricultural Process Engineering, Mahatma Phule Krishi Vidyapeeth, Rahuri, India

Kalyani Kenghe
Department of Mechanical Engineering,Sanjivani College of Engineering, Kopargaon, India

Index

www.ingramcontent.com/pod-product-compliance
Lightning Source LLC
Chambersburg PA
CBHW080515200326
41458CB00012B/4221